Excitations in simple liquids, liquid metals and superfluids

Excitations in simple liquids, liquid metals and superfluids

Wouter Montfrooij

*Department of Physics and Astronomy and the Missouri Research Reactor,
University of Missouri, Columbia, Missouri, United States*

Ignatz de Schepper

Reactor Institute Delft, Technical University Delft, The Netherlands

OXFORD
UNIVERSITY PRESS

OXFORD
UNIVERSITY PRESS

Great Clarendon Street, Oxford OX2 6DP

Oxford University Press is a department of the University of Oxford.
It furthers the University's objective of excellence in research, scholarship,
and education by publishing worldwide in

Oxford New York

Auckland Cape Town Dar es Salaam Hong Kong Karachi
Kuala Lumpur Madrid Melbourne Mexico City Nairobi
New Delhi Shanghai Taipei Toronto

With offices in

Argentina Austria Brazil Chile Czech Republic France Greece
Guatemala Hungary Italy Japan Poland Portugal Singapore
South Korea Switzerland Thailand Turkey Ukraine Vietnam

Oxford is a registered trade mark of Oxford University Press
in the UK and in certain other countries

Published in the United States
by Oxford University Press Inc., New York

British Library Cataloguing in Publication Data

Data available

Library of Congress Cataloging in Publication Data

Data available

Typeset by SPI Publisher Services, Pondicherry, India
Printed in Great Britain
on acid-free paper by
CPI Antony Rowe, Chippenham, Wiltshire

ISBN 978–0–19–956412–5

10 9 8 7 6 5 4 3 2 1

This book is dedicated to E.G.D. Cohen and E.C. Svensson

Preface

This book is written with the experimentalist in the field of liquids in mind. The past two decades have seen a remarkable increase in experimental capabilities at neutron sources and even more so, at X-ray sources. This has resulted in a wealth of highly accurate data on the dynamics of the liquid state. For instance, it is now possible to probe the microscopic dynamics of ^{3}He by means of X-ray scattering, something unheard of until very recently. Given the continuing development of new and higher-powered sources for doing neutron- and X-ray-scattering experiments, we can look forward to increasingly more accurate experimental data that will allow us to measure even the smallest changes in atomic motions in liquids.

Our book will aid the experimentalist in ensuring that these small changes can indeed be teased out of the experimental data. First, it provides an intuitive framework for modelling the excitations in liquids using a minimum number of adjustable parameters, while satisfying all known theoretical constraints. For instance, the built-in sum rules link the amplitudes of the excitations to their decay rates. This ensures that the parameters that are inferred from the experimental data are as accurate as possible and that the maximum amount of information will be extracted from the data. Secondly, by using this framework the interpretation of the scattering data becomes very transparent. In particular, by treating on an equal footing classical liquids and quantum liquids as well as dense and dilute gases, the comparison between different thermodynamic states of the same liquid as well as comparisons between different liquids will be straightforward to carry out.

Other than some fairly basic knowledge of statistical physics and quantum mechanics the reader is not required to already have a deep understanding of classical liquids or of quantum fluids. As such, this self-contained textbook should be useful to anyone who does research on liquids, from the graduate student level and up.

Notwithstanding our practical rather than theoretical focus, this book is more than a mere practical guide; it serves as a complement to existing textbooks on normal liquids and superfluids. Our emphasis on excitations augments classic textbooks that mainly concentrate on models for and calculations of the averaged structure in classical liquids (Pryde 1966; Croxton 1975).

In addition, it also fully complements more recent efforts that tackle the theory of liquid dynamics, both in classical liquids (Balucani and Zoppi, 1994; Hansen and McDonald 2006) as well as in superfluids (Griffin, 1993; Glyde, 1994). Essentially, we start where these other books leave off. Our approach is to use one overarching formalism for the description of the dynamics in both classical and quantum liquids, and then to compare in detail how various couplings and decay mechanisms are similar and/or subtly different between various liquids, and how these parameters evolve as a function of density and temperature. We use very simple models to aid the interpretation of the data. Thus, our aim is to probe the overall similarities between all liquids on a microscopic scale, and to learn from the small differences. In fact, by expanding the viewpoint to now also include quantum liquids we found it possible to develop a better general understanding of excitations (and their fate) in both classical and quantum liquids.

Moreover, we believe that this book will also be of interest to researchers in the field of quantum liquids. Superfluid helium textbooks tend to focus on the origin of superfluidity and its relationship to the formation of a Bose-condensate. They discuss in detail the role of the Bose-condensate, and its potential implications for the excitations in superfluid helium. By and large, we do not deal with any of this. We simply take the onset of superfluidity for granted as something that must eventually happen in a Bose-liquid because of the formation of a Bose-condensate. Instead, we scrutinize how excitations that are also present in the normal fluid phase are affected by the onset of superfluidity, and how these standard excitations get modified in the superfluid phase to take on the role of elementary excitations. In doing so, we show the very close similarity between high-temperature normal fluids, low-temperature normal fluids, and low-temperature superfluids. Combined with our discussion on the use of perturbation theory to reproduce the elementary excitation curve in superfluid helium from scratch, we believe that our book constitutes a worthwhile addition to standard texts on superfluidity.

Finally, this book does not purport to be a review of the state of the field. In fact, far from it. Our intention is to provide as much useful information as we can in order to aid in the analysis of experiments, and to present this information in a logical manner. As a consequence, we will not refer to every single outstanding paper in the field, or necessarily touch upon the (very) latest findings. No slight was intended in any of our choices.

The Liquid State, J.A. Pryde, Hutchinson University Library, London, 1966.

Introduction to Liquid State Physics, Clive A. Croxton, John Wiley & Sons, London, 1975.

Theory of Simple Liquids, J.-P. Hansen and I.R. McDonald, Academic Press, 3rd edn, London, 2006.

The Dynamics of the Liquid State, Umberto Balucani and Marco Zoppi, Clarendon Press, Oxford, 1994.

Excitations in a Bose-condensed Liquid, Alan Griffin, Cambridge Studies in Low Temperature Physics, 1993.

Excitations in Liquid and Solid Helium, Henry R. Glyde, Clarendon Press, Oxford, 1994.

Acknowledgements

Foremost, we would like to thank our colleagues Ubaldo Bafile and Kunimasa Miyazaki for their reading of this book, and for their invaluable help in correcting mistakes and in clarifying obtuse passages, especially concerning the memory function formalism in general and the subtle difference between this formalism and the effective eigenmode formalism when applied to real systems in particular (UB), and concerning our brief description of mode-coupling effects (KM).

A great many figures have been reproduced, with permission, from the literature. We thank all our colleagues who made these figures which made our work much easier. We also thank, in no particular order, the Physical Review, the Journal of Chemical Physics, the Journal of Physics: Condensed Matter, the Journal of Low Temperature Physics, Condensed Matter Physics, Europhysics Letters, the Czechoslovak Journal of Physics, the Review of Modern Physics, Il Nuovo Cimento, Nature, and the Journal of Non-Crystalline Solids for their kind permission. Credit is given in the appropriate figure captions by listing the lead author(s) and the publication year; the full references are written out on p. 262 and beyond. We would also like to thank Ilse Koudijs for obtaining all these permissions.

We are also indebted to Dylan Moore for his work on Fig. 4.8, to Alexander Schmets for his visualizations (Figs. 4.9 and 4.10) of a simple model we use to help with the interpretation of short-wavelength fluctuations, and to Mark Patty for his work on some of the gallium experiments presented in this book. Lastly, WM is indebted to Peter Pfeifer for freeing up his teaching schedule during the writing of this book.

Contents

1 Introduction   1

2 Excitations, relaxation and effective eigenmode formalism   6
2.1   Excitations and correlation functions   9
2.2   Approximate theories for the relaxation of disturbances   15
2.3   Effective eigenmode formalism   32
2.4   Memory function formalism   54

3 Experiments and computer simulations   64
3.1   Graphical representation of scattering techniques   64
3.2   Corrections to the data   71
3.3   Computer simulations   81

4 Simple liquids   88
4.1   Density fluctuations: general behavior of the extended modes   88
4.2   The hydrodynamic limit and the approach thereof   92
4.3   Beyond hydrodynamics   95
4.4   The response for $q\sigma \simeq 2\pi$ 100
4.5   Cage diffusion   110
4.6   Temperature fluctuations and oscillations   113

5 Colloidal suspensions   115
5.1   Charged colloidal suspensions   118
5.2   Neutral colloidal suspensions   122

6 Binary mixtures   126
6.1   The experimental signature of fast and slow sound   126
6.2   Effective eigenmode formalism for mixtures   129
6.3   The hydrodynamic limit   131
6.4   Optical modes and missing modes in a mixture   134
6.5   Fast sound versus viscoelastic behavior   138

7 Liquid metals   143
7.1   Extended sound modes in liquid metals   144
7.2   The hydrodynamic limit   151
7.3   Excitations with $q\sigma \simeq 2\pi$ 160
7.4   New excitations and old warnings   162

8 Very cold liquids   172
8.1   Prominent sound modes   172
8.2   Lineshape distortion   174

| 8.3 | Low-temperature versus high-temperature excitations | 177 |
| 8.4 | ^{4}He versus ^{3}He | 181 |

9 Superfluids | 185 |
9.1	Superfluidity and Bose–Einstein condensation	185
9.2	Why Bose-liquids must become superfluids	190
9.3	Density fluctuations in very cold superfluid ^{4}He	193
9.4	The normal-fluid to superfluid transition	196
9.5	Perturbation theory for the extended sound modes in ^{4}He	202

10 Summary and outlook | 216 |
| 10.1 | Summary | 216 |
| 10.2 | Outlook | 219 |

Appendix A   Conversions | 226 |

Appendix B   Derivation of the effective eigenmode formalism | 227 |
B.1	Projection formalism	227
B.2	Coupling constants	230
B.3	Frequency-sum rules	233

Appendix C   (Almost) exact results for hard-sphere fluids | 236 |

Appendix D   Detailed behavior of the eigenmodes | 240 |

Appendix E   Memory functions and effective eigenmodes | 248 |

Appendix F   Effective eigenmode formalism for mixtures | 252 |

Appendix G   Glory oscillations | 256 |

References | 262 |

Index | 271 |

1
Introduction

The solid state is dominated by the lattice vibrations of the atoms, whereas the gas state is dominated by the ballistic motion of the atoms. Both these extremes have allowed for the construction of theories that describe the microscopic behavior of the atoms in these states and how in turn this microscopic behavior determines the macroscopic properties.

The state of affairs for the liquid phase is very different. Not only are both vibrational and diffusive motions important, they couple so strongly to each other that it is pointless to try to develop a theory where both are not treated on an equal footing. Needless to say, the lack of a rigid lattice structure does not help since it prevents us from using the perturbation techniques that have been developed for the solid state. On the other hand, this level of difficulty is probably what attracts people to the field of liquids in the first place.

Yet, we recognize a liquid when we see one. While we may not have a microscopic theory that yields details about liquid behavior such as how the viscosity of a liquid is related to the collision between two atoms, it is obvious that liquids have many features in common. With this book we aim to extend this general notion to the microscopic realm. We will be looking at how small-scale perturbations relax back to equilibrium, how long this takes, and whether the way this happens is subtly different or rather similar between a variety of liquids, ranging from the simplest monoatomic liquids to the spectacular superfluid systems.

In order to describe the relaxation of small-scale perturbations, we shall employ the effective eigenmode formalism. The premise behind this formalism is very straightforward: it simply separates processes that take place on a very short timescale from processes that take a while. Fast processes are replaced by their time-integrated contributions. In this way, the fate of a perturbation that takes a while to decay back to equilibrium will be expressed in terms of coupling parameters to processes that decay rapidly, and the decay rate of such a fast process will be given by an adjustable parameter to be determined from a model fit to experiment. Thus, the formalism effectively reduces the some 10^{23} degrees of freedom in a liquid to a few by only looking at those modes of decay that take the longest time.

To be more precise, let us look at how a density disturbance would be described in this formalism. In Fig. 1.1 we have sketched a small-scale density disturbance (denoted 'n'), that is, a departure from the average density. Clearly, this is a sketch. First, it is an exaggeration, and secondly, we assumed that we are in the classical limit where atoms behave nicely as particles. We can imagine that this density disturbance arose

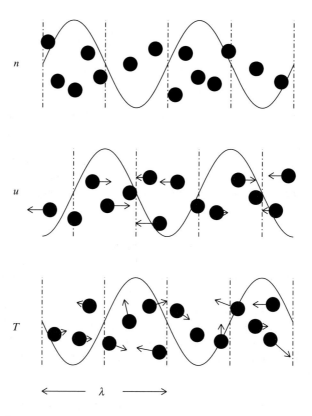

n

u

T

λ

Fig. 1.1 A sketch representing microscopic fluctuations of wavelength λ of the density 'n' (top panel), velocity 'u' (middle panel) and temperature 'T' (bottom panel).

spontaneously, or perhaps it was created when the liquid was struck by a neutron in a scattering experiment. For this disturbance to decay, atoms must move from areas of high density to areas of low density. Therefore, when we look at the same situation a little while later we find that this density disturbance has created a (longitudinal) velocity disturbance, denoted 'u' (Fig 1.1). In our formalism we say that the density couples to the longitudinal component of the velocity. This velocity disturbance can relax through collisions, but it can also result in a kinetic-energy disturbance, or equivalently, a temperature disturbance (denoted 'T') (Fig. 1.1).

Thus, the velocity couples to the temperature, however, there is no direct coupling of the density to the temperature. The temperature disturbance will even out through collisions. This collisional process will be much faster than the transition from a density disturbance to a velocity disturbance. If we are interested in the relatively slow decay of the density disturbance, we would simply replace the underlying decay mechanisms of the velocity and temperature disturbances by time-integrated decay constants (denoted z_u and z_T, respectively), and cut off our decay tree at this point. This is depicted in Fig. 1.2.

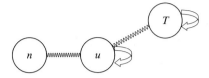

Fig. 1.2 The forces and decay rates for the microscopic fluctuations sketched in Fig. 1.1. The springs denote the couplings between the variables, implying that, for instance, a density fluctuation 'n' can give rise to a velocity disturbance 'u'. The looping arrows indicate that disturbances such as the temperature 'T' disturbance shown in Fig. 1.1 will relax back to equilibrium; the detailed mechanism through which a disturbance at the end of the branch decays is not important in modelling the decay of the density fluctuation that we are measuring.

In fact, Fig. 1.2 could well depict the familiar hydrodynamics equations (Hansen and McDonald, 2006), with the decay constants determined by the shear and bulk viscosity and by the thermal conductivity. Or it could depict the decay of disturbances such as the one shown in Fig. 1.1 that take place on such small length scales that hydrodynamics is no longer valid. Nonetheless, in this case we can experimentally determine the decay constants through a fitting procedure, and we infer how hydrodynamics can be extended to smaller length scales, even if we can no longer calculate the decay constants – that have now acquired an intrinsic dependence on the wavelength of the fluctuation – from scratch.

The situation becomes especially interesting when we look at disturbances on length scales comparable to the interatomic spacing d. Then, we could be looking at the collision between two atoms, or at the motion of an atom between collisions. Thus, on this length scale we are looking directly at those processes that determine how larger-scale disturbances eventually decay. For instance, we can use the effective eigenmode formalism to determine how long the liquid maintains a memory of an inter-atomic collision and we can see how this time compares to the decay time relevant to the larger length scales. In this way, we can assess whether the main decay channel for a particular perturbation is related to interatomic collisions, or not. Not only will we find that the effective eigenmode formalism provides us with a transparent way of characterizing all of these types of processes, we will find upon comparing various types of liquids such as liquid metals and dense noble gases that the microscopic processes in all these liquids are more similar than dissimilar.

Of course, there are many more types of excitations possible in liquids than the ones described so far, and quite frequently a process has to be broken up into various timescales. For instance, the diffusion of an atom through the rest of the liquid on very short timescales is determined by the frequent collisions of an atom with its neighbors. Here, we see a particle trapped in the cage formed by those neighbors, and it behavior resembles that of a vibrational mode in a solid. In binary mixtures such as a helium–neon mixture, we find that the lighter atoms can vibrate at a frequency that the heavier atoms cannot follow, and vice versa. These so-called fast- and slow-sound modes are very similar to the density disturbance depicted in Fig. 1.1. In liquid metals we find that the decay mechanism of density disturbances is slightly less effective, resulting in the ability of a liquid metal to sustain well-defined propagating density fluctuations on short length scales. That is, well defined compared to a simple fluid, where such disturbances are fully relaxed back to equilibrium before they have propagated by much more than a full wavelength.

Quantum fluids are capable of much more. In ^{3}He one observes particle–hole excitations directly related to the fermionic nature of these atoms, whereas in superfluid ^{4}He one finds that small-scale density disturbances do not decay at all because of the Bose nature of the atoms and the ensuing presence of a Bose-condensate. In this case, the density disturbances take on the role of elementary excitations. In addition, superfluid ^{4}He exhibits excitations that are the direct equivalent of multiphonon modes in solids. Remarkably, it turns out that the excitations in a superfluid are virtually identical to the modes in normal fluids. In fact, the persistence of density disturbances in superfluids allows us to calculate from scratch parameters such as propagation speeds that can only be determined through a fitting procedure in simple fluids. In other words, it turns out that superfluid ^{4}He is not only very similar to simple liquids, it might even be the simplest liquid of them all. Superfluid helium provides us with the equivalent of an ideal gas in the description of real gases, and with the equivalent of a perfectly ordered solid in the description of real solids.

Our approach is the following. In Chapter 2 we introduce the effective eigenmode formalism, and discuss how it relates to the various types of excitations. In Chapter 3 we briefly discuss the experimental techniques that probe departures from equilibrium, as well as computer simulations. In Chapter 4 we apply the effective eigenmode formalism to simple liquids, and we will follow the change in behavior of fluctuations when probed on smaller and smaller length scales. We will pay special attention to density fluctuations on the same length scale as the interatomic distances, and we will discuss approximate models that relate the main parameters of interest in this region to average quantities such as thermal velocities and nearest-neighbor distances.

We will find that three modes, out of all possible modes of excitation and decay, are of particular importance in the description of simple liquids when describing the decay of density fluctuations; this holds true for fluctuations ranging from hydrodynamic length scales down to the very short length scales characteristic of the mean-free path between collisions. These three modes have been dubbed the (extended) heat and sound modes, their names signifying that they are the continuous extensions down to smaller length scales of the well-known hydrodynamic Rayleigh and Brillouin triplet of modes that describe the diffusion of heat and the propagation of sound waves, respectively (Hansen and McDonald, 2006). It is a most remarkable experimental fact that in simple liquids the full decay process of any density fluctuation on any length scale (larger than the mean-free path) can be very well described by just taking into account these three modes. It is not known why simple fluids would exhibit this type of behavior. After all, one would expect that upon investigating the fluid on shorter and shorter wavelengths, that the number of modes needed to accurately describe the decay of a density fluctuation would increase. This increase is anticipated since on the shorter length scales the finer details of the interparticle interactions are expected to come into play, in contrast to the averaged view of hydrodynamic scale disturbances where the details of the interparticle interactions are washed out because of the great many collisions that take place between the particles before a density fluctuation has relaxed back to equilibrium.

We will scrutinize the behavior of these three modes in particular, and try to unravel what decay mechanisms they reflect in fluids on short length scales. In Chapter 5 we look at the behavior of colloidal suspensions whose dynamics are dominated by cage diffusion. This gives us insight in the behavior of the (extended) heat mode in the absence of the (extended) sound modes. In Chapter 6 we will extend the eigenmode formalism to binary mixtures and scrutinize the changes linked to short length-scale concentration gradients. In these mixtures we will see that the extended sound modes become more prominent and that they do not necessarily involve both species of atoms. In Chapter 7 we apply the formalism to liquid metals and discuss the transition from ordinary hydrodynamic behavior to the non-hydrodynamic character that the heat and sound modes acquire, and we discuss – and even speculate on – the presence of excitations that are unique to the liquid metals.

In Chapter 8 we will focus on the changes that become apparent when quantum effects start to be important at low temperatures. In particular, we will see that the extended sound modes will start to dominate the spectra. In Chapter 9 we look at the behavior of density fluctuations when they take longer and longer to decay on going from the normal fluid phase upon entering the superfluid phase. We will see that there is a one-to-one correspondence between the elementary excitations in superfluid ^{4}He and the extended sound modes discussed in simple liquids and liquid metals. We conclude Chapter 9, and the main part of the book, with a calculation using standard perturbation theory that allows us to predict the energies of the elementary excitations in superfluid ^{4}He. In doing so, we hope to have shown that deep down – on a microscopic scale – all fluids are very similar indeed.

Our final chapter contains a brief summary, and a debriefing of what we consider to be the main questions that have been raised by the experiments presented in this book as well as by the comparison between the various types of liquids. Our final introductory note pertains to the notation adopted in this book. When the probing radiation interacts with the sample, it transfers $p = \hbar q$ in momentum, and $E = \hbar \omega$ in energy. Here, $\hbar$ is Planck's constant h divided by 2π, q is the wave number and ω the radial frequency. We will use q in units of Å^{-1}, while ω is measured in units of ps^{-1} throughout the written text, and wherever possible in the figures. However, most of our figures have been reproduced from the literature and have a notation (such as Q instead of q) as well as units different from ours. In Appendix A we list the symbols for and the conversions between the different units for energy transfer. Reduced momentum transfer $p/\hbar$ is measured in Å^{-1} or nm^{-1}, with $1 \text{ Å}^{-1} = 10 \text{ nm}^{-1}$.

2

Excitations, relaxation and effective eigenmode formalism

We do not have a microscopic theory for liquids. While this is stating the obvious, it is good to briefly enumerate what we do know, and what we (still) do not know (Brush, 1976; Cohen, 1993b). Thanks to Maxwell, we do know – since 1859 – the distribution function of the velocities of the individual particles of a classical fluid in thermal equilibrium (Maxwell, 1902), but we do not know how to compute the macroscopic transport coefficients that determine how a fluid goes back to thermal equilibrium after it has been disturbed. These days we can, of course, use a computer to solve the classical equations of motion – once we know the interaction potential – and determine the transport coefficients in that way, but this does not provide us with a nice warm feeling since it is essentially the same as simply measuring them. We have also learned that the once cherished belief that the transport coefficients could be expressed in terms of a virial expansion, similar to what is done for the pressure of a system, is in fact not true; non-analytical terms will appear (Cohen, 1966). For instance, the density expansion for the viscosity η reads (Cohen, 1967)

$$\eta(n, T) = \eta_0(T) + \eta_1(T)n + \eta_2(T)n^2 + \eta_2'(T)n^2 \ln n + \dots, \qquad (2.1)$$

where η_0 represents the contribution due to all binary collisions, η_1 all contributions due to ternary collisions, etc. The term $\sim \ln n$ represents the end of the simple, and intuitively appealing idea that one could calculate how the non-equilibrium velocity distribution function (resulting from a disturbance) would relax back to the Maxwellian equilibrium distribution by including the free-streaming of particles between collisions, as well as the effects of binary, ternary, quarternary (and so forth) collisions. We now know that this simple idea fails because such a scheme does not take into account that the mean-free path between collisions decreases rapidly with increased density. As a result, too many contributions are incorporated into the collision terms, and the higher-order terms in the virial expansion diverge. Regrouping various sets of collisions removes (Cohen, 1993b) the divergences, but it leads to non-analytical terms such as $n^2 \ln n$.

So why would anyone care whether non-analytical terms appear? For one thing, it sets up a firm roadblock in the attempts to generalize (Résibois and de Leener, 1977) the Boltzmann equation,[1] which only deals with uncorrelated binary collisions, to the

---

[1] The Boltzmann equation describes the dynamics in dilute gases. In Section 2.2.4 we touch upon a generalization to higher densities.

case of dense fluids. After all, how does one classify and group all possible (correlated) collision sequences? How many of these sequences are needed to correctly reproduce the long-term behavior of the correlation functions? These questions are pertinent not only to the calculation of the macroscopic transport coefficients, but also to the decay of disturbances on any length scale since the physical processes underlying this decay are identical.

Of all the efforts during the last 50 years that have been put into developing a better understanding of liquids on all length scales – efforts involving doing scattering experiments, computer simulations and endless shuffling of equations – two major findings stand out in our opinion. On the one hand, there is the realization that out of all the possible collision sequences, two have been shown to play a particularly important role (Cohen, 1993*b*). They are commonly referred to as cage diffusion and vortex diffusion, and we will discuss those collision sequences in the subsequent sections of this chapter.

On the other hand, there is the finding that when a fluid is disturbed on a length scale that is not smaller than the distance between collisions its decay back to equilibrium is characterized by a very limited number of exponential functions (Alley and Alder, 1983; de Schepper *et al.*, 1988). For instance, when a density disturbance is imposed upon a monoatomic liquid (such as the one depicted in Fig. 1.1), then three exponentials are all that is needed to describe how this disturbance will propagate away while being damped out, and how it will give rise to a change in local temperature, and how this change will be undone. The remarkable thing is that irrespective of whether such a disturbance involves hundreds of particles, or only very few, three is all that is needed. This observation is of course what makes the effective eigenmode description so successful: had increasingly more exponentials been required for disturbances of increasingly shorter wavelength, then the effective eigenmode formalism would not have been more than a rewriting of the equations of motion. In this and subsequent chapters we will scrutinize these exponentials, in particular as to what determines their amplitudes and how the arguments of these exponentials are related to the transport coefficients of the fluid. We will derive the master equation of the effective eigenmode formalism, which gives us the relationship between the measured dynamic structure factor $S_{nn}(q,\omega)$ and the behavior of the various decay channels through which fluctuations relax back to equilibrium. This can all be captured in terms of a matrix $G(q)$ that consists of all the coupling parameters and decay rates associated with these channels:

$$S_{nn}(q,\omega) = \frac{\beta\hbar\omega}{1 - e^{-\beta\hbar\omega}} \frac{S_{\text{sym}}(q)}{\pi} \text{Re} \left[\frac{1}{i\omega 1 + G(q)} \right]_{11}. \tag{2.2}$$

In this equation, β is the inverse temperature of the fluid $\beta = 1/k_{\text{B}}T$ with k_{B} Boltzmann's constant, and $S_{\text{sym}}(q)$ is a measure of the instantaneous order present in the liquid, to be formally defined in subsequent sections.

Before we look into details and discuss the meaning, shape and form of the matrix $G(q)$, we should place the above equation in the context of analyzing experimental results. This is the equation that experimental results will be fitted to. The matrix embodies everything that can be learned from experiment. The elements of the matrix $G(q)$ are determined from experiment through a fitting procedure. The size of the matrix, and the number of non-zero matrix elements are dependent on the liquid that is being modelled, and on the length scale λ of the fluctuations that are being probed. The diagonal elements of the matrix are decay rates, and they always have to be determined through a fitting procedure. In virtually all cases, there are exactly two such decay rates that have to do with momentum transfer and energy transfer. The off-diagonal elements represent the coupling parameters, and they can be expressed as equal-time correlation functions (see next section), implying that they represent instantaneous, structure-like properties of the liquid. In principle, these off-diagonal elements can be determined through the sum rules that govern the scattering spectra; in practice, they are adjusted as free parameters because of experimental limitations. In computer simulations however, one has direct access to these coupling constants, and as such, they do not have to be determined through a fitting procedure.

Thus, the dynamics of a liquid can be expressed in terms of static parameters – the off-diagonal elements of $G(q)$ – combined with a few (~ 2) decay rates – the diagonal elements of $G(q)$. In words, this means that the relaxation back to equilibrium of any type of fluctuation of a certain wavelength λ is completely determined by two decay rates. For a given wavelength, the relaxation of a density fluctuation, or of a temperature fluctuation, or of a fluctuation in the local stress level are all determined by the very same two (adjustable) parameters. This book explains how to determine these two parameters from experiment, which turns out to be the easy part. The hard part is to ascribe physical meaning to these two parameters.

The advantage of fitting the experimental results to eqn 2.2 is that all the sum rules are automatically taken care of, ensuring that the data are fitted using a minimal number of adjustable parameters, with a maximum number of physical constraints. In particular, the matrix $G(q)$ not only determines the decay rates of the excitations; the amplitudes of the excitations in the scattering spectra are also determined by the matrix elements of $G(q)$, and they cannot (and should not) be adjusted as free parameters independently of the decay rates of these excitations. The latter is of particular importance given the accuracy of present-day scattering data: it imposes the correct relative strengths (amplitudes in the scattering spectra) onto the various decay channels given their decay rates. This provides a very stringent test on models. Frequently, one encounters in the literature that the amplitudes of the features in the scattering spectra are being fitted independently of their decay rates, whereas the two carry the same amount of information. Consequently, as a result of having too many adjustable parameters at one's disposal, some models are accepted, whereas they should have been rejected. This is something that should be avoided at all cost. Using eqn 2.2 provides a fail-safe method for avoiding such pitfalls.

As a final note in this preamble, in practice one convolutes eqn 2.2 with the experimental resolution function $R(\omega)$ in order to be able to compare the model directly to the measured data $S_{\exp}(q, \omega)$ without having to resort to some deconvolution procedure:

$$S_{\text{exp}}(q,\omega) = \int_{\infty}^{\infty} d\omega' R(\omega - \omega') S_{nn}(q,\omega'). \tag{2.3}$$

We give the procedure for how to fit to the results of molecular dynamics computer simulations (in the time domain) in Section 3.3.1.

2.1 Excitations and correlation functions

Excitations in liquids are not as snazzy as their counterparts in solids. When a fluid is disturbed from equilibrium, a deviation from average in the quantity A is created at some position $\vec{r}$ in the fluid at time t. This can be a deviation from the average density, from the average temperature etc. This deviation represents a locally excited state of the liquid. The decay of this disturbance will be noticeable in other parts of the liquid at position $\vec{r}'$ and at later times t' in either the same quantity A or in some other quantity B. So, in general, we will follow the decay of the disturbance by following the correlations between $A(\vec{r},t)$ and $B(\vec{r}',t')$. When we do experiments, we will actually average over the entire fluid, and we will repeat the measurement many times. Thus, we measure an averaged correlation function $< A(\vec{r},t)B(\vec{r}',t') >_{\text{eq}}$, where the brackets $< \ldots >_{\text{eq}}$ denote the appropriate ensemble average.

In most scattering experiments we do not actually measure correlation functions as a function of position and time, rather we measure them as a function of their Fourier variables $\vec{q}$ and ω. In these scattering experiments the probing radiation transfers an amount of momentum $\hbar\vec{q}$ to the sample as well as an amount of $\hbar\omega$ in energy. If the combination of momentum and energy transferred is just right to excite the sample (by perhaps creating a sound wave in it), then we are likely to observe a scattering event. If the combination is not very special, then we are far less likely to see the radiation scattered by the sample. The ins and outs of scattering experiments are discussed in Chapter 3, in this section we give a qualitative description of the various correlation functions of interest, both in real space and in reciprocal space, both as a function of time and as a function of frequency.

The upcoming discussion of the various correlation functions brings us directly to a semantics problem. All of our definitions will be valid for the quantum-mechanical case (and we shall give the proper classical limits when needed), and therefore we must acknowledge that the quantity $A(t)$ will not necessarily commute with the same quantity at a later time. Thus, strictly speaking we cannot say that we observe a density fluctuation at some point in the fluid, and observe a change in density at some other point in the fluid at some later time since the two do not commute. However, we shall ignore this in our written text, while of course ensuring that the commutation relationships are included in the equations.

The microscopic number density $n(\vec{r},t)$ for a system of N particles is defined in real space as a function of time as (Balucani and Zoppi, 1994)

$$n(\vec{r},t) = \sum_{i}^{N} \delta(\vec{r} - \vec{r}_i(t))/\sqrt{N}, \tag{2.4}$$

with $\vec{r}_i(t)$ the position of particle i at time t. The density–density correlation function $G_{nn}(\vec{r} - \vec{r}', t - t')$ measures what effect a density disturbance at position $\vec{r}$ and time t has on the density of the fluid at position $\vec{r}'$ at a later time t' (van Hove, 1954):

$$G_{nn}(\vec{r} - \vec{r}', t - t') \equiv\, < n(\vec{r}, t) n(\vec{r}', t') >_{\text{eq}}. \tag{2.5}$$

The ensuing density disturbance $n(\vec{r}', t')$ can take on various guises. For instance, we could observe that at time t' the atoms alternately compress and rarefy, until they return back to their average equilibrium state. This of course would have happened because a sound wave that originated at $\vec{r}$ passed through the region around $\vec{r}'$ with a propagation speed of $|\vec{r} - \vec{r}'|/(t' - t)$. Alternatively, we could observe that some new atoms streamed into the region around $\vec{r}'$, pushing out some of the atoms already there. This would correspond to the density disturbance at $\vec{r}$ relaxing back to equilibrium through the collective diffusion mechanism with the diffusion constant for this process given by $D = |\vec{r} - \vec{r}'|^2/(t' - t)$. Another possibility is that some of the original atoms that were part of the density fluctuation at $\vec{r}$ made it over to $\vec{r}'$. This process is called self-diffusion and the self-diffusion constant D_{s} for this process would be given by $D_{\text{s}} = |\vec{r} - \vec{r}'|^2/(t' - t)$.

The effects of the density disturbance $n(\vec{r}, t)$ can be even more varied. For example, one could observe that the average speed of the atoms at $\vec{r}'$ at time t' increases, even though nothing noteworthy necessarily happens to the density at this point. In this case, the original density disturbance has created a temperature disturbance at $\vec{r}'$ at time t' such as the one depicted in Fig. 1.1. In order to capture this effect in a correlation function we generalize eqn 2.5 to capture all correlations between all microscopic quantities. Thus, the generalized (van Hove) correlation functions are defined (van Hove, 1954) as

$$G_{AB}(\vec{r} - \vec{r}', t - t') \equiv\, < A(\vec{r}, t) B(\vec{r}', t') >_{\text{eq}}. \tag{2.6}$$

Frequently, it is more convenient to restate these quantities in reciprocal space; not only does this capture the wave nature of the disturbances as shown in Fig. 1.1, it also allows for a more convenient comparison to experiments and computer simulations where correlations are measured in reciprocal space. This conversion to reciprocal space is achieved through a Fourier transform of the real-space quantities; the microscopic density in reciprocal space is given by $n(\vec{q}, t) = \sum_{i=1}^{N} e^{i\vec{q}\cdot\vec{r}_i(t)}/\sqrt{N}$. For future reference, in Table 2.1 we list the most commonly used microscopic quantities for which correlations functions are measured through scattering experiments and through computer simulations.[2]

The transform from the real-space correlation function $G_{AB}(\vec{r} - \vec{r}', t - t')$ to the reciprocal-space correlation function $F_{AB}(\vec{q}, t - t')$ is achieved through a Fourier transform (Balucani and Zoppi, 1994). This reciprocal-space correlation function is most commonly referred to as the intermediate scattering function when $A = B = n$.

---

[2] In this book we have chosen the variables A in such a way that $< A(t) >_{\text{eq}}=< B(t) >_{\text{eq}}= 0$, which can be done without loss of generality by redefining $A(t)$ as $A(t) = A(t)- < A(t) >_{\text{eq}}$. We do not always explicitly mention that we have subtracted the equilibrium value from the quantity of interest.

Table 2.1 The five atomic quantities $A_j^{(l)}$ that determine the five microscopic variables needed for the description of the decay of collective excitations (de Schepper *et al.*, 1988). In this table, these quantities are written out for the classical case. Given these $A_j^{(l)}$, the full microscopic wavevector-dependent variables $a_j(q)$ can be constructed through $a_j(q) = (1/\sqrt{N}) \sum_{l=1}^{N} A_j^{(l)} e^{i\vec{q}\cdot\vec{r}_l}$. In order of appearance these quantities are the microscopic density, longitudinal velocity, energy density, longitudinal momentum flux and longitudinal energy flux. The summations run over all the particles in the fluid, and ϕ denotes the interparticle potential.

identification	$A_j^{(l)}$
$j = 1$; 'n'	1
$j = 2$; 'u'	$\vec{v}_l.\vec{q}/q$
$j = 3$; 'e'	$\frac{1}{2}mv_l^2 + \frac{1}{2}\sum_{i\neq l}^{N} \phi(r_{li})$
$j = 4$; 'σ'	$(\vec{v}_l.\vec{q}/q)^2 + \frac{i}{2mq^2}\sum_{i\neq l}^{N} \vec{q}.\frac{\partial\phi(r_{li})}{\partial\vec{r}_{li}}(e^{i\vec{q}.\vec{r}_{li}} - 1)$
$j = 5$; 'q'	$[\frac{1}{2}mv_l^2 + \frac{1}{2}\sum_{i\neq l}^{N} \phi(r_{li})](\vec{v}_l.\vec{q}/q) + \frac{i}{2q}\sum_{i\neq l}^{N} \vec{v}_l.\frac{\partial\phi(r_{li})}{\partial\vec{r}_{li}}(e^{i\vec{q}.\vec{r}_{li}} - 1)$

We shall refer to all correlation functions $F_{AB}(\vec{q}, t - t')$ as intermediate scattering functions. In the purely classical case where all variables commute, these intermediate scattering functions are determined through computer simulations. Note that in general the $F_{AB}(\vec{q}, t - t')$ are complex functions, so standard computer simulations (see Section 3.3) can only approximate reality. For systems where the dynamics are very slow, the intermediate scattering functions can be obtained directly from light-scattering experiments. Examples of the latter are given in Chapter 5.

The correlation function measured in most scattering experiments is the dynamic structure factor $S_{nn}(\vec{q}, \omega)$, and it is related to the intermediate scattering function $F_{nn}(q, t)$ through a transform from the time coordinate t to the radial frequency ω. The details of all the transformations are given in Appendix B.1. The dynamic structure factor is a measure of how spontaneous density fluctuations in a fluid propagate and decay. This function satisfies the detailed balance condition (Balucani and Zoppi, 1994)

$$S_{nn}(\vec{q}, -\omega) = e^{-\beta\hbar\omega} S_{nn}(\vec{q}, \omega). \tag{2.7}$$

The detailed balance condition essentially states that the number of spontaneous fluctuations depends on the temperature of the system; at $T = 0$, no spontaneous fluctuations – whose energy can be absorbed by scattering radiation – occur in the liquid. We can also generalize the dynamic structure factor to include spontaneous temperature fluctuations, etc. Here, we will also refer to these generalized functions $S_{AB}(\vec{q}, \omega)$ as dynamic structure factors. As mentioned, the density–density dynamic structure factor $S_{nn}(\vec{q}, \omega)$ is measured in scattering experiments, other dynamic structure factors such as for instance $S_{nT}(\vec{q}, \omega)$, relating to the density and temperature, can be determined by means of computer simulations.

Table 2.2 Overview of the various correlation functions, and their interconnectedness. This list is incomplete, we refer the reader to standard textbooks (e.g., Balucani and Zoppi 1994) for additional functions. $L[...]$ stands for the Laplace transform (eqn B.5), and $Ft[...]$ for the Fourier transform (both from $\vec{r} \to \vec{q}$ and from $t \to \omega$).

name	notation	definition	
van Hove correlation function	$G_{AB}(\vec{r} - \vec{r}', t - t')$	$=$	$<A(\vec{r}, t)B(\vec{r}', t')>_{\text{eq}}$
response function	$\chi"_{AB}(\vec{r} - \vec{r}', t - t')$	$=$	$<[A(\vec{r}, t), B(\vec{r}', t')]>_{\text{eq}}$
relaxation function	$C_{AB}(\vec{r} - \vec{r}', t - t')$	$=$	$<[A(\vec{r}, t)]^* B(\vec{r}', t')>$
	with		$i\beta\partial_t C_{AB}(\vec{r}, t)/2 = \chi"(\vec{r}, t)$
intermediate scattering function	$F_{AB}(q, t)$	$=$	$Ft[G_{AB}(\vec{r}, t)]$
susceptibility	$\chi_{AB}(q, z)$	$=$	$L[\chi"_{AB}(\vec{r}, t)]$
		$=$	$\frac{1}{\pi}\int_{-\infty}^{+\infty} d\omega \chi"_{AB}(q, \omega)\frac{1}{\omega - z}$
dynamic susceptibility	$\chi_{AB}(q, \omega)$	$=$	$\chi_{AB}(q, z = \omega + i0^+)$
imaginary part of	$\chi"_{AB}(q, \omega)$	$=$	$\frac{1 - e^{-\beta\hbar\omega}}{2\hbar}S_{AB}(q, \omega)$
dynamic susceptibility		$=$	$Ft[\chi"_{AB}(\vec{r}, t)]$
dynamic structure factor	$S_{AB}(q, \omega)$	$=$	$Ft[F_{AB}(q, t)]$
symmetrized dynamic structure factor	$S_{AB}^{\text{sym}}(q, \omega)$	$=$	$Ft[C_{AB}(\vec{r}, t)]$
		$=$	$\frac{1 - e^{-\beta\hbar\omega}}{\beta\hbar\omega}S_{AB}(q, \omega)$
current–current correlation function	$C_L(q, \omega)$	$=$	$\frac{\omega^2}{q^2}S_{nn}(q, \omega)$
static structure factor	$S_{AB}(q)$	$=$	$F_{AB}(q, t = 0)$
static susceptibility	$\chi_{AB}(q)$	$=$	$\chi_{AB}(q, z = 0)$
		$=$	$\int_{-\infty}^{+\infty} \frac{d\omega}{\pi} \frac{\chi"_{AB}(q, \omega)}{\omega}$

In analyzing and modelling the data from scattering experiments one resorts to two additional sets of correlation functions, namely the response functions $\chi_{AB}(\vec{r} - \vec{r}', t - t')$ and the relaxation functions $C_{AB}(\vec{r} - \vec{r}', t - t')$. We list the interconnectedness of all these correlation functions in Table 2.2. The double transform of the response function is called the dynamic susceptibility $\chi_{AB}(\vec{q}, \omega)$, whose imaginary part $\chi"_{AB}(\vec{q}, \omega)$ is a measure of the dissipation that takes place in a fluid. In Appendix B.1 we discuss how to transform one correlation function into another, and we detail their precise definitions in terms of commutators and ensemble averages. We moved those details to the appendix as they do not add much to what is easily interpretable:

the van Hove correlation functions in the classical limit. The response functions measure how a fluid responds to being disturbed by a weak probe, such as a neutron. The dynamic susceptibility measures how susceptible the fluid is to absorbing a certain amount of momentum and energy from the probing radiation. The imaginary part of the dynamic susceptibility and the dynamic structure factor are related through the fluctuation-dissipation theorem (Balucani and Zoppi, 1994):

$$\chi"_{AB}(\vec{q},\omega) = \frac{1 - e^{-\beta\hbar\omega}}{2\hbar} S_{AB}(\vec{q},\omega). \tag{2.8}$$

Finally, the double transform of the relaxation functions $C_{AB}(\vec{r} - \vec{r}', t - t')$ do not go by any particular name; here, we will refer to them as the symmetrized (in frequency) dynamic structure factors $S_{AB}^{\text{sym}}(\vec{q},\omega)$ since they are related to the dynamic structure factors as

$$S_{AB}^{\text{sym}}(\vec{q},\omega) = \frac{1 - e^{-\beta\hbar\omega}}{\beta\hbar\omega} S_{AB}(\vec{q},\omega) = \frac{2}{\beta\omega}\chi"_{AB}(\vec{q},\omega). \tag{2.9}$$

These symmetrized dynamic structure factors, which can be directly determined from the experimental $S_{AB}(\vec{q},\omega)$, will play a central role in the effective eigenmode formalism.

For the effective eigenmode formalism we will also need the static counterparts to the dynamic structure factor and the response functions. These are defined through integrals over all frequencies, and thus respond to the equal-time, or instantaneous correlation functions. The static structure factors are defined as

$$S_{AB}(\vec{q}) = F_{AB}(\vec{q}, t = t') = \int_{-\infty}^{+\infty} S_{AB}(\vec{q},\omega)\mathrm{d}\omega, \tag{2.10}$$

and the static susceptibilities by

$$\chi_{AB}(\vec{q}) = \int_{-\infty}^{+\infty} \frac{\chi"_{AB}(\vec{q},\omega)}{\omega} \frac{\mathrm{d}\omega}{\pi}. \tag{2.11}$$

These quantities represent a measure of the instantaneous structure that is present in a fluid. As such, the nomenclature 'static' is somewhat misleading since there is nothing in a fluid that is static; rather they signify the average structure, averaged over many snapshots of the atomic positions. Similarly to the above two equations, the static counterparts to $S_{AB}^{\text{sym}}(\vec{q},\omega)$ are given by

$$S_{AB}^{\text{sym}}(\vec{q}) = \int_{-\infty}^{+\infty} S_{AB}^{sym}(\vec{q},\omega)\mathrm{d}\omega. \tag{2.12}$$

We will frequently use the short-hand notation $S_{\text{sym}}(q)$ to refer to $S_{nn}^{\text{sym}}(|\vec{q}|)$, $S(q)$ to refer to $S_{nn}(|\vec{q}|)$ and $\chi(q)$ to refer to $\chi_{nn}(|\vec{q}|)$, provided that there is no room for confusion. The relationship between the static structure factor $S(q)$ and the pair correlation function $g(r)$ reads (Balucani and Zoppi, 1994):

$$S(q) - 1 = 4\pi n \int_{0}^{\infty} \mathrm{d}r r^2 [g(r) - 1]\frac{\sin(qr)}{qr}. \tag{2.13}$$

Thus, the pair correlation function contains information about the average distance between neighboring atoms, but it does not yield information on the trajectories of the particles. For that, we need the full van Hove correlation functions.

If we want to model a correlation function, we need an expression for the time evolution of a specific quantity, such as the microscopic density. Of course, the time evolution of a quantity $A(t)$ is determined by the Hamiltonian of the system, and can be expressed in terms of the Liouville operator L as

$$\frac{\partial}{\partial t}A(t) = \frac{1}{i\hbar}[A(t), H] \equiv iLA(t). \tag{2.14}$$

In here, the square brackets [...] denote the Poisson brackets for the classical case (other than a factor $1/i\hbar$), and the standard commutator in the quantum-mechanical case. The reader will also notice that our equations differ in some subtle aspects from those presented in textbooks that treat liquids as purely classical liquids (Balucani and Zoppi, 1994; Hansen and McDonald, 2006). In instances where the static structure factor $S(q)$ – corresponding to the positions of the liquid particles as seen in a snapshot – shows up in classical models, it will be replaced by the static susceptibility $\chi(q)$ in our quantum-mechanical models. Normally, this does not lead to any confusion, although the reader might find it somewhat disturbing that a quantum-mechanical description will end up yielding models for the dynamic structure factor $S_{nn}(q, \omega)$ in terms of a sum over a limited number of terms, but that it does not provide equally compact expressions for $S(q)$.

The effective eigenmode description models the response functions $\chi''_{AB}(\vec{q}, \omega)$ and the (transform of the) relaxation functions $S^{\text{sym}}_{AB}(\vec{q}, \omega)$. In the classical limit ($\beta \to 0$) the relaxation function and the dynamic structure factor are trivially connected by $S^{\text{sym}}_{AB}(\vec{q}, \omega) = S_{AB}(\vec{q}, \omega)$, but at finite temperatures we can expect some lineshape distortion to appear in $S_{AB}(\vec{q}, \omega)$. What we mean by this is that excitations that decay exponentially in time (yielding Lorentzian lineshapes[3] in $S^{\text{sym}}_{AB}(\vec{q}, \omega)$) no longer appear to do so in $S_{AB}(\vec{q}, \omega)$. At very low temperatures, this lineshape distortion can become so strong that it can easily lead to a misinterpretation of the excitations in a very cold liquid (see Chapters 3 and 8). Thus, one should always bear in mind that models deal with the relaxation of disturbances back to equilibrium, whereas what is measured are the spontaneous fluctuations that arise in liquids. The two are intimately related, but they are not the same. Second to last, when the correlation functions refer to the correlations between one particle with itself, then these functions are referred to as self-correlation functions, and we denote them by the subscript s. For instance, $F_s(\vec{q}, t)$ denotes the self-intermediate scattering function, describing the wanderings of an individual particle through the rest of the fluid. And lastly, since fluids are isotropic, we will replace $\vec{q}$ by $q = |\vec{q}|$ in the remainder of this book.

---

[3] We define a Lorentzian line shape in ω by $\text{Re}[A/(i\omega + b)]$, with *both* A and b complex numbers. This definition, which is dictated by the theory behind the eigenmode formalism, sometimes leads to confusion with the convention of a Lorentzian line being given by $A\text{Re}[b]/((\omega - \text{Im}[b])^2 + (\text{Re}(b))^2)$. For complex A, the two definitions represent very different lineshapes.

2.2 Approximate theories for the relaxation of disturbances

Before we set up the effective eigenmode formalism, we discuss some limiting cases of disturbances in a liquid, namely the limits where the wavelength of the fluctuation is either very large, or very small, representing the hydrodynamic and ideal-gas limit, respectively. We will also discuss the special case of a hard-sphere fluid as described by the Enskog theory. This section is intended to give the reader some background that will be useful in the interpretation of the effective eigenmode formalism; however, this section can be skipped.

2.2.1 The hydrodynamic limit

In the hydrodynamic limit the wavelength of the disturbance is so large that a great many particles are involved. Many collisions between the particles will take place before the disturbance is evened out, and as a result we can treat the fluid as a continuum; the details of the particle spacings and interactions are not important. The hydrodynamic region has been dealt with extensively in many textbooks (Balucani and Zoppi, 1994; Hansen and McDonald, 2006), so here we just wave our hands and in addition, we will treat the excitations in the classical limit.[4] In addition, we do not discuss transverse correlation functions, solely the longitudinal ones as observable in scattering experiments.

Imagine increasing the density locally at some point in the liquid with the result that a deviation $n(\vec{r}, t)$ from the average equilibrium density n_{eq} has been created. For this excess density to disappear, it must be that more particles are leaving the region than are entering it, so we must have that $\partial/\partial t\, n(\vec{r}, t) \sim -\vec{\nabla}.\vec{u}(\vec{r}, t)$. Here, $\vec{u}(\vec{r}, t)$ is the velocity of the fluid near $\vec{r}$ at time t. This is, of course, the continuity equation (Balucani and Zoppi, 1994). Furthermore, the velocity itself must have arisen in the first place because of a gradient in the density; after all, particles move from places of high density to places of low density. Thus, $\vec{u}(\vec{r}, t) \sim -\vec{\nabla} n(\vec{r}, t)$. After eliminating the velocity, we get $\partial/\partial t\, n(\vec{r}, t) \sim \nabla^2 n(\vec{r}, t) \equiv D_s \nabla^2 n(\vec{r}, t)$. This is Fick's law for diffusion (Hansen and McDonald, 2006). This equation can be solved directly, or it can be solved by Fourier or Laplace transforming. It is easily verified that the solution is $G_s(r, t) = 1/\sqrt{4\pi D_s t}\, e^{-r^2/4D_s t}$, $F_s(q, t) = e^{-D_s q^2 t} \equiv e^{-t/\tau}$, and the associated dynamic self-scattering function is obtained through a Laplace transform to read:

$$S_s(q, \omega) = \frac{1}{\pi}\frac{D_s q^2}{\omega^2 + (D_s q^2)^2}. \tag{2.15}$$

This equation shows us that what we would call the relaxation rate Γ is given by $\Gamma = D_s q^2$, or equivalently, a characteristic decay time τ of $\tau = 1/D_s q^2$. This $\sim q^2$ dependence is characteristic of all damping processes in hydrodynamics, whether it is the damping of sound waves or the thermal diffusion associated with the decay of temperature disturbances.

---

[4] This simplification is justified since the hydrodynamic damping rates are so small that the excitations show up as very sharp features in $S_{nn}(q, \omega)$; as a result, lineshape distortion is unnoticeable.

Table 2.3 Results for the hydrodynamical modes that are visible in scattering experiments. The propagation speed, damping rates and amplitudes of the modes are determined by the two thermodynamic quantities γ and χ_T, with γ the ratio of the specific heats $\gamma = c_p/c_V$ and χ_T the isothermal compressibility, as well as by the transport coefficients a and ϕ through $\Gamma = (\gamma - 1)a/2 + \phi/2$. The thermal diffusivity a is related to the thermal conductivity κ by $a = \kappa/mnc_p$ and ϕ is the longitudinal kinematic viscosity (Hansen and McDonald, 2006). $C_T(q, \omega)$ is the transverse current–current correlation function for which an expression can be found in Hansen and McDonald (2006).

mode	propagation speed	damping rate	amplitude	structure factor
self-diffusion	diffusive	$D_s q^2$	1	$S_s(q, \omega)$
heat mode	diffusive	aq^2	$(\gamma-1)/\gamma$	$S_{nn}(q, \omega)$
sound mode	$\pm cq = \pm q\sqrt{\gamma/m\chi_T}$	Γq^2	$1/(2\gamma)$	$S_{nn}(q, \omega)$
shear mode	diffusive	$(\eta/nm)q^2$	1	$C_T(q, \omega)$

The excess energy generated by increasing the density locally will diffuse away in a similar manner, with the relaxation rate given by aq^2, in which a is the coefficient of thermal diffusivity. The excess density can also propagate itself though the fluid as a pair of damped sound waves. Note that we need a pair of sound waves propagating in opposite directions in order to conserve momentum. As can be anticipated, the speed of propagation of these sound waves c depends on the compressibility of the fluid. We summarize these results in Table 2.3. Note that the ratio of the specific heats γ plays a major role in hydrodynamics; it determines the relative amplitudes of the modes, and it determines to a large extent the propagation speed and damping rate of the sound modes.

2.2.2   Mode-coupling theory

Mode-coupling theory extends the validity of the results obtained in the hydrodynamic region to slightly larger values of q, or equivalently, to slightly shorter length scales (Bedeaux and Mazur, 1974; de Schepper *et al.*, 1974). In liquids at intermediate densities it takes into account that correlations in liquids have a way of persisting due to large-scale backflow. When a particle (or a group of them) is moving through a liquid it will push other particles in the liquid out of the way, while more particles will stream in to fill the void it leaves behind. This collective motion, involving pushing particles out of the way that will in turn push others out of the way, leads to a vortex type of flow pattern. This type of backflow is actually very common, and it can even be seen by the naked eye when a smoke ring moves through the air. In liquids at very high densities a backflow pattern will be unable to establish itself, but we can still find persisting memory effects resulting from correlated collision sequences. The backflow patterns are referred to as current–current modes in mode-coupling theory, while the correlated collision sequences are denoted by density–density modes

(Hansen and McDonald, 2006). Our discussion focuses on the current–current backflow patterns since they are most easy to visualize. We refer the reader to the original work by Sjögren and Sjölander (1979) and to the subsequent literature for details on the density–density modes and their effects on the dynamics.

The net effect of this backflow pattern is that it gives the moving particle a push from behind. This ensures that it will take longer for a moving particle to slow down; it will lead to what is called long-time tails in the correlation functions. Since correlations will persist for longer times τ, we will observe such tails as a decrease in characteristic widths $\omega_H = 1/\tau$ of the scattering spectra. We can also expect an increase in the speed of propagation c_s of sound waves since it will be (slightly) easier for particles to propagate.[5] These expectations have indeed been borne out by experiments. The most studied effects, experimentally, are those that pertain to the motion of individual particles, as measured in the self-scattering functions. We give examples of the results of extensive theoretical work in determining the latter, and we discuss the comparison to experiments. Note that the backflow pattern is only expected (Pilgrim and Morkel, 2002) to emerge at intermediate densities. At low density the coupling to the shear modes would be too weak, while at very high densities it is not anticipated that a backflow pattern can establish itself since the particles do not diffuse past each other fast enough (here, the density–density modes dominate). Therefore, while the mode-coupling predictions themselves do not come with a region of validity as a function of density, we should anticipate that these predictions will deviate from experiment at low and high densities.

The motion of individual particles is determined by the coefficient of self-diffusion D_s. The backflow pattern involves a pushing out of the way of other particles, and therefore, we can expect the shear viscosity η to come into play since this is a measure of how easy it is to maintain fluid flow through a channel of non-moving 'wall' particles. We can also expect this coupling of the forward, longitudinal motion to the transverse, shear modes to depend on the density of the fluid, its temperature and on the mass on the particles. In addition, it should depend on the length scale of the flow pattern. All this is captured in the following equation for the dynamic structure factor for individual particles (Bedeaux and Mazur, 1974; de Schepper *et al.*, 1974):

$$S_s(q, \omega) = \frac{1}{\pi} \frac{D_s q^2}{\omega^2 + (D_s q^2)^2} + \frac{1}{\pi} \frac{1}{D_s q q^*} \text{Re} \left[G \left(\frac{1 + i\omega/D_s q^2}{\delta} \right) \right]. \tag{2.16}$$

Here, $q^* = 16\pi m n D_s^2/(k_B T)$ and $\delta = D_s/(D_s + \eta/nm)$, while the mode-coupling function $G(z)$ is given by

$$G(z) = \arctan \left[\frac{1}{\sqrt{z-1}} \right] - \frac{(z-2)\sqrt{z-1}}{z^2} \approx \frac{8}{3z\sqrt{z}}[1 + O(1/z)]. \tag{2.17}$$

---

[5] Mode-coupling theory predicts that the leading correction to the propagation frequency of sound waves is given by $c_s q + a q^{5/2}$ (Balucani and Zoppi, 1994).

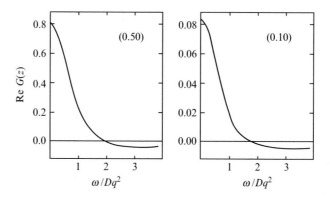

Fig. 2.1 Figure reproduced with permission from Montfrooij and de Schepper (1989). The mode-coupling correction function $G(z)$ leads to an increased value of $S_s(q, \omega = 0)$ both for a dilute gas ($\delta = 0.5$, left panel) as well as for a dense gas ($\delta = 0.1$, right panel). The negative behavior of Re $G(z)$ for $\omega > 2D_s q^2$ is responsible for the long-time tails in the velocity autocorrelation function $C_s(t)$. Note that the shape of the function is almost independent of density, but experiments have shown that the mode-coupling theory fails at very low and very high densities.

The numerical dependencies are not easy to derive (Balucani and Zoppi, 1994; Hansen and McDonald, 2006), however, it should be intuitively clear that dimensionless numbers like $D_s/(\eta/nm)$ that measure the relative strengths of the couplings should come into play. In addition, the only way to create a length scale like $1/q^*$ out of the variables D_s, $k_B T$, n and m through dimensional analysis is the combination given above (apart from a numerical factor). So the mode-coupling prediction in terms of the variables q^* and δ stands to reason.

We plot Re$[G(z)]$ in Fig. 2.1 for two values of δ, corresponding to a dilute gas (0.5) and to a dense gas (0.1) (Montfrooy *et al.*, 1986). It can be seen in this figure that the shape of Re$[G(z)]$ is very similar in both cases, with the function going negative for $\omega/Dq^2 > 2$. It is this negative part of the function that actually produces the long-time tails in the velocity autocorrelation function $C_s(t)$.[6] We can also see from its shape that the correction will lead to an increase in $S_s(q, \omega = 0)$, while at the same time it will lead to a decrease in the characteristic width $\omega_H(q) = D_s q^2$. This is in agreement with our qualitative expectations outlined earlier in this section. The explicit expressions for $\omega_H(q)$ and $S_s(q, \omega = 0)$ are (Bedeaux and Mazur, 1974; de Schepper *et al.*, 1974)

$$\omega_H(q) = Dq^2[1 - bq + O(q^{3/2})] \tag{2.18}$$

and

$$S_s(q, \omega = 0) = \frac{1}{\pi D_s q^2}[1 + aq + O(q^{3/2})]. \tag{2.19}$$

---

[6] Classically, this function is defined as $C_s(t) = < \vec{v}_i(0).\vec{v}_i(t) >_{eq}$.

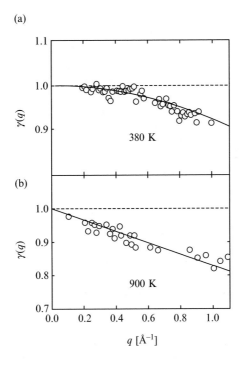

Fig. 2.2 Figure reproduced with permission from Pilgrim and Morkel (2002). The reduced half-width $\gamma(q) = \omega_{\mathrm{H}}^{\mathrm{exp}}/(D_s q^2)$ for liquid sodium at high density (top half, close to melting), and at intermediate density (bottom half). The mode-coupling prediction (solid line in bottom panel) of eqn 2.18 gives a very satisfactory description at the intermediate density, but at the high density eqn 2.18 breaks down. Instead, a quadratic departure is observed (solid line in top panel), which can be understood based on the dominance of cage diffusion at these densities (Pilgrim and Morkel, 2002).

Here, $a = G(1/\delta)/q^*$, $b = H(\delta)/q^*$ and $H(\delta) = 1.4531\delta^{3/2}[1 - 0.7276\delta - 0.1523\delta^2 - O(\delta^3)]$.

We show the agreement between theory and experiment for neutron-scattering data on liquid sodium in Fig. 2.2 (Pilgrim and Morkel, 2002). This figure demonstrates that mode-coupling theory based on current–current modes does indeed give a satisfactory prediction, but not at the highest densities. At the very high densities, the vortex-like backflow is not effective since cage diffusion dominates; it is simply too hard for a particle to escape the cage of its neighbors at solid-like densities, and a flow pattern including backflow cannot be established. Instead the density–density modes dominate the long-time behavior, leading to a quadratic correction to the characteristic decay time.

At this point we take a little digression to discuss the meaning of the $\sim q^2$ departure of the characteristic half-width at high densities as caused by the density–density modes and to make the connection to correlated collision sequences. The reason for this is that the experiments by Pilgrim and Morkel (2002) are so accurate that we can distinguish between two scenarios of how self-diffusion actually takes place at these very high densities. In the first scenario, the macroscopic coefficient for self-diffusion D_s in dense fluids is determined by repeated sequences in which a particle bounces around in its cage before it escapes after a number of collisions that take a total time τ_{escape}. In the second scenario, self-diffusion is determined by a particle bouncing around in its cage, but making its way through the rest of the liquid because the cage as a whole slowly moves around. The first scenario comes with a quantitative prediction that is not borne out by the experiments shown in the top panel of Fig. 2.2.

Let us suppose that the first scenario is valid. Then, when we observe the particle on timescales $t < \tau_{escape}$ we would overestimate how much it wandered away from its initial position based on the decay time $\tau = (D_s q^2)^{-1}$ since the particle has not gone anywhere much yet. Therefore, when we correct for this we expect a negative correction to the characteristic width of the scattering spectra reflecting that in fact it takes a little longer to cover the distance determined by the self-diffusion coefficient. A reasonable measure for this would be obtained by replacing $\tau = (D_s q^2)^{-1}$ by $\tau_{new} = (D_s q^2)^{-1} + \tau_{escape}$, yielding a characteristic width $\omega_H(q)$ of

$$\omega_H(q) = \frac{1}{\tau_{new}} = \frac{D_s q^2}{1 + D_s q^2 \tau_{escape}}. \tag{2.20}$$

In Fig. 2.2 we would then expect $\gamma(q) \approx 1 - D_s q^2 \tau_{escape}$, yielding the $\sim q^2$ dependence and identifying the level of departure from the hydrodynamic prediction with a crude measure of the characteristic time it takes a particle to escape its cage. While the functional form and the sign of the correction are reproduced by the data, the magnitude of the effect is not. The self-diffusion coefficient for liquid Na at 380 K is 0.423 Å^2/ps, and hence, the departure shown in Fig. 2.2 would correspond to an escape time of $\tau_{escape} = 0.2$ ps. This is an unrealistically short time, given that the collision time is of the order of 0.1 ps (Pilgrim and Morkel, 2002), thereby ruling out this scenario.

In the second scenario, the movement of a particle along with its immediate surroundings would imply that the most important mode-coupling effect would be the coupling to longitudinal collective modes of wavelength q_{max} (with q_{max} the position of the main peak in $S(q)$). The mode-coupling prediction based on this scenario is shown by the solid line through the data points in the top panel of Fig. 2.2. Thus, at these high densities the data shows that the cage-diffusion mechanism in liquid Na consists of the particle being locked up in a cage that slowly makes its way through the liquid.

Going back now to our discussion on current–current modes and their effects on the behavior of fluids at intermediate densities, from the mode-coupling expression for $S_s(q, \omega)$ we can determine the Fourier transform $z(\omega)$ of the velocity autocorrelation function $C_s(t)$ through (Montfrooij and de Schepper, 1989)

$$z(\omega) = \lim_{q \to 0} z(q, \omega) = \lim_{q \to 0} \left[\frac{\omega^2}{q^2} + D_s^2 q^2 \right] S_s(q, \omega). \tag{2.21}$$

Combining eqns 2.16 and 2.21 we find

$$z_{\mathrm{M}}(q, \omega) = \frac{D_s}{\pi} \left\{ 1 + \frac{q}{q^*} \left[1 + \left(\frac{\omega}{D_s q^2} \right)^2 \right] \mathrm{Re}(G \left[\frac{1 + i\omega/D_s q^2}{\delta} \right]) \right\}, \tag{2.22}$$

so that the mode-coupling prediction $z_{\mathrm{M}}(\omega)$ for the leading correction to the velocity autocorrelation function $z(\omega)$ reads

$$z_{\mathrm{M}}(\omega) = \frac{D_s}{\pi} - \frac{\sqrt{2}k_{\mathrm{B}}T}{12(D_s + \eta/nm)^{3/2}\pi^2 nm} \sqrt{|\omega|} + \dots . \tag{2.23}$$

The square root cusp in $|\omega|$ leads to the long-time tail $\sim t^{-3/2}$ in $C_s(t)$, as can be directly verified by writing out the Fourier transform and making the transition to the dimensionless variable $x = \omega t$. In Fig. 2.3 we plot $z(\omega)$ determined from a molecular dynamics computer simulation (Montfrooij and de Schepper, 1989) representing hydrogen gas at $T = 119$ K and $n = 0.0212$ Å^{-3}. It can be seen from this figure that eqn 2.23 gives a reasonable description of $z(\omega)$, especially considering the fact that there are no adjustable parameters. It is also clear from this figure that the difference between a $\sim \sqrt{|\omega|}$ and a $\sim \omega$ dependence of $z(\omega)$ for small ω is very subtle, and that it would be difficult to ascertain any particular functional form from a scattering experiment.

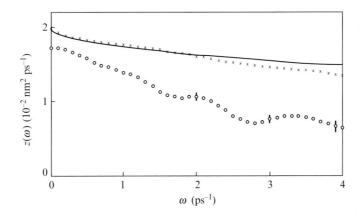

Fig. 2.3 Figure reproduced with permission from Montfrooij and de Schepper (1989). The Fourier transform $z(\omega)$ (crosses) of the velocity autocorrelation function $C_s(t)$ determined from a molecular dynamics computer simulation representative of hydrogen at $T = 119$ K and $n = 0.0212$ Å^{-3}. The solid line is the mode-coupling prediction of eqn 2.23 with $D_s = 6.22$ Å^2/ps determined from the simulation. Also shown (circles) are the results of an extrapolation $q \to 0$ based on a potential misinterpretation of the scattering data (see text and Fig. 2.4).

The velocity autocorrelation function has been determined from experiment in sodium (Morkel *et al.*, 1987) and hydrogen (Verkerk *et al.*, 1989). Since this function involves the single-particle dynamics, it can only be obtained through neutron-scattering experiments that are sensitive to single-particle dynamics through the incoherent cross-section. However, its determination is by no means straightforward, even in liquids that have a large incoherent cross-section (see Chapter 3), let alone that verification of the mode-coupling prediction is a foregone conclusion. The basic premise is clear, one needs to extrapolate the experimental results at finite wave number q down to $q \to 0$. As it turns out, the smallest q values experimentally accessible are typically of the order of $q \sim 0.2$ Å^{-1}, and this value is still so large that the $q \to 0$ limit cannot be carried out unambiguously. In fact, in order to prove the validity of the mode-coupling theory, one has to assume its validity. We clarify this in the following, and give a recipe for how to determine the velocity autocorrelation function from experiment.

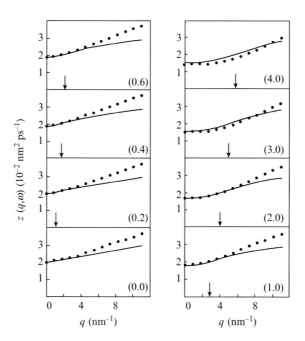

Fig. 2.4 Figure reproduced with permission from Montfrooij and de Schepper (1989). Shown (circles) are the $z(q, \omega)$ as they can be determined from neutron-scattering experiments. The data are plotted versus q at constant ω (given in ps^{-1} in parentheses in the figure). The $z(q, \omega)$ shown here are actually determined from the computer simulation described in the caption of Fig. 2.3. The solid line is the prediction of the mode-coupling theory (see eqn 2.22). The arrows are given by $\omega = 2D_s q^2$, and they mark the upper limit of the range where the long-time tails of $C(t)$ make their presence felt in scattering experiments. Note that $z(q, \omega)$ behaves linearly in q for most of the neutron-scattering range $q > 0.2$ Å^{-1}, and that a linear extrapolation from this range down to $q = 0$ would yield the wrong limit $z(\omega)$ (as shown in Fig. 2.3).

The neutron-scattering results for $S_s(q, \omega)$ can be transformed into $z(q, \omega)$ without any problem. However, it is in taking the $q \to 0$ limit that the limitations of the kinematic range accessible to neutrons are playing a role. As mentioned, the long-time tail, leading to the cusp $\sim \sqrt{|\omega|}$ in $z(\omega)$, can only be accessed for $\omega > 2D_s q^2$ (see Fig. 2.1). We plot the relevant range in Fig. 2.4. As is clear form this plot, the relevant q range for a given value of ω is mostly outside of the range available to neutrons: $q > 0.2$ Å^{-1}, and $z(\omega)$ can only be determined through extrapolation. From the shape of the curves shown in Fig. 2.4 in conjunction with the mode-coupling prediction, it is clear that while the curves show an almost linear behavior in the neutron-scattering range $q > 0.2$ Å^{-1}, a linear extrapolation would yield the wrong result. For comparison, we show the results of such a linear extrapolation in Fig. 2.3; obviously a linear extrapolation fails to reproduce the details of $z(\omega)$. However, a quadratic extrapolation in q would work very well, albeit that such an extrapolation requires the assumption a priori that the mode-coupling theory is correct, the very premise that one wants to investigate. We will encounter more $q \to 0$ extrapolation issues in other sections in this book. Lastly, this brief discussion on mode-coupling effects has centered on the self-correlation functions; for the predictions regarding the full correlation functions describing the collective dynamics we refer the reader to the literature (Balucani and Zoppi, 1994; Hansen and McDonald, 2006).

2.2.3 The ideal-gas limit and the approach thereof

The ideal gas consists of particles that do not interact with each other through an interaction potential, and therefore the dynamical behavior is given by the free movement of the atoms. Given a distribution of velocities of the gas particles, the dynamic structure factor $S_{nn}(q, \omega)$ can be calculated exactly for the ideal gas, both in the classical and in the quantum-mechanical limit (Balucani and Zoppi, 1994). The idea is that this ideal-gas limit will be reached in liquids at very large momentum transfers q. Interestingly enough, experiments have shown (Schimmel *et al.*, 2002) that the ideal gas limit is only approached, but that it cannot be reached. In this section we give the results for the ideal-gas limit, as well as how it is approached in some cases.

It is intuitively clear how the ideal-gas limit should be approached. For the sake of explanation, it is easiest to use the intermediate scattering function for density–density correlations $F_{nn}^{cl}(q, t)$. We use the superscript 'cl' for classical to denote that we are taking the easy way out in making a qualitative argument:

$$F_{nn}^{cl}(q, t) = \frac{1}{N} < \sum_i e^{i\vec{q}\cdot\vec{r}_i(0)} \sum_j e^{-i\vec{q}\cdot\vec{r}_j(t)} >_{eq} = \frac{1}{N} < \sum_{i,j} e^{i\vec{q}\cdot(\vec{r}_i(0)-\vec{r}_j(t))} >_{eq} . \quad (2.24)$$

When q is very large, the only terms that will contribute to the double summation in eqn 2.24 or those for which $|\vec{r}_j(t) - \vec{r}_i(0)|$ is small; large values would lead to random phases that would cancel each other in the summation of eqn 2.24. Small values are possible provided that $i = j$, in which case F_{nn}^{cl} reduces to the self-scattering function $F_s(q, t)$ describing the behavior of individual particles. In addition, we will only see contributions for short times t, since a particle will have wandered away too far from its initial position for large t and $|\vec{r}_i(t) - \vec{r}_i(0)| = v_i t$ would be too large to yield a significant contribution. Thus, at large q we are (mainly) seeing the movement of individual particles over very short times. Since for short enough times we can assume that the particles trajectory is determined mainly by free-streaming (that is, the translational transfer of momentum occurs mostly through free flight and not through collisions), the behavior in this limit should be that of an ideal gas of non-interacting particles. Given a particular distribution of particle speeds, we can then calculate $F_s(q,t)$ and its Fourier transform $S_s(q,\omega)$. For the case of a monoatomic, classical system we have a purely Maxwellian velocity distribution and we find

$$S_s(q,\omega) = C e^{-m\omega^2/(2k_B T q^2)}, \tag{2.25}$$

with the normalization constant C determined by

$$\int_{-\infty}^{+\infty} S_s(q,\omega)\mathrm{d}\omega = 1. \tag{2.26}$$

When incorporating quantum mechanics, we also have the f-sum rule (see Appendix B)

$$\int_{-\infty}^{+\infty} \omega S_{nn}(q,\omega)\mathrm{d}\omega = \omega_{\mathrm{recoil}}(q) = \hbar q^2/2m. \tag{2.27}$$

This will lead to a shift in the maximum of $S_{nn}(q,\omega)$ compared to $S_s(q,\omega)$, so that now the function will be centered at $\omega_{\mathrm{recoil}}(q)$. The spectral width of $S_{nn}(q,\omega)$ is determined by the average kinetic energy, which does not necessarily have to be $3k_B T/2$ as in the monoatomic classical case, but it can also include the zero-point energy of the atoms. Thus, in general, the spectral shape at very high momentum transfers reflects the velocity-distribution function of the individual atoms in the gas. Moreover, in the case of an ideal Bose-condensed gas, the spectral shape should reflect the fact that the velocity distribution now includes a macroscopic number of particles in the zero momentum state. In practice, many neutron-scattering experiments have been performed on superfluid ^{4}He at very high momentum transfers (Sears *et al.*, 1982; Sokol, 1995) in order to measure the velocity-distribution function with the aim of determining what fraction of the atoms have Bose-condensed in the superfluid phase.

So how fast will the ideal-gas limit be reached, how high does one have to go in q to find that the spectral weight is centered around $\omega_{\mathrm{recoil}}(q)$? The answer is somewhat surprising (Schimmel *et al.*, 2002): if we do a quantum-mechanical treatment for a liquid at low temperature, we find that the peak position $\omega_{\max}(q)$ of $S_{nn}(q \to \infty, \omega)$ is always located below $\omega_{\mathrm{recoil}}(q)$, and that the shape of the spectral distribution is asymmetric around $\omega_{\max}(q)$. In fact, not only is the ideal-gas limit not reached, the difference between $\omega_{\mathrm{recoil}}(q)$ and $\omega_{\max}(q)$ approaches a constant value as $q \to \infty$. We show this behavior in Fig. 2.5 where we plot the calculated frequency shifts $\Delta\omega_M(q) = \omega_{\max}(q) - \omega_{\mathrm{recoil}}(q)$ for an atom that is subject to a harmonic potential.

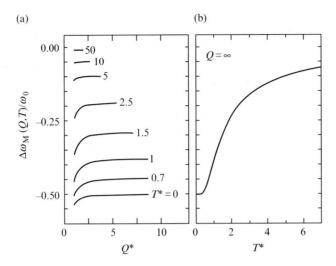

Fig. 2.5 Figure reproduced with permission from Schimmel *et al.* (2002). (a) Calculated frequency shifts $\Delta\omega_M(q) = \omega_{max}(q) - \omega_{recoil}(q)$ as a function of $Q^* = \sqrt{\omega_{recoil}(q)/\omega_0}$ for the harmonic-oscillator model. Here, $3\hbar\omega_0/2$ is the zero-point energy, and the results are shown for various reduced temperatures $T^* = 2k_BT/(\hbar\omega_0)$ given in the figure. (b) Same as in (a), but now plotted as a function of temperature in the ideal-gas limit $q = \infty$. Note that for the quantum case $T^* \to 0$ the frequency shift approaches a constant value $\Delta\omega_M(q) = -\omega_0/2$.

It can be seen in this figure that the frequency shift, compared to the expected ideal-gas limit, indeed approaches a constant value. This is a pure quantum effect and it is not restricted to a harmonic potential; it is valid for any potential that locks a particle up in a cage formed by its neighbors. Whenever a particle is confined, its energy levels will be quantized. In the case of the harmonic oscillator shown in Fig. 2.5, the spacing in energy between the levels is $\hbar\omega_0$. When energy is transferred to this atom in a scattering experiment, it can only absorb $n\hbar\omega_0$ (with n an integer), rather than the full amount E that would be expected in the ideal-gas limit in the absence of a confining potential. The average mismatch between the ideal gas result and the actual amount of absorbed energy would correspond to $1/2$ the spacing between levels, or $\hbar\omega_0/2$ for the case of a harmonic potential. In general, the frequency shift for any confining potential should be equal to half the spacing between energy levels. In principle, the spacing between energy levels in any type of system could be determined in this way from a scattering experiment by measuring the response at sufficiently high q. In Fig. 2.6 we show the frequency shift for ^{4}He, and it can be seen from this figure that this shift does indeed approach a constant value.

In the absence of quantum effects, we can calculate the approach to ideal-gas behavior in some detail, in particular for a fluid consisting of hard spheres. The following will be valid for a hard-sphere fluid, for realistic fluids we refer the reader to Moraldi *et al.* (1992) and Miyazaki and de Schepper (2001). For times not infinitely short, it is possible that some collisions take place in a fluid. A measure of how

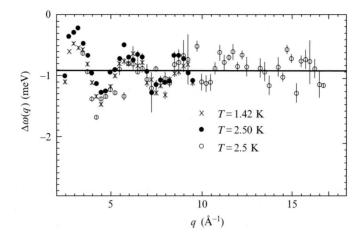

Fig. 2.6 Figure adapted with permission from Andersen *et al.* (1997) and from Schimmel *et al.* (2002). The measured difference $\Delta\omega(q)$ between the peak position $\omega_{\max}(q)$ of the dynamic structure factor of ^{4}He and the recoil frequency $\omega_{\mathrm{recoil}}(q)$, as determined by neutron scattering as a function of q. The solid line is the predicted shift $\Delta\hbar\omega(q) = -\hbar\omega_0/2 = -0.92$ meV, with the value for ω_0 inferred from the measured width $\omega_{\mathrm{H}}(q)$ (in energy) of $S_{nn}(q,\omega)$. This inference was accomplished by setting $\omega_{\mathrm{H}}(q)$ equal to the kinetic energy E_{kin} of a particle locked up in a harmonic potential well: $E_{\mathrm{kin}} = 3\hbar\omega_0/4$. The oscillations in peak position are caused by quantum diffraction effects, and they are referred to as Glory oscillations (see Chapter 10).

important collisions are compared to the probing wavelength λ is provided by the mean-free path between collisions l_{free}, and leading corrections can be calculated in terms of the dimensionless parameter $1/(ql_{\mathrm{free}})$. We can expect that the leading correction will produce a reduction of the characteristic width $\omega_{\mathrm{H}}(q)$ compared to the ideal-gas result $\omega_{\mathrm{H}}(q) = (2k_{\mathrm{B}}T\ln 2/m)^{1/2}q$. In an ideal gas we only have translational transfer of momentum (free-streaming); when we include the instantaneous transfer of momentum that occurs during a collision, we effectively get an increased rate of momentum transfer. To compare this to an ideal gas, we would have to wait a little longer in the latter case for momentum to have been transferred across a similar distance; it is as if the characteristic time τ has become longer, and therefore, the characteristic width $\omega_{\mathrm{H}} \sim 1/\tau$ should get smaller. This is indeed what is calculated (Sears, 1973; Montfrooij *et al.*, 1986):

$$\omega_{\mathrm{H}}(q) = (2k_{\mathrm{B}}T\ln 2/m)^{1/2}q[1 - \frac{0.4486}{ql_{\mathrm{free}}} + O(q^{-2})]. \tag{2.28}$$

The correction to the dynamic structure factor $S_{nn}(q,\omega)$ can similarly be expressed as $[1+s_1(q,\omega)/(ql_{\mathrm{free}}) + \ldots]$. This correction is given explicitly in Appendix C and is plotted in Fig. 2.7. Note that this function yields a change in the shape of $S_{nn}(q,\omega)$, but unlike the quantum-mechanical case, it does not produce a shift in the position of the peak of $S_{nn}(q,\omega)$. From the correction function s_1 it is also possible to calculate the

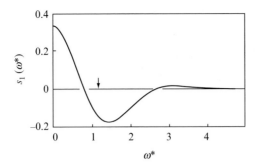

Fig. 2.7 Figure reproduced with permission from Montfrooij *et al.* (1986). The first correction term $s_1(\omega^*)$ to the dynamic structure factor of an ideal gas, calculated for a fluid made up of hard spheres. Here, ω^* is the scaled frequency $\omega^* = (\beta m)^{1/2}\omega/q$. The arrow points to where the ideal gas has its characteristic width.

increased maximum height of $S_{nn}(q, \omega = 0)$. Again, an explicit expression is given in Appendix C. In fact, the correction to ideal-gas behavior works over a remarkably large range in q space, as we illustrate in Fig. 2.8 for the case of liquid sodium (Montfrooij *et al.*, 1986). The data shown in this figure pertain to the self-part of the dynamic structure factor, so strictly speaking we are not testing the approach to ideal-gas behavior for the intermediate scattering function in this figure; rather we are testing whether the calculated expressions for hard-sphere fluids can describe a real fluid to some degree of accuracy. The figure clearly shows that this is the case and therefore, we

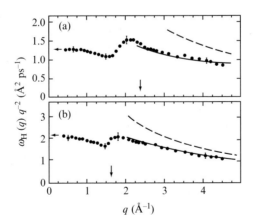

Fig. 2.8 Figure reproduced with permission from Montfrooij *et al.* (1986). The characteristic width $\omega_H(q)/q^2$ for liquid sodium (Gläser and Morkel, 1984) at 602 K (a) and 803 K (b). The ideal-gas prediction $\omega_H(q) = (2\ln 2/\beta m)^{1/2}q$ is given by the dashed line, the prediction of eqn 2.28 by the solid line. The vertical arrow points to $q l_{\text{free}} = 1$, indicative of the lower range of validity of eqn 2.28. The results for the maximum value of $S_s(q, \omega)$ give an equally good agreement (Montfrooij *et al.*, 1986).

can expect that the hard-sphere expressions will also properly describe the full dynamic structure factor, provided we are in the limit where this function is dominated by the motion of individual particles.

2.2.4 The Enskog theory

Given how well the prediction of the behavior of a hard-sphere fluid reproduces the measured data of a real fluid like sodium (see Fig. 2.8), it makes sense to present the predictions of the Enskog theory for hard spheres. We will do this in a qualitative manner; in Section 2.3 we show a quantitative prediction (Kamgar-Parsi *et al.*, 1987) for the full dynamics. The Enskog theory is an approximate solution for the dynamics of a fluid consisting of impenetrable hard spheres (Chapman and Cowling, 1970; van Beijeren and Ernst, 1973). The predictions of this theory are in remarkably good agreement with the measurements and computer simulations for a range of classical fluids, even fluids that are not considered to be good candidates for resembling hard-sphere fluids. In this section, we discuss some qualitative features of the Enskog theory, features that we will also encounter in the dynamics of real fluids.

The articles by Cohen (1993*a*; 1993*b*) on kinetic theory provide a very good overview of the history of the development of the Enskog theory, and here we draw heavily upon these papers. The Enskog theory started as a generalization of the Boltzmann equation (Brush, 1976) which is applicable to a dilute gas. The Boltzmann equation describes how a non-equilibrium velocity distribution $f(\vec{r}, \vec{v}, t)$ relaxes back to the equilibrium distribution:

$$\frac{\partial f}{\partial t} = -\vec{v}.\frac{\partial f}{\partial \vec{r}} + J_{\mathrm{B}}(ff). \tag{2.29}$$

The first term on the right-hand side is of course the free-streaming of the particles, and the second term describes how the velocities of two particles change upon a binary collision. All collisions are assumed to be uncorrelated, and the particles are point particles. Clearly, these assumptions are not tenable in a real fluid, but they work well for a dilute gas. Enskog generalized the Boltzmann equation (Chapman and Cowling, 1970) in order to try to incorporate the effects of higher-order collisions. In this generalization only the $J_{\mathrm{B}}(ff)$ term is modified, no additional terms specifically dealing with ternary or higher-order collisions are introduced. However, the binary collisions, though still assumed to be uncorrelated, now occur with greater frequency. Enskog also included the fact that the particles are not point particles by taking into account the difference in position between two colliding particles. This allowed for the transfer of momentum through collisions, which is a much more effective way of transferring momentum in a dense fluid. The predictions of the Enskog theory proved to be fairly successful in describing non-dilute gases, even though the predictions were getting increasingly less accurate with increased density.

Starting in the 1970s, the Enskog theory has been generalized (van Beijeren and Ernst, 1973) in order to compensate for its shortfalls in describing very dense fluids. This generalization was achieved by going beyond binary uncorrelated collisions through the inclusion of two correlated collision sequences that were shown, through computer simulations and through theoretical work, to play a major role in the time

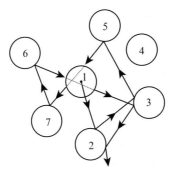

Fig. 2.9 Figure reproduced with permission from Cohen (1993*b*). The figure depicts how a particle (labelled '1') escapes the cage formed by its neighbors through a correlated collision sequence. This represents the initial part of the cage-diffusion process, the second stage of the process is a repetition of the first stage. Through this repeated escape sequence the particle wanders further and further away from its original position. For clarity, the distances between collisions (the mean-free path) have been exaggerated; in dense fluids this mean-free path is only a fraction of the diameter of the particles.

evolution of f. These two collision sequences go by the name of cage diffusion and vortex diffusion and through inclusion of these processes the range of validity of the Enskog theory was pushed to longer timescales (Cohen, 1993*b*). As a result, it led to a significant improvement in the calculated values for the transport coefficients, especially at high densities.

The cage-diffusion process is especially important in dense liquids. In order for a particle, surrounded by the 'cage' of its neighbors to diffuse to another part of the fluid, it must either escape its confinement or the cage as a whole must diffuse through the fluid. The escape will only be effectted after many, repeated collisions. This is illustrated in Fig. 2.9. Clearly, this process cannot be described as a sequence of uncorrelated, binary collisions, since repeated collisions between the same particles will occur frequently. In dense fluids, the typical time between collisions is less than 1 ps, and the length scales of the cage-diffusion process are such that the mean-free path l_{free} is much smaller than the size of the particles. Since this process determines how quickly particles can slide past each other, it should be the most important mechanism determining the speed at which the fluid back relaxes back to its equilibrium state. The associated cage-diffusion coefficient is determined by the density of the fluid since this determines the degree of order of the neighbors, by the collision frequency since this determines how frequently a particles will try to escape its cage, and by the amount of momentum transferred during collisions, since this determines how easy it is for a particle to push its neighbor out of the way. The explicit prediction for the characteristic decay time of density fluctuations based upon cage diffusion is given in eqn 4.9. We will encounter further links between the cage-diffusion mechanism and the values for microscopic decay rates in later sections of this chapter and in subsequent chapters.

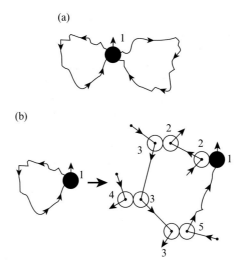

Fig. 2.10 Figure reproduced with permission from Cohen (1993*b*). This figure represents the large-scale hydrodynamic backflow generated by a moving particle (labelled '1'). Panel (a) depicts the moving particle shedding a vortex ring, which in this two-dimensional projection yields two vortices denoted by the wiggly lines. Panel (b) gives an example of a correlated collision sequence that could constitute the details of the vortex-diffusion event and that would lead to the moving particle being accelerated from behind.

The inclusion of vortex diffusion extends the validity of the Enskog theory to longer timescales. The vortex-diffusion mechanism is the same mechanism that produces the long-time tails in the mode-coupling theory at intermediate densities. The main idea is that when a particle moves forward, it pushes other particles out of the way. These particles in turn collide with others while imparting some momentum and energy. This process will repeat itself until a particle in this collision sequence hits the original forward-moving particle from behind, accelerating it. This is illustrated in Fig. 2.10. The name vortex diffusion originates from the similarity between this process, and that of a macroscopic sphere moving through a fluid, shedding vortex rings during its forward motion. Thus, what is referred to as current–current modes in mode-coupling theory is referred to as correlated collision sequences in kinetic theory. Whatever the name, we see a backflow phenomenon that takes place over large, almost hydrodynamic, length scales, with the net effect that it increases the time over which correlations exist. These long-time tails have indeed been observed in computer simulations (Alley and Alder, 1983; Balucani and Zoppi, 1994; Hansen and McDonald, 2006). Since the time and length scales of vortex diffusion are much larger than those of cage diffusion, both processes contribute almost separately to the transport coefficients.

The only parameters that enter the predictions by the Enskog theory are the diameter of the hard spheres and the density of the fluid. In order to be able to apply the Enskog theory to a real fluid, the real fluid needs to behave to some degree like a fluid of hard-sphere particles, and we need to come up with an effective hard-sphere diameter. The former appears to be the case for a surprisingly wide range of fluids, and the latter turns out to be a matter of making some educated guesses to be justified (a posteriori) by the agreement between theory and experiment.

The fact that many dense fluids behave like hard-sphere fluids is not entirely surprising since dense fluids are dominated by collisions, and collisions are dominated by the 'hard' core of the interatomic potential. For systems in which this core is steeply repulsive, we can expect that most of its dynamics will be captured by those of a hard-sphere system. Conversely, if the repulsive part of the interaction potential is not very steep, then we should see significant deviations from hard-sphere dynamics. Because of this, the noble gases and alkali metals are considered to be good 'hard-sphere' candidates, soft systems like gallium are considered to be bad candidates. As we will see in Chapter 7, the latter is not necessarily the case.

Since the temperature of the system influences the distance of closest approach of two fluid particles during a collision, the effective hard-sphere diameter for a real fluid is expected to be temperature dependent, and weakly density dependent. There are a few standard ways in practice to determine the equivalent hard-sphere diameter σ_{eq}^{hs}. One way is to calculate the static structure factor of a hard-sphere fluid, and compare this to the measured $S(q)$. The hard-sphere diameter can then be adjusted until the peak position of the main peak for the calculated and measured structure factors coincide. This ensures that the equivalent hard-sphere diameter correctly reproduces the average separation between the particles.

An alternative method is to ensure instead that the height of the main peak of $S(q)$ is correctly reproduced, even though this tends to lead to discrepancies in the calculated position of this peak. Another procedure one can apply is to take the interparticle potential (when known), and to calculate what the closest distance of approach would be, given the average amount of kinetic energy, and to equate this distance to the hard-sphere diameter. Yet another established method (Ross, 1969; van Loef, 1974) is to compare the density of a liquid at the melting point to the density at which a hard-sphere fluid melts, and deduce a hard-sphere diameter from this. This is known as Lindemann's law, which simply states that a solid melts when the amplitude of the vibrational motion of the atoms exceeds a certain fraction of the lattice parameters. For hard spheres this relationship reads $n(\sigma_{eq}^{hs})^3 = 0.93 \pm 0.02$. The fact that this relationship holds (van Loef, 1974) for inert gases, metals, diatomic molecules as well as a range of pseudo-spherical molecules (see Fig. C.1 in Appendix C) illustrates that solidification is essentially a packing problem.

All these methods – and the list is not exhaustive – will yield a slightly different hard-sphere diameter. The best approach in practice is to choose the hard-sphere diameter from this range of possibilities that produces the best agreement with experiment regarding the dynamics. Hence, in practice, one has one adjustable parameter at one's disposal in order to try to reproduce the full range of observed properties, from transport coefficients to the details of the dynamics, such as the damping of the various modes at all measured probing wavelengths. As a final note in this section, the Enskog theory methodology can also be applied to more realistic potentials as long as the Enskog assumption of instantaneous collisions is reconsidered. We encourage the reader to peruse the more recent literature on this (Leegwater, 1991; Sharma *et al.*, 1998; Miyazaki *et al.*, 2001).

2.3 Effective eigenmode formalism

We will now formally set up the effective eigenmode formalism. We will push most of the equations into Appendix B.1, and we refer the reader to this appendix for a step-by-step derivation. Here, we touch upon the salient points, and present the models that are used most in the description of liquids. The effective eigenmode description presented here is based on the projection formalism introduced by Zwanzig (1961) and further developed by Mori (1965). The quantum-mechanical aspects of the effective eigenmode description have been worked out by Forster (1975).

What the effective eigenmode formalism accomplishes in a nutshell is the following. One not only looks at the time evolution of the microscopic quantities of interest, but also at the time evolution of the time derivatives of these quantities, and one then determines whether correlations between these derivative quantities decay rapidly, or not. If they do not decay rapidly, then one takes the time derivative of the derivative quantity and repeats the procedure. In words, one acknowledges that a microscopic density fluctuation gives rise to a microscopic disturbance of the velocity, as shown in Fig. 1.1. In turn, the velocity fluctuation can give rise to a microscopic temperature disturbance (Fig. 1.1), and to a microscopic stress field.[7] Then, we scrutinize how the stress relaxes back to equilibrium by studying the evolution of the microscopic quantity given by the time derivative of the microscopic stress. If this derivative quantity (that might be called stress flux or that might remain unnamed) relaxes very rapidly, then the procedure for this part of the decay is cut off here and we simply capture the very rapid decay associated with the time derivative of the microscopic stress in a single number. That is, we would not look into the details of exactly how the generated stress decays, but rather we would just characterize the decay by one number, the decay rate z_σ. It is important to realize that the fact that we cut off the decay tree (in this instance) at the microscopic stress variable does not imply that the microscopic stress relaxes back to equilibrium rapidly; the decay rate z_σ can be a small number, or it can be a large number. The only thing it implies is that we no longer need to scrutinize whether z_σ changes on the timescales of the decay of the correlation functions that are part of our decay tree; instead, we can simply ignore any time dependence and take it as a number to be determined from experiment.

Similarly, if the microscopic temperature disturbance relaxes equally rapidly, then one also cuts off the decay tree here (as would be the case in hydrodynamics). Conversely, if microscopic temperature disturbances persist for quite a while, one goes a step further and one incorporates how the heat flux generated by the temperature disturbance will relax. In this way, one ends up with a number of couplings between microscopic variables representing that one disturbance generates another, and with a number of decay rates, representing those microscopic variables whose time derivatives decay very rapidly (de Schepper *et al.*, 1988). The power of the formalism is that

---

[7] We refer to the time derivative of the longitudinal microscopic velocity both as longitudinal momentum flux as well as microscopic stress ($j = 4$ in Table 2.1).

one will be able to describe the experimental data with a minimum number of adjustable parameters, resulting in maximum information that can be gleaned from the experiments. This holds particularly true for computer simulations (see Section 3.3) where all the couplings can be determined directly from the equal-time correlation functions, without having to resort to a fitting procedure. Thus, the time evolution of all independent correlation functions would be described by a set of static, fixed parameters and the smallest number of adjustable variables: the decay rates.

Our formalism is valid for quantum fluids as well as for classical fluids (Montfrooij *et al.*, 1997), so technically we cannot use the classical microscopic quantities that we listed in Table 2.1 since we have to replace them by their quantum-mechanical operator equivalent. For instance, the microscopic density and longitudinal velocity become the following operators, respectively:

$$\hat{n}(\vec{q},t) = \frac{1}{\sqrt{N}} \sum_{i=1}^{N} (e^{i\vec{q}.\hat{r}_i(t)} - n_{\text{eq}}),\tag{2.30}$$

with n_{eq} the equilibrium density and

$$\hat{u}(\vec{q},t) = \frac{1}{\sqrt{N}} \sum_{i=1}^{N} \left(\frac{\vec{q}.\hat{p}_i(t)}{2qm} e^{i\vec{q}.\hat{r}_i(t)} + e^{i\vec{q}.\hat{r}_i(t)} \frac{\vec{q}.\hat{p}_i(t)}{2qm} \right).\tag{2.31}$$

We could do the same for the microscopic energy density, momentum flux, and energy flux, however, this would be a pointless exercise. After all, it is $S_{nn}(q, \omega)$ that is measured in scattering experiments, and this function will be modelled using a number of decay channels. So, in the end, the precise form of these microscopic quantities does not matter since their only purpose is to provide a decay channel. The way these decay channels show up in the formalism is as a number of coupling parameters f between the various microscopic quantities, and as a limited number of decay rates z. The decay rates will be associated with those quantities whose time derivatives undergo a rapid decay. Thus, power put into these decay channels will be lost (almost) immediately; oscillations will only occur between the 'conserved' variables. We use the word 'conserved' very loosely here, it just implies that their decay takes much longer than the variables, whose decay is fast.

The set of microscopic variables listed in Table 2.1 is a somewhat inconvenient choice since we would rather work with a set where the variables are orthonormal to each other. So, we are seeking to change $\{a_1, a_2, ..., a_N\} \rightarrow \{b_1, b_2, ..., b_N\}$, with $b_i.b_j = \Delta_{ij}$, the Kronecker delta. This can easily be implemented once a scalar product has been defined (see Appendix B.1), and the result will be that all couplings will be symmetric; for instance $f_{un} = f_{nu}$. Explicit expressions for the classical b_i are given in de Schepper *et al.* (1988). For the quantum-mechanical case pertinent to the microscopic density and microscopic longitudinal velocity, the change between the two sets is minimal, only resulting in a renormalization (by $\sqrt{\chi_{nn}(q)}$ and $1/\sqrt{\beta m}$, respectively). In the remainder of this book we will refer to the five microscopic quantities b_i as derived from the five a_i listed in Table 2.1 as the density ($i = 1$, 'n'), longitudinal momentum ($i = 2$, 'u'), temperature ($i = 3$, 'T'), longitudinal momentum flux or microscopic stress ($i = 4$, 'σ') and heat flux ($i = 5$, 'q').

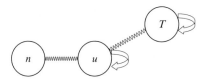

Fig. 2.11 The coupling between the microscopic variables as observed in the hydrodynamic region. This decay scheme has also been observed to work well for some liquids for shorter-wavelength disturbances than those dealt with in hydrodynamics. The density fluctuation 'n' relaxes by creating a velocity disturbance 'u'. This velocity disturbance can give rise to some stress 'σ' that relaxes so quickly that its influence on the decay scheme can be captured by one number, the damping rate z_u. This is indicated by the looping arrow at 'u'. In addition, the relaxation of 'u' gives rise to a temperature disturbance 'T'. In turn, this temperature disturbance gives rise to the transport of heat (heat flux 'q') that decays so quickly that its effect can also be captured in a damping rate z_T. Here, the springs denote the coupling constants f_{ij} between the microscopic variables, and the arrows looping back denote their decay rates. Corresponding to this decay tree, there would be 2 coupling constants (f_{un} and f_{uT}), as well as two decay rates (z_u and z_T).

Employing this orthonormal set of microscopic variables b_i, we can calculate the coupling constants between them. For instance, the coupling $f_{un}(q)$ between 'n' and 'u' is given by

$$f_{un}(q) = \frac{q}{\sqrt{S_{\text{sym}}(q)\beta m}} = \frac{q}{\sqrt{2\pi m \chi_{nn}(q)}}, \tag{2.32}$$

and the coupling between momentum and temperature $f_{uT}(q)$ is given by

$$f_{uT}(q) = f_{un}(q)\sqrt{\gamma(q) - 1}, \tag{2.33}$$

where $\gamma(q)$ is the generalized ratio of the specific heats. Expressions for $f_{u\sigma}(q)$ and $f_{Tq}(q)$ are given in Appendix B.1. There is nothing mysterious about these coupling constants, or forces. They just represent the fact that the Liouville operator L connects the variables since velocity is the derivative of position, etc. Thus, $Lb_1(q) = -if_{un}(q)b_2(q)$, telling us that a density fluctuation will only decay because more particles enter a region than are leaving it. This was already shown schematically in Fig. 1.1. Similarly, we have that $Lb_2(q) = -if_{un}(q)b_1(q) - if_{uT}(q)b_3(q) - if_{u\sigma}(q)b_4(q)$. Not only does this show the new symmetry of the variables, it also tells us that a velocity fluctuation can give rise to a temperature fluctuation 'T', or that it can set up the microscopic equivalent of stress 'σ'. Exactly how many of the variables b_i are needed depends on the wavelength of the fluctuations, and the number must be determined by comparison to experiment.

Let us look at a concrete example to make the above clear, namely the case valid for disturbances of very long wavelength as one encounters in hydrodynamics. After that, we shall generalize this case to disturbances of shorter wavelength. In hydrodynamics, we deal with only three microscopic variables: 'n', 'u' and 'T'. In Fig. 2.11 we sketch the interconnectedness of the tree of these three variables. Physically, this sketch represents the following. A density fluctuation 'n', similar to the one depicted in Fig. 1.1 but on a slightly larger length scale, can only decay by giving rise to a velocity disturbance 'u' (Fig. 1.1). Thus, 'n' and 'u' couple to each other, as indicated by the spring – representing the coupling constant f_{un} – in Fig. 2.11. The velocity disturbance relaxes back to equilibrium through two decay channels. On the one hand, a velocity disturbance involves fluid particles streaming past other particles, which will generate the equivalent of stress in a liquid. This stress does not last very long, at least not very long compared to how long it takes for the density fluctuation to decay. This very rapid relaxation of the generated stress can be described by a single number, the damping rate z_u of the velocity disturbance (depicted by a looping arrow in Fig. 2.11).

On the other hand, a velocity disturbance can also set up a temperature disturbance 'T' (see Fig. 1.1). Once there is a temperature disturbance, then we will find that heat will be transported. In other words, the temperature disturbance sets up a heat flux 'q'. Like the generated stress, the heat flux will decay very rapidly (in this example representing hydrodynamics), and its effects on the temperature disturbance can be captured by a number, the damping rate z_T. From this damping rate z_T one can calculate the transport coefficient related to thermal conductivity. Similarly, the transport coefficient ϕ (longitudinal kinematic viscosity) can be inferred from the damping rate z_u since this number tells us how quickly the generated stress dissipates. The details for these transport coefficients are given in Section 4.2.

We can also express the above in mathematics. The formal procedure for doing so is given in Appendix B.1, but we can summarize it as follows. If we were *not* to replace the time dependence of any variable by its time-integrated behavior, which is what we do by saying that some disturbances relax back to equilibrium so quickly that we can capture that by a single number, then all our equations would be exact. In this case, our efforts would just have amounted to a rewriting of the laws of motion (even though we would have managed to distinguish the slower timescales from the faster timescales). We can capture all this in a matrix $G(q, t)$ corresponding to Fig. 2.11. This matrix describes the time evolution of all possible correlation functions between all possible combinations of the three variables b_i, including the density–density correlation function. The change from exact equations to approximate equations can mathematically be captured by replacing the time dependence of the fast-decaying channels by their time-integrated values. Mathematically, it is easier to express this by first doing a Laplace transform on the time dependence of the equations of motion in order to go from $G(q, t) \rightarrow G(q, z)$, with z the Laplace variable, and $z = 0$ corresponding to performing a time integration $\int_0^\infty dt$. Thus, for the case corresponding to Fig. 2.11 we have that we will replace (de Schepper *et al.*, 1988)

$$G(q, z) = \begin{pmatrix} 0 & if_{un}(q) & 0 \\ if_{un}(q) & z_u(q, z) & if_{uT}(q) \\ 0 & if_{uT}(q) & z_T(q, z) \end{pmatrix} \tag{2.34}$$

by

$$G(q) = \begin{pmatrix} 0 & if_{un}(q) & 0 \\ if_{un}(q) & z_u(q) & if_{uT}(q) \\ 0 & if_{uT}(q) & z_T(q) \end{pmatrix}. \tag{2.35}$$

If the above replacement is justified, then we would have the situation that the transport coefficients would be determined by $z_u(q)$ and $z_T(q)$, and $z_u(q)$ would be given by $z_u(q) = \phi q^2$ and $z_T(q)$ by $z_T(q) = \gamma a q^2$ in this example pertaining to the hydrodynamics region (de Schepper *et al.*, 1988). If the above replacement would also be justified outside of the hydrodynamics region then we would most likely observe departures from the hydrodynamic q dependence of $z_u(q)$ and $z_T(q)$. Note that the procedure outlined above does not imply that the damping rates z are necessarily much larger than the forces f, it just implies that, e.g., a fluctuation of the momentum flux ('σ') that arose from a fluctuation of the microscopic longitudinal momentum ('u') decays almost as soon as it forms. Whether the damping rate z_u is small or large under these circumstances also depends on the strength of the coupling between 'u' and 'σ', in other words, how effective is setting up a stress field in making velocity disturbances decay back to equilibrium?

When we look at the liquid on shorter and shorter length scales, we find that we have to take more microscopic variables into account besides the three hydrodynamic ones ('n'), ('u') and ('T'). By and large, the shorter the wavelength of the disturbance, the more quickly it will decay. Thus, we cannot necessarily say that the rate of decay of the generated stress is much faster than that of the generated velocity disturbance. In this case, we will have to expand our decay tree, and incorporate the time dependence of the decay of the generated stress. An example of such an expanded tree is shown in Fig. 2.12 where we sketch the interconnectedness of the tree of variables ('n'), ('u'), ('T'), ('σ') and ('q'). For this particular case, the tree is cut off after five variables because (presumably) it was found that the decay rates of the disturbances that are being generated by 'σ' and 'q' were so substantial that it took only a fraction of the time that a microscopic density disturbance (or a microscopic temperature disturbance) would persist for. As before, we would replace the time dependence of these quickly decaying channels by their time-integrated values. In Section 2.3.3 we discuss in more detail (from a parameter point of view) when the transition from one set of variables to an extended set of variables is warranted.

The matrix of all possible correlations between the 5 microscopic variables discussed in the preceding paragraph is given by (de Schepper *et al.*, 1988)

$$
G(q) = \begin{pmatrix}
0 & if_{un}(q) & 0 & 0 & 0 \\
if_{un}(q) & 0 & if_{uT}(q) & if_{u\sigma}(q) & 0 \\
0 & if_{uT}(q) & 0 & 0 & if_{Tq}(q) \\
0 & if_{u\sigma}(q) & 0 & z_\sigma(q) & iz_{q\sigma}(q) \\
0 & 0 & if_{Tq}(q) & iz_{q\sigma}(q) & z_q(q)
\end{pmatrix}. \tag{2.36}
$$

Should this decay scheme (Fig. 2.12) be appropriate for a particular liquid when describing the decay of a density fluctuation of a particular wavelength, then we would have the situation that the transport coefficients would be determined by z_q, z_σ and $z_{q\sigma}$. Should this decay scheme not be appropriate, then we would have to expand the set of microscopic variables by including more time derivatives, thereby increasing the dimension of the matrix $G(q)$.

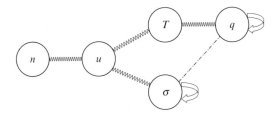

Fig. 2.12 Schematic diagram of how a density fluctuation 'n' gives rise to a velocity disturbance 'u'. The velocity disturbance relaxes back to equilibrium by causing a temperature disturbance 'T' and by setting up microscopic stress 'σ'. The microscopic stress gives rise to another microscopic disturbance that relaxes so rapidly that its influence can be captured by a decay rate (arrow). The temperature disturbance creates a microscopic heat flux 'q', which also gives rise to a microscopic disturbance that decays very quickly. Here, the springs denote the coupling constants f_{ij} between the microscopic variables, and the arrows looping back denote their decay rates. Corresponding to this decay tree, there would be 4 coupling constants (f_{un}, f_{uT}, $f_{u\sigma}$ and f_{Tq}), as well as two decay rates (z_σ and z_q). The dotted line between 'q' and 'σ' shows another allowed connection ($z_{q\sigma}$), even though in practice (de Schepper *et al.*, 1988) this connection has been found to be very weak in the case of a Lennard-Jones fluid.

So this is the long and the short of the effective eigenmode formalism. One looks at the time evolution of the microscopic quantities in successive steps, and one then determines whether correlations between those derived quantities decay rapidly, or not. If they do, then one will not include any more variables b_i, and the very rapid decay is captured in a time-integrated number, the decay rate. In this way, all the couplings between the microscopic variables are taken into account, which results in a very stringent restriction on the number of fit parameters that one has at one's disposal.[8] Conversely, it implies that one will be able to get the maximum amount of information out of an experiment, with minimal error bars on the derived quantities. The number of microscopic variables that are required depends on the wavelength of the fluctuation. Of course, the decay tree can be expanded ad absurdum, but in general no more than five microscopic variables are required, although not necessarily the ones shown in Fig. 2.12. When more variables are required, one is likely looking at fluctuations of such short wavelengths that a description based on ideal-gas behavior is more appropriate.

Since an appropriately chosen matrix $G(q)$ gives a full description of the dynamics, and since a matrix is determined by its eigenvalues and eigenfunctions, the dynamics of the liquid are fully specified by giving the eigenvalues and eigenfunctions of $G(q)$. We refer to an eigenvalue and corresponding eigenfunction as an eigenmode (or shortly, mode) of the matrix $G(q)$. An m by m matrix will have m eigenmodes. Each mode corresponds to a Lorentzian line in the correlation function of interest; the position and characteristic width of such a Lorentzian line is given by the eigenvalue of the

---

[8] The coupling constants f can always be written as static (equal-time) correlation functions (see Appendix B.2), and they are directly available in computer simulations.

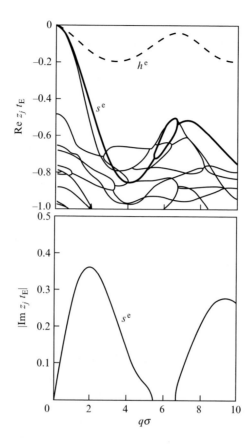

Fig. 2.13 Figure adapted with kind permission of Societa Italiana di Fisica from Cohen and de Schepper (1990). Shown are the real parts (top panel, thinner continuous lines) of the eigenvalues z_j that describe the dynamics between 55 microscopic variables as calculated using the Enskog theory for hard spheres for a fluid at reduced density $n\sigma^3 = 0.88$. The data have been plotted as a function of reduced wave number $q\sigma$ and they have been scaled using t_E, the average time between collisions for a hard-sphere fluid at this density. The thick lines (labeled h^e and s^e) are the result of describing $S_{nn}(q,\omega)$ with three effective eigenmodes. The imaginary parts (propagation frequencies) of these effective eigenmodes are shown in the bottom panel. The extended heat-mode eigenvalue (dashed curve) is indistinguishable from one of the 55 modes (not plotted since it is hidden under the dashed curve); in contrast, the extended sound modes harbor the characteristics of many modes. Note that the real parts of z_j have been assigned a minus sign in this figure, unlike our convention in this book.

mode, the amplitude (or strength) of the mode is given by the eigenfunction. The details on how to determine these amplitudes are given in the Section 2.3.1.

We illustrate the discussion above in Fig. 2.13 where we show the results for 55 modes calculated (Cohen and de Schepper, 1990) using the Enskog theory for hard spheres. The top panel of the figure depicts the real parts of the eigenvalues that correspond to the longest-living modes. Note that these lines merely depict the eigenvalues of the modes as calculated for 55 microscopic quantities; this figure does not reveal what the contributions (amplitudes) of these modes are in the dynamic structure factor, or whether they even show up at all in this correlation function. These amplitudes can also be calculated (see eqn 2.44) but they are not shown in this figure.

In practice, one observes that the amplitudes of the (55) modes in the dynamic structure factor $S_{nn}(q, \omega)$ are such that for any q value only three *combinations of modes* will have a large enough amplitude to be observable. This is a phenomenological rule, and it is one of the deepest mysteries in fluid dynamics on these length scales. These three combinations go by the name of extended heat mode and the two extended sound modes. These special combinations, whose amplitudes are such that together they faithfully reproduce the dynamic structure factor, are also shown in Fig. 2.13 as thick curves. Thus, of all the modes that represent the dynamics of 55 microscopic variables, only three combinations are relevant to understand the dynamic structure factor for density–density correlations. As can be seen in this figure, at small momentum transfers these three effective eigenmodes are easily identifiable as (extensions of) the hydrodynamic heat and sound modes, and hence the names extended heat and sound modes. With increasing q, more of the other modes are mixed in and their character changes.

Since we are talking about two distinct types of modes, we should take an inventory at this point. On the one hand, we have the dynamic matrix $G(q)$ that describes the dynamics of all possible correlation functions between all possible variables. The eigenvalues of this matrix we call the extended eigenmodes of the fluid. On the other hand, we have the experimental observation that for the density–density correlation function only three combinations of these eigenmodes actually show up. We call these special combinations the extended heat mode and extended sound modes. We do not have an explanation or prediction which eigenmodes will combine to form the extended heat and sound modes; this is something we determine from experiment and computer simulations. Note that more than one eigenmode can be involved in a combination to form the extended sound mode. Also note that one eigenmode does not correspond to the dynamics of one particular microscopic variable, an eigenmode is a reflection of the interplay between the various microscopic variables.

There are two avenues to infer the behavior of the extended heat and sound mode from experiment. The first one is simply to fit $S_{nn}^{\text{sym}}(q, \omega)$ to the sum of three Lorentzian lines:

$$S_{nn}^{\text{sym}}(q, \omega) = \frac{1}{\pi} \text{Re} \left[\frac{A_h}{i\omega + z_h} + \frac{A_{s,1}}{i\omega + z_{s,1}} + \frac{A_{s,2}}{i\omega + z_{s,2}} \right]. \tag{2.37}$$

The extended heat mode has a real amplitude A_h and a real value for its eigenvalue z_h. The two sound modes form a conjugate pair when they are propagating ($A_{s,1} = A_{s,2}^*$; $z_{s,1} = \Gamma_s + i\omega_s = z_{s,2}^*$), or they have real amplitudes and real $z_{s,1}$ and $z_{s,2}$ when they are overdamped (non-propagating). The fit only has three independent parameters since the amplitudes A are determined by z_h, $z_{s,1}$ and $z_{s,2}$ through the following three frequency-sum rules ($m = 0$, 1, 2):

$$\sum_{i=h,s1,s2} A_i(q) z_i(q)^m = R_m(q). \tag{2.38}$$

Here, $R_0(q) = S_{\text{sym}}(q)$; $R_1(q) = 0$ and $R_2(q) = -q^2/\beta m$. This avenue is the one that was historically employed in describing scattering data on dense fluids (Alley and Alder, 1983; de Schepper *et al.*, 1983; Bruin *et al.*, 1985).

It was also found in these earlier studies that adding more Lorentzian lines (with the appropriate restrictions on the amplitudes) to the description did not improve the quality of the fit. This three-modes-are-sufficient scenario has since been borne out by scattering studies on a range of liquids. Note that this description does not tell us anything about the character of the extended heat and sound modes. It merely yields their characteristic decay times and propagation frequencies as a function of wave number q. It does not tell us, for instance, through which particular set of couplings an extended sound mode decays back to equilibrium.

The second avenue open to the interpretation of the scattering data is to fit the data by employing a dynamic matrix $G(q)$ of sufficient dimension to be able to satisfactorily describe the density–density dynamic structure factor for all q values probed. In this way one (experimentally) finds that still only three combinations of eigenmodes are needed to describe the fate of density fluctuations, however, these three combinations are now expressed in terms of coupling parameters between, and damping rates of, microscopic variables. This yields information on how (through which channels) density fluctuations decay. What it does not do is give us an explanation as to why only three combinations are needed. In addition, as (will be) mentioned repeatedly in this book, knowledge of the density–density fluctuation spectra alone is in general not sufficient to determine unambiguously which microscopic variables are actually involved in the decay of density fluctuations. The latter can only be determined when we also have information on other correlation functions. Since the description of the data in terms of the dynamic matrix $G(q)$ still yields the extended heat and extended sound modes (expressed in combinations of the eigenmodes of $G(q)$), we refer to this particular avenue as the extended eigenmode formalism.

In short, the reason that the extended effective eigenmode formalism works over a large q range is that only certain combinations of the many modes (eigenvalues and eigenfunctions of the matrix $G(q)$) lead to processes that are observable in $S_{nn}(q,\omega)$. The other processes merely affect the decay rates, but they do not add to the number of Lorentzians visible in $S_{nn}(q,\omega)$. At the same time, the spaghetti shown in Fig. 2.13 clearly illustrates the problems associated with interpreting the meaning of the decay parameters that describe the time evolution of the density fluctuations. This interpretation can only be done unambiguously when one has access to more correlation functions than just the density–density correlation function.

The number of microscopic variables that one requires for a satisfactory description of the experimental data over the entire q range of interest corresponds to the dimension of the dynamic matrix $G(q)$. It does not correspond to the number of Lorentzian lines that are required to describe the measured dynamic structure factor $S_{nn}(q, \omega)$, or equivalently, to the number of exponentials that one needs for the intermediate scattering function $F_{nn}(q, t)$. This is captured in the master equation for the dynamic structure factor, and is worked out in detail in Appendix B.1 (Forster, 1975; Montfrooij *et al.*, 1997):

$$S_{nn}(q, \omega) = \frac{S_{\text{sym}}(q)}{\pi} \frac{\beta \hbar \omega}{1 - e^{-\beta \hbar \omega}} \text{Re} \left[\frac{1}{i\omega 1 + G(q)} \right]_{11}. \qquad (2.39)$$

The pre-factor $S_{\text{sym}}(q)$ is needed to go from the orthonormal set of microscopic variables to the density fluctuations measured in scattering experiments. The subscript '11' in this equation implies that one has to take the '11' element of the matrix on the right-hand side, where the numbering corresponds to Table 2.1 in which '1' denotes the microscopic density n so that '11' corresponds to the density–density correlation function. Similarly, the '13' element would correspond to the density–temperature correlation function $S_{nT}(q, \omega)$ (with $S_{\text{sym}}(q)$ replaced by $S_{nT}^{\text{sym}}(q)$), and the '22' element to the velocity–velocity correlation function $S_{uu}(q, \omega)$ (with $S_{\text{sym}}(q)$ replaced by $S_{uu}^{\text{sym}}(q)$), which is directly related to the momentum–momentum correlation function $C_{\text{L}}(q, \omega)$ (Table 2.2).

If, for example, the matrix $G(q)$ is a 5×5 matrix, then eqn 2.39 would yield five poles in the complex plane corresponding to the five eigenvalues of the matrix $G(q)$, or to five Lorentzian lines in $S_{nn}(q, \omega)$ after we take the real part. However, as mentioned above through some unknown mechanism only three of those Lorentzian lines have an observable amplitude in scattering experiments. As we will see in the following chapters, exactly which three Lorentzian lines show up, and exactly what combination of eigenmodes they represent, changes very rapidly as a function of q (or equivalently, as a function of the wavelength of the density fluctuation).

2.3.1   Details

Before we look at more simple decay schemes like the viscoelastic model and the damped harmonic oscillator, and before we make the connection to other successful descriptions of liquids, we must discuss some more generalities, such as the pre-factor in eqn 2.39 and how the differences between a purely classical and a quantum-mechanical treatment show up.

The poles of the dynamic susceptibility are given by the eigenvalues of $G(q)$, as they should be in any self-respecting formalism. This is implicit in eqn 2.39, and written out explicitly in Appendix B.1. The effective eigenmode formalism distinguishes between long-lasting correlations, and rapidly decaying ones. To do this mathematically, the rapidly decaying variables are projected out. Thus, one has to introduce some projection operator, which can only be done once a scalar product has been defined. Of course, there are many possible scalar products that one can define in the Hilbert space that is spanned by the microscopic variables a_i. The one that we have chosen (see Appendix B.1) was the one that made the eigenmode formalism the most transparent in the following sense. We wanted to ensure that the quantum to classical limit would be visible as a pre-factor in the frequency dependence of correlation functions. Also, we wanted to ensure that the frequency-sum rules (see Appendix B), such as the f-sum rule, were transparent. And we wanted to ensure that it would be straightforward to analyze the data using the effective eigenmode formalism, both in the almost classical case as well as for those liquids where quantum effects are important.

Our choice of scalar product $< \ldots >$ accomplished the following. While the intermediate scattering functions are given by $F_{AB}(q,t) =< A(t)B(0) >_{eq}$, we have that the relaxation functions $C_{AB}(q,t)$ are given by $C_{AB}(q,t) =< A(t)B(0) >$. With our choice of scalar product, the time derivative of these relaxation functions is proportional to the imaginary part of the susceptibility (see eqn B.11). The consequence of all this is that the Fourier transform $S_{AB}^{\mathrm{sym}}(q,\omega)$ of the relaxation function is given by (expressed in terms of matrices as opposed to matrix elements)

$$S^{\mathrm{sym}}(q,\omega) = \frac{1 - \mathrm{e}^{-\beta\hbar\omega}}{\beta\hbar\omega}S(q,\omega) = \frac{2}{\beta}\mathrm{Re}\left[\frac{\chi(q)}{i\omega\mathbf{1} + G(q)}\right] = \frac{1}{\pi}\mathrm{Re}\left[\frac{S_{\mathrm{sym}}(q)}{i\omega\mathbf{1} + G(q)}\right]. \quad (2.40)$$

The label 'sym' indicates that the $S_{AB}^{\mathrm{sym}}(q,\omega)$ are even functions in ω. The eigenvalues of $G(q)$ that determine the poles of the dynamic susceptibility will be prominent features in $S^{\mathrm{sym}}(q,\omega)$; this is why we have chosen $S_{nn}^{\mathrm{sym}}(q,\omega)$ over $S_{nn}(q,\omega)$ in our effective eigenmode description. Also, this function can be determined directly from experiment by a simple multiplication of $S_{nn}(q,\omega)$. Moreover, in the classical limit $\beta \to 0$, $S_{nn}(q,\omega)$ and $S_{nn}^{\mathrm{sym}}(q,\omega)$ coincide, while the frequency-sum rules of $S_{nn}(q,\omega)$ and $S_{nn}^{\mathrm{sym}}(q,\omega)$ are also directly related (see Appendix B):

$$\int_{-\infty}^{\infty} \omega^{2k} S_{nn}^{\mathrm{sym}}(q,\omega)\mathrm{d}\omega = \frac{2}{\hbar\beta}\int_{-\infty}^{\infty}\omega^{2k-1}S_{nn}(q,\omega)\mathrm{d}\omega. \quad (2.41)$$

Thus, the first frequency moment of $S_{nn}(q,\omega)$, which is exactly equal to $\hbar q^2/(2m)$ yields the second frequency moment for $S_{nn}^{\mathrm{sym}}(q,\omega)$. Also, note that the zeroth frequency moment of $S_{nn}(q,\omega)$, which of course is the static structure factor $S(q)$, plays no role. The role of $S(q)$ has now been taken over by the static susceptibility $\chi_{nn}(q)$, which is essentially the (-1)st frequency moment of $S_{nn}(q,\omega)$.

The m eigenvalues of the $m \times m$ matrix $G(q)$ yield the time evolution of any disturbance. The m eigenvalues and eigenvectors are formally defined by

$$G(q)\psi^{(i)}(q) = z_i(q)\psi^{(i)}(q), \quad (2.42)$$

where $z_i(q)$ is an eigenvalue corresponding to the eigenvector $\psi^{(i)}(q)$ $(i = 1, \ldots, m)$, with components $\psi_l^{(i)}(q)$ $(l = 1, \ldots, m)$. The eigenvectors are orthonormal to each other:

$$\sum_{k=1}^{m}\psi_k^{(i)}\psi_k^{(j)} = \delta_{ij}. \quad (2.43)$$

Note that the orthonormalization condition does not involve taking the complex conjugate of the eigenvectors.

The dynamic structure factor as measured in experiments can then be expressed in terms of a sum of Lorentzians:

$$S_{nn}^{\text{sym}}(q,\omega) = \frac{1 - e^{-\beta\hbar\omega}}{\beta\hbar\omega} S_{nn}(q,\omega) = \frac{S_{\text{sym}}(q)}{\pi} \text{Re} \sum_{i=1}^{m} \frac{[\psi_1^{(i)}(q)]^2}{i\omega + z_i(q)}. \tag{2.44}$$

Thus, the amplitudes of the m effective eigenmodes that make up the dynamic structure factor are based on $[\psi_1^{(i)}(q)]^2$. These quadratic quantities are complex for propagating modes, and real for diffusive modes. Complex values lead to asymmetric Lorentzian lines, an explicit example of which is written out in Section 3.2.4 for the damped harmonic oscillator case. The amplitudes are fully determined by the matrix elements of $G(q)$, just as these matrix elements determine the eigenvalues. Therefore, the amplitudes of the Lorentzian lines corresponding to the individual modes in $S_{nn}(q,\omega)$ should not be adjusted as free parameters during a fitting procedure in which the eigenvalues are already allowed to vary freely.

Expressions similar to eqn 2.44 hold for the other correlation functions of interest. For instance, the density–temperature correlation function would be given by:

$$S_{nT}^{\text{sym}}(q,\omega) = \frac{1 - e^{-\beta\hbar\omega}}{\beta\hbar\omega} S_{nT}(q,\omega) = \frac{2\pi\chi_{nT}(q)}{\beta\pi} \text{Re} \sum_{i=1}^{m} \frac{\psi_1^{(i)}(q)\psi_3^{(i)}(q)}{i\omega + z_i(q)}. \tag{2.45}$$

Note that, in general, the amplitudes of the Lorentzians in eqn 2.44 as determined by the components of the eigenvalues are complex, there is no requirement that the amplitudes should be real. In fact, there is no requirement that the amplitudes squared of the individual modes should be positive. The only sum rule that controls the amplitudes is the sum over all of them. As an example of what these eigenvectors look like, in hydrodynamics the three eigenvectors corresponding to the heat and the sound modes are

$$\psi^{\text{heat}}(q) = (\sqrt{(\gamma-1)/\gamma}, 0, -\sqrt{1/\gamma}) \tag{2.46}$$

for the heat mode, and

$$\psi^{\text{sound}}(q) = \frac{1}{\sqrt{2}}(\sqrt{1/\gamma}, \pm 1, \sqrt{(\gamma-1)/\gamma}) \tag{2.47}$$

for the two sound modes. It is easy to verify that these eigenvectors produce the amplitudes given in Table 2.3, and that the eigenvectors form an orthonormal set. The sound-mode eigenvectors technically also have an imaginary part to them, however, this part is proportional to the ratio of damping rate and propagation frequency. In the hydrodynamics region, this ratio is negligible.

2.3.2   Commonly used models for $G(q)$

Next, we discuss the viscoelastic model, as well as the damped harmonic oscillator. Other than the extended hydrodynamics model (Fig. 2.12) discussed thus far, these two models are the ones most frequently employed in the description of liquids. The

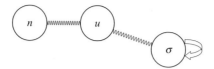

Fig. 2.14 The decay-tree representative of the simple viscoelastic model. This decay scheme is valid for the cases where a microscopic velocity disturbance 'u' gives rise to a microscopic stress field 'σ', while there is barely any coupling between the microscopic longitudinal velocity and the microscopic temperature 'T'. This would be the case if the generalized specific heat ratio $\gamma(q) \approx 1$ since $f_{uT} \sim \sqrt{\gamma(q) - 1}$.

simple viscoelastic model is sketched in Fig. 2.14 and its time-evolution matrix $G(q)$ is given by (Lovesey, 1971; Balucani and Zoppi, 1994)

$$G(q) = \begin{pmatrix} 0 & if_{un}(q) & 0 \\ if_{un}(q) & 0 & if_{u\sigma}(q) \\ 0 & if_{u\sigma}(q) & z_\sigma(q) \end{pmatrix}. \tag{2.48}$$

Balucani and Zoppi (1994) give a very thorough discussion of when the viscoelastic model is expected to be valid, to which we have very little to add. The viscoelastic model clearly cannot be used to describe the dynamics over the entire wavelength range of fluctuations, as it does not take into account the coupling between velocity 'u' and temperature 'T' and therefore would be unable to reproduce the adiabatic speed of sound.[9] On the other hand, the viscoelastic model can be expected to yield a good description of $S_{nn}(q,\omega)$ when the atoms cannot follow the oscillations of the probe; in this case, their response to the probe is an instantaneous response, largely determined by the interparticle forces. This situation is expected to be encountered when probing the liquid on the length scale of the interparticle separation, that is, for wave number corresponding to the main peak in $S(q)$. On these length scales, the dynamics are expected to be dominated by binary collisions, and the long-time tails are not expected to play a vital role in the decay of such density fluctuations. However, the applicability of this basic viscoelastic model is not necessarily limited to this q region. As an example, we mention the case of a dense helium gas at 39 K, for which it was found (Montfrooij *et al.*, 1991) that the dynamics over a large q range ($0.3 < q < 2.4$ Å^{-1}) could be satisfactorily described by this model that only requires one damping rate.

We give the equivalent expressions for the viscoelastic model in terms of the memory function (see Section 2.4) and the continued fraction expressions in Appendix E. The two forces that determine the response to the external probe are f_{un} and $f_{u\sigma}$ (Fig. 2.14). Both these forces are given by the ratio of frequency moments when $\gamma(q) = 1$ (which implies that $f_{uT} = 0$), and these ratios reflect the local structure present in

[9] The extension(s) of the viscoelastic model to be discussed later on in this section and in Section 2.4 do incorporate the hydrodynamic limit.

the liquid. In particular, in the absence of thermal fluctuations arising from velocity disturbances $f_{u\sigma}$ is given by (see eqns B.32 and B.33 in Appendix B.2)

$$f_{u\sigma}(q) = q\sqrt{\frac{M^{(3)}}{M^{(1)}} - \frac{M^{(1)}}{M^{(-1)}}}, \tag{2.49}$$

with $M^{(n)}$ the nth reduced frequency moment of $S_{nn}(q,\omega)$ defined by

$$M^{(n)} = \int_{-\infty}^{+\infty} S_{nn}(q,\omega)\frac{\omega^n}{q^n}\mathrm{d}\omega. \tag{2.50}$$

Note that as before only the odd frequency moments of $S_{nn}(q,\omega)$ play a role. The conditions on f_{un}, $f_{u\sigma}$ and z_σ that lead to propagating modes, or even to discernible side peaks in $S_{nn}(q,\omega)$ have been discussed extensively by Balucani and Zoppi (1994), and we refer the reader to their book and to Appendix D. Finally, for the reader who might be more familiar with the viscoelastic model using the memory function formalism (see Section 2.4 and Appendix E), we mention that the decay time τ_q encountered in the latter formalism is given by $\tau_q = 1/z_\sigma(q)$ and that the parameter Δ_q is given by $\Delta_q = f_{u\sigma}(q)^2$.

The viscoelastic model has also been extended so that it does include the proper hydrodynamic limit. We sketch this model in Fig. 2.15, and we give its equivalent memory function expression in Appendix E. With the inclusion of the coupling parameter $f_{uT}(q)$ the model will now be able to properly reproduce the hydrodynamic limit. The fact that the decay tree is still cut off at the hydrodynamic level at the microscopic variable 'T' implies that we expect when using this model that the decay of the heat flux will be much faster than the decay of the longitudinal momentum flux. In practice though, this is not the real reason why the decay tree is cut off at the microscopic variable 'T'. We do not have obvious arguments that tell us that the heat flux should decay faster than the longitudinal momentum flux, however, the systems for which the extended viscoelastic model is being used tend to have a very weak coupling between 'u' and 'T' so that the details of the decay of the microscopic temperature disturbances cannot be inferred from experiment.

The damped harmonic oscillator model, sketched in Fig. 2.16, is very restricted in its applications. Not only does it require the coupling between 'u' and 'T' to be

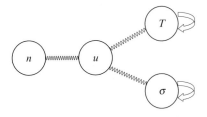

Fig. 2.15 The decay-tree representation of the extended viscoelastic model. This is a straightforward modification of Fig. 2.14, and it can succesfully account for those situations where $\gamma(q) \neq 1$, thereby ensuring that the hydrodynamic limit is properly reproduced.

Fig. 2.16 The decay tree for the damped harmonic oscillator. This decay scheme has been found to work well in the description of quantum fluids. Here, a density fluctuation gives rise to a velocity disturbance. The velocity disturbance sets up a microscopic stress that decays so rapidly that its effects can be captured in the damping rate z_u (indicated by the arrow) of the velocity disturbance. Note that z_u is not necessarily a large number compared to the coupling constant f_{un}. For this decay tree to be valid we must have that the (generalized) specific heat ratio $\gamma(q) \approx 1$.

negligible (that is, $\gamma(q) \approx 1$), for the model to be valid we must also have that the stress fluctuations generated by the velocity disturbance relax back to equilibrium very rapidly. The damped harmonic oscillator can thus be viewed as a special case of the (simple) viscoelastic model (see also Appendix E). Stringent as they are, these conditions are actually met in quantum fluids over a surprisingly large q range (see Chapters 8 and 9). Its dynamic matrix $G(q)$ is given by

$$G(q) = \begin{pmatrix} 0 & i f_{un}(q) \\ i f_{un}(q) & z_u(q) \end{pmatrix}. \tag{2.51}$$

This matrix is readily evaluated for its two eigenvalues, the poles of the dynamic susceptibility: $z_{\pm} = z_u/2 \pm i\sqrt{f_{un}^2 - z_u^2/4}$. If the force $f_{un} > z_u/2$ then the two modes are propagating with propagation frequency $\pm\sqrt{f_{un}^2 - z_u^2/4}$, and damping rate $\Gamma = z_u/2$. When $z_u/2 > f_{un}$ the modes become overdamped resulting in two diffusive modes with decay rates $\Gamma_{\pm} = z_u/2 \mp \sqrt{z_u^2/4 - f_{un}^2}$. Even though the damped harmonic oscillator is as basic a model as they come with a clearcut prediction of what the poles of the dynamics susceptibility are, there appears to be quite some room for misinterpreting its results. We list some common pitfalls in Section 3.2.4. Of particular interest in the study of liquids has been whether the extended sound modes remain propagating ($f_{un} > z_u/2$)) for wavelengths corresponding to the interparticle spacing, or whether they become overdamped ($f_{un} < z_u/2$)). This issue is tackled in Chapters 4 and 9.

2.3.3   How many parameters are needed?

In the previous section, we have discussed how the simple hydrodynamics decay scheme shown in Fig. 2.11 can be generalized to shorter length scales of fluctuations by including shorter timescales. This resulted in the decay tree shown in Fig. 2.12. We also discussed some variations on this decay scheme, such as the simple viscoelastic model shown in Fig. 2.14. By and large, when the wavelength of the density fluctuations gets shorter their damping rates increase. When the damping rate becomes appreciable compared to the propagation frequency upon leaving the hydrodynamic region, we witness a transition from hydrodynamic behavior to finite-q behavior. One of the most common features accompanying such a transition is an increase in the speed of propagation of the extended sound modes. Such an increase has been observed, amongst others, in simple liquids and in liquid metals. In liquid metals in particular we see that the speed of propagation of sound waves becomes dependent of the coupling parameter $f_{u\sigma}$ (Scopigno and Ruocco, 2005). The explanation that is given for this phenomenon is that for fluctuations of very short wavelengths the fluid behaves almost elastically; the frequencies are so high that the fluid no longer has time to relax as it does for fluctuations on a hydrodynamic length scale. Sometimes, this type of sound propagation is referred to as zero sound, with the zero meaning that essentially zero collisions are involved. This is in contrast to the hydrodynamics region (referred to as first sound) where many collisions between the atoms that make up the density fluctuation take place.

The question is how, and under what conditions, this transition from hydrodynamic behavior to an increased speed of propagation of the sound waves manifests itself in the decay trees of the effective eigenmode formalism. In particular, what is the range of values that the matrix elements of $G(q)$ should exhibit before the non-hydrodynamical elements (such as $f_{u\sigma}$ and f_{Tq}) of the matrix come into play. We will revisit this issue in various places in this book; in this section we illustrate the general behavior signalling the transition from one region to another.

Our starting point is the extended decay tree shown in Fig. 2.12. There are many aspects to this discussion as the extended hydrodynamics scheme involves four coupling parameters f and three transport coefficients z. The transition from hydrodynamic behavior to extended hydrodynamics is marked by a change in the speed of propagation of sound waves and by an increase in damping rate of the density fluctuations. Since the full decay tree also describes hydrodynamic behavior, albeit in an overparameterized form, one might expect that this increase in damping is reflected in an increase of the decay rates $z_\sigma(q)$ and $z_q(q)$. However, by and large this is not the case. As an example, we show the values for z_σ, z_q and $z_{q\sigma}$ in Fig. 2.17 as determined through a molecular dynamics simulation for a dense fluid. As can be seen in this figure, the variation with q of these matrix elements is much slower than the variation with q of the damping rates of the extended modes that typically vary as $\sim q^2$ in the hydrodynamic region and as $\sim q^2$ or $\sim q$ for larger q values up to $q_{max}/2$ (q_{max} is the position of the main peak of the static structure factor $S(q)$). In addition, the matrix elements z_σ and z_q reach finite values when $q \to 0$, whereas the damping rate of the extended heat and sound modes vanishes as $\sim q^2$ for $q \to 0$. Thus, one should let go of the intuitively appealing notion that the q dependence of the damping rates of density fluctuations is a reflection of the q dependence of the underlying microscopic decay rates (in this case $z_\sigma(q)$ and $z_q(q)$).

The explanation for this behavior of the extended modes is that the q dependence of the damping rates of the heat and sound modes is determined by the q dependence of the coupling constants f. In addition, the q dependence of these coupling constants also determines the transition from one region to another when it comes to the speed of propagation of the extended sound modes. In order to see this mathematically, we need to look closer at the interconnectedness of the matrix elements of $G(q)$.

In mathematical terms, the transition from ordinary hydrodynamics (Fig. 2.11) to extended hydrodynamics (Fig. 2.12) becomes a little clearer when we write out the

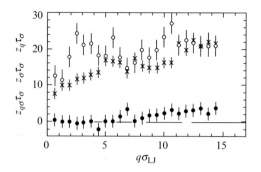

Fig. 2.17 Figure adapted with permission from de Schepper *et al.* (1988). Shown are the three matrix elements of $G(q)$ that relate to the transport coefficients (see Fig. 2.12) for a Lennard-Jones fluid corresponding to liquid argon at its triple point. The data have been plotted in reduced units, with $\sigma_{\mathrm{LJ}} = 3.36 \text{ Å}^{-1}$ and τ_σ the time it takes to traverse the particle's radius $\sigma/2$ with thermal velocity $\tau_\sigma = \sigma/2\sqrt{\beta m}$. The q dependence of these matrix elements is fairly minor in the sense that they are close to constant over most of the range, and stay finite for $q = 0$ at a value that is of the same order of magnitude as the maximum value. Also note that the damping rates of the momentum flux and of the heat flux are very similar in magnitude, and that they can be associated with the time it would take a particle to travel across a distance of $\sigma/40 - \sigma/20$ (the collisional timescale).

matrix expressions for $S_{nn}^{\mathrm{sym}}(q, \omega)$ (eqn 2.40) for both decay trees. The expression for the hydrodynamic case then reads (see Appendix E):

$$S_{nn}^{\mathrm{sym}}(q, \omega) = \frac{S_{\mathrm{sym}}(q)}{\pi}\mathrm{Re}\left[i\omega - \frac{f_{un}^2}{i\omega + \frac{f_{uT}^2}{i\omega + z_T} + z_u}\right]^{-1}, \tag{2.52}$$

and the expression for the extended hydrodynamic case is given by:

$$S_{nn}^{\mathrm{sym}}(q, \omega) = \frac{S_{\mathrm{sym}}(q)}{\pi}\mathrm{Re}\left[i\omega + \frac{f_{un}^2}{i\omega + \frac{f_{uT}^2}{i\omega + \frac{f_{Tq}^2}{i\omega + z_q}} + \frac{f_{u\sigma}^2}{i\omega + z_\sigma}}\right]^{-1}. \tag{2.53}$$

In eqn 2.53 we have taken $z_{q\sigma}$ to be identical to zero to make our life easier in this qualitative section; however, this appears to be justified based upon the results of Fig. 2.17. Inspecting the two above equations it becomes clear in what limit they become equivalent. First, the damping rates z_q and z_σ have to be large enough so that we can ignore the term $i\omega$ in the expressions $i\omega + z_q$ and $i\omega + z_\sigma$. As a rule of thumb, this tends to be the case when $z_q/2 > f_{Tq}$ and $z_\sigma/2 > f_{u\sigma}$. Since both damping rates approach finite values for $q \to 0$ and since the forces f are proportional to q in the hydrodynamic limit, the above condition will end up being satisfied upon going to lower and lower q values. Once we reach this limit, we make the substitution of $f_{u\sigma}^2/z_\sigma \to z_u$ and $f_{Tq}^2/z_q \to z_T$ to recover eqn 2.52. This limiting behavior also directly demonstrates why the decay rates in the hydrodynamic region are proportional to q^2: the forces f are linear in q, the decay constants z_q and z_σ are independent of q for $q \to 0$, so that the hydrodynamic damping rates will vary as f^2/z, which produces the observed $\sim q^2$ dependence.

The above type of reasoning holds for any transition. Provided that the pertinent force f is larger than the corresponding damping rate $z/2$ in expressions such as eqn 2.53, then the propagation frequencies will become dependent upon that force. In the opposite case, the force will not feature in the expression for the propagation frequencies. In fact, the combination f^2/z will now act as a single decay constant. For a numerical example we refer the reader to Appendix E where we discuss the transition from the simple viscoelastic model to the damped harmonic oscillator in some detail and where we illustrate that the transition from one model to the other takes place around $f \approx z/2$. In Appendix D we show the behavior of the eigenmodes as a function of the elements of the matrix $G(q)$ for the decay tree shown in Fig. 2.12.

When one applies a fitting procedure to the experimental data one notices near a transition that the two fitting parameters f and z become cross-correlated and that an equally good fit can be obtained using a less-elaborate decay tree. This discussion puts into mathematical perspective what we mean by cutting off a decay tree and replacing it by time-integrated constants. When we say that the frequency (or force f) is so high that the liquid can no longer follow the rapid oscillations, what we actually mean is that the decay constant z is now so small compared to the driving force that the oscillation is no longer completely damped out within one oscillation. This part of the branch is then no longer overdamped (the overdamping condition given by $z/2 > f$), and we have to take the oscillatory behavior into account, which is achieved by inclusion of the frequency ω in the expression $i\omega + z$. A crude way of saying this is that some of the power that is put into this channel comes back so that oscillatory behavior can result, and the speed of propagation will now (also) depend on this additional coupling constant f. An example of a transition related to temperature fluctuations is discussed in Section 4.6.

This limiting behavior explains the overall behavior of the linewidths and speeds of propagation. In the hydrodynamic region we have that the combinations f_{qT}^2/z_q and $f_{u\sigma}^2/z_\sigma$ yield the hydrodynamic damping rates z_T and z_u, respectively (see eqn 4.2 for the explicit expressions). In the hydrodynamic limit the two coupling constants f_{un} and f_{uT} combine to give the speed of propagation of sound waves $c = [f_{un}^2 + f_{uT}^2]^{1/2}/q$. Then, with increasing q the forces f_{Tq} and $f_{u\sigma}$ grow beyond $z_q/2$ and $z_\sigma/2$, respectively. At this point the damping rate of the fluctuations will cease to be proportional to q^2 but instead increase more slowly with increasing q. With ever-increasing q, the forces f tend to reach a maximum near $q \approx q_{max}/2$, after which the damping rates of the fluctuations vary weakly with q. So, the bottom line is that the relationship between the coupling constants f and the damping rates z determines where the decay tree will be cut off. The damping rates of the heat flux and the momentum flux are only weakly dependent on the wavelength of the fluctuation, hence it is the coupling constants that essentially determine the extent of the decay tree. Of course, this is not a surprise as it is well known that higher frequencies of the probing radiation require the inclusion of shorter timescales, timescales on which one has to take into account more details of the decay mechanism than can be captured in the hydrodynamic damping rates z_u and z_T.

What might be slightly counterintuitive, however, is that the damping rates of the heat flux and the momentum flux play such a small role in determining the extent of the decay tree. After all, we are somewhat used to thinking in terms of the damping rates determining whether we are still in the hydrodynamic region or not because of criteria such as that the propagation frequency should be much larger than the linewidth (that is, the damping rate). There is no real controversy here, the only difference is whether one thinks of damping rates as z_u ($= f_{u\sigma}^2/z_\sigma$) and z_T ($= f_{Tq}^2/z_q$), or in terms of z_σ and z_q. If one thinks in terms of the latter, then the hydrodynamic limit is crossed once $f_{u\sigma} \approx z_\sigma/2$ or $f_{Tq} \approx z_q/2$.

In Figs. 2.18 and 2.19 we show two cases for various combinations of the forces f demonstrating how this transition from hydrodynamics to extended hydrodynamics turns out in terms of propagation frequencies. Depending on the relative sizes of the forces we can have various transitions; for instance, we can see an increase in propagation speed from $[f_{un}^2 + f_{uT}^2]^{1/2}/q$ to $[f_{un}^2 + f_{u\sigma}^2]^{1/2}/q$. There also appears to be the possibility, in principle, that we can observe a decrease in propagation speed, from the adiabatic propagation speed $[f_{un}^2 + f_{uT}^2]^{1/2}/q$ to the isothermal speed $f_{un}(q)/q$. For this to happen, we need that the damping z_T becomes very large compared to f_{uT} *before* the force $f_{u\sigma}$ becomes comparable to z_σ. If this were the case then we would expect to see the speed of propagation change from $[f_{un}^2 + f_{uT}^2]^{1/2}/q$ to $[f_{un}^2]^{1/2}/q$ upon leaving the hydrodynamic region. In addition, it might even be possible for this speed to subsequently increase to $[f_{un}^2 + f_{u\sigma}^2]^{1/2}/q$ with increasing q, provided the ratio of $f_{u\sigma}^2/z_\sigma$ becomes favorable. We discuss these more esoteric transitions in Chapter 7. One system that might show this type of behavior is liquid Ni. We show the results for this system in Chapter 7 where we discuss whether such a type of transition is actually possible.

The speed of propagation of the extended sound modes is only part of the overall behavior of the effective eigenmodes. We also have to consider what happens to their damping rates and what the relative amplitudes are of the modes in the dynamic structure factor $S_{nn}(q, \omega)$. We show the results for all five modes in Figs. 2.20 and 2.21. The damping rates and amplitudes shown are calculated for the same set of parameters as the set used in the calculation of Figs. 2.18 and 2.19.

We first discuss Fig. 2.20. The damping rates Γ^* of the extended heat and sound modes start off at their hydrodynamic equivalents $f_{Tq}^2/z_q\gamma$ and $f_{u\sigma}^2/z_\sigma$, respectively.[10] The two other (non-hydrodynamic or kinetic) modes of the matrix $G(q)$ start off at a

---

[10] We have opted to show, in this and in similar figures, f_{Tq}^2/z_q rather than $f_{Tq}^2/z_q\gamma$ since the former relates to the hydrodynamic matrix element z_T in the matrix $G(q)$. Thus, only one of the quadratic curves in these figures (and in the figures of Appendix D) directly relates to hydrodynamic damping of density fluctuations, the other curve has to be divided by γ first.

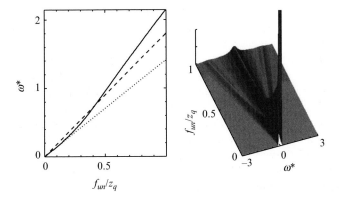

Fig. 2.18 The peak positions (solid line, left panel) of the current–current correlation function ($\sim \omega^2 S^{\mathrm{sym}}(q,\omega)$) mark the transition from hydrodynamic behavior to extended hydrodynamic behavior. This figure is representative of a simple liquid with specific heat ratio $\gamma = 2$. In this q range where the damping rates of the extended sound modes are still very small compared to their propagation frequencies, the peaks of the current–current correlation function virtually coincide with the eigenvalues of the propagating mode of the matrix $G(q)$. The right panel shows a surface plot of $S^{\mathrm{sym}}(q,\omega)$, convoluted with the energy-resolution function of a very good spectrometer. This figure has been calculated based on the 5×5 matrix $G(q)$ corresponding to Fig. 2.12 using the following combination of forces f: $f_{un} = f_{uT} = f_{u\sigma}/2 = f_{Tq}/1.5$, and hence $\gamma = 2$. The two damping rates have been kept constant at $z_q = z_\sigma = 1$. The label ω^* refers to the fact that these frequencies are not those of any particular system; the overall frequency scale has not been set, rather this figure is solely based on the relative strength of the various forces. The dotted line in the left panel corresponds to the adiabatic propagation of sound modes ($[f_{un}^2 + f_{uT}^2]^{1/2}$), the dashed line that underestimates the propagation speed corresponds to $[f_{un}^2 + f_{Tq}^2]^{1/2}$. Other possible combinations of forces would correspond to straight lines that intersect the vertical axis ($f_{un}/z_q = 1$) at $\omega^* = 2.23$ (for $[f_{un}^2 + f_{u\sigma}^2]^{1/2}$) and at $\omega^* = 2.45$ (for $[f_{un}^2 + f_{uT}^2 + f_{u\sigma}^2]^{1/2}$).

finite value. The heat-mode amplitude is $(\gamma - 1)/\gamma = 0.5$ for the smallest values of f, the sound-mode amplitudes start off at $1/2\gamma = 0.25$. The amplitudes of the two kinetic modes in $S_{nn}^{\mathrm{sym}}(q,\omega)$ are identical zero for the smallest f values. With increasing f – which is equivalent to increasing q in scattering experiments – and upon leaving the hydrodynamic region we observe gradual changes in the damping rates and in the amplitudes. The two sound-mode amplitudes diminish, while the two kinetic modes acquire a very small amplitude. Note that amplitudes can be negative since they are given by the real part of a complex number (see eqn 2.44). There is no restriction on the amplitudes of individual modes, just on the sum of the amplitudes of all the modes. Overall, the five modes are very well behaved in the sense that there is barely any mixing of them. Nonetheless, the changes are quite noticeable, in particular in the region where the speed of propagation starts to depart from the hydrodynamic value of $[f_{un}^2 + f_{uT}^2]^{1/2}/q$ (see Fig. 2.18) at $f_{un}/z_q \approx 0.2 - 0.3$. In addition, in this example $S_{nn}^{\mathrm{sym}}(q,\omega)$ technically consists of five modes for $f_{un}/z_q > 0.2$; however,

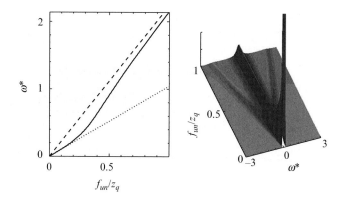

Fig. 2.19 This figure is similar to Fig. 2.18 with the parameters values for f and z of Fig. 2.12 chosen so as to resemble a liquid metal with $\gamma = 1.1$. The dotted line in the left panel corresponds to the adiabatic propagation of sound modes ($[f_{un}^2 + f_{uT}^2]^{1/2}$), the dashed line to $[f_{un}^2 + f_{u\sigma}^2]^{1/2}$. This figure has been calculated using $f_{un} = f_{uT}/0.3 = f_{u\sigma}/2 = f_{Tq}/0.6$. The two damping rates have been kept constant at $z_q = z_\sigma = 1$. Note that the transition away from hydrodynamic speeds of propagation takes place near $f_{un}/z_q = 0.25$, corresponding to $f_{u\sigma} \approx z_\sigma/2$.

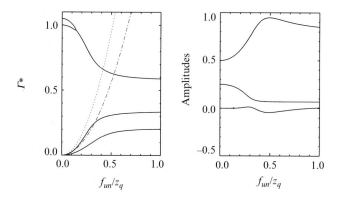

Fig. 2.20 This figure shows the damping rates Γ^* (left panel) and the relative amplitudes (right panel) for the five modes in $S_{nn}^{sym}(q, \omega)$ whose spectra are shown in Fig. 2.18. The dotted line in the left panel corresponds to the hydrodynamic damping of the sound modes $f_{u\sigma}^2/z_\sigma$, the dashed line to the equivalent rate for the heat mode: f_{Tq}^2/z_q, other than a factor γ. The parameters in this figure are identical to these in Fig. 2.18 except the rate z_σ has been increased slightly to 1.05 to make the hydrodynamic behavior of the two kinetic modes better visible.

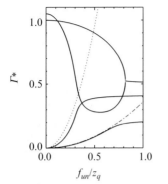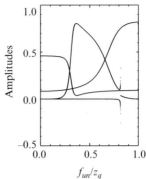

Fig. 2.21 This figure shows the damping rates (left panel) and the relative amplitudes (right panel) for the five modes whose spectra are shown in Fig. 2.19. The dotted line in the left panel corresponds to the hydrodynamic damping of the sound modes $f_{u\sigma}^2/z_\sigma$, the dashed line forms the basis of the damping of the heat mode: f_{Tq}^2/z_q, other than a factor γ. The parameters in this figure are identical to these in Fig. 2.19 except the rate z_σ has been increased slightly to 1.05 to make the hydrodynamic behavior of the two kinetic modes better visible.

the amplitude of the two additional modes (the curve close to zero in the right panel of Fig. 2.20) would probably be too weak to actually be detected in a scattering experiment.

The situation depicted in Fig. 2.21 corresponds to the case where we saw an increase in the speed of propagation from $[f_{un}^2 + f_{uT}^2]^{1/2}/q$ to $[f_{un}^2 + f_{u\sigma}^2]^{1/2}/q$ (see Fig. 2.19). The damping rates and amplitudes of the five modes show the very complex behavior that takes place during such transitions. At the smallest f values, the heat-mode amplitude starts off at its hydrodynamic value of $(\gamma - 1)/\gamma = 0.09$ and the sound modes at $1/2\gamma = 0.46$ for the γ value of 1.09 used to calculate the dispersion of Fig. 2.19. What is seen to happen at the transition around $f_{un}/z_q = 0.3$ is that the kinetic mode corresponding to the longitudinal momentum flux 'σ' is being mixed in. When this happens, the character of the extended eigenmode changes markedly, as can be seen by studying the behavior of the amplitudes (right panel of Fig. 2.21). One can see from this figure that the kinetic mode 'σ' is now fully mixed with the standard hydrodynamic modes. This is the region of q where one would expect the viscoelastic model to yield a good description of the data. Also note the discontinuity in the amplitudes of the modes near $f_{un}/z_q = 0.8$ where two of the modes change in character. Here, two diffusive modes become propagating modes, as can be seen by the fact that two damping rates merge and acquire the same value (left panel of Fig. 2.21). Finally, the extended heat mode that starts off at a very modest value at the smallest f values dominates the spectra for the higher f values. It appears to have switched roles with the extended sound modes that now make only a modest contribution to the spectra. Yet, even such a modest contribution is clearly visible as a propagating mode in Fig. 2.19.

The preceding discussion merely shows the eigenmodes of the matrix $G(q)$ under various conditions. It does not predict when the switch from hydrodynamics to extended hydrodynamic behavior will occur in a particular liquid. These figures also illustrate the difficulty in extrapolating experimental results from finite q down to $q \to 0$. It is quite possible that the lowest q value accessible to experiment would correspond to a relatively large value of f/z in these figures. For instance, if the lowest q value were to correspond to $f_{un}/z_q = 0.4$ in Fig. 2.21, then clearly such an extrapolation is not possible, even though the damping rates of the modes do not appear to deviate that much from their hydrodynamic equivalent (dotted and dashed lines in the left panel of Fig. 2.21). We show more examples of the behavior of the effective eigenmodes in Section 7.2, and we show the generic changes in the behavior of the modes as a function of the parameters f and z – including the parameter $z_{q\sigma}$ – in Appendix D.

We have centered the discussion in this section on the damping rates of the longitudinal momentum flux and of the heat flux. This does not imply that these two rates are more fundamental rates than for example, z_T, or the damping rates associated with even shorter-lived correlation functions given by the derivatives of the heat and momentum flux. Rather, these two damping rates are just somewhat special; they represent a convenient timescale that covers most of the fluctuations studied in experiments and they provide insight into how the transition between hydrodynamic behavior and the decay of shorter-wavelength fluctuations takes place. These two decay rates are directly related to the two quantities that originate in the conservation laws for (longitudinal) momentum and energy. First, the fact that energy and momentum are not conserved is of course a consequence of looking at the system on small length scales; at these length scales the entities that we are studying are not isolated and hence we should not expect the conservation laws to hold. Secondly, the fact that the damping rates z_q and z_σ vary so little as a function of probing wavelength already tells us that the decay process of density fluctuations ultimately involves collisions between the particles. The two decay rates describe the characteristics of these individual collisions when it comes to momentum and energy transfer. As such, the effective eigenmode formalism allows us to scrutinize how these collisional decay rates determine the dynamics both at the hydrodynamic length scales as well as the atomic length scale.

2.4 Memory function formalism

The memory function formalism is a phenomenological description of the dynamic structure factor, just as the effective eigenmode formalism is a phenomenological description. In fact, the two descriptions contain the same amount of information, provided they are applied correctly (Bafile *et al.*, 2006). In practice, some (small) differences show up between the two formalisms. Given the widespread and successful use of the memory function formalism (Balucani and Zoppi, 1994) in the description of dense liquids in general and of liquid metals (Scopigno and Ruocco, 2005) in particular (see Chapter 7), we detail the connection between the two in this section. We pay particular attention to the sum rules on the amplitudes of the various modes, and to the rules regarding the relaxation times that show up in both formalisms. We refer the reader to the literature (Balucani and Zoppi, 1994; Bafile *et al.*, 2006) for details not covered in this section.

Upon inspection of the general decay tree in the effective eigenmode formalism (such as the one shown in Fig. 2.12), or from inspection of the general structure of the dynamic matrix $G(q)$, it is evident that the expression for the density–density dynamic structure factor can be rewritten in the following form:

$$S_{nn}^{\mathrm{sym}}(q,\omega) = \frac{S_{\mathrm{sym}}(q)}{\pi}\mathrm{Re}\left[\frac{1}{i\omega + G(q)}\right]_{11} = \frac{S_{\mathrm{sym}}(q)}{\pi}\mathrm{Re}\left[\frac{1}{i\omega + \dfrac{f_{un}(q)^2}{i\omega + M(q,\omega)}}\right]$$

$$= \frac{1}{\pi}\frac{q^2/(\beta m)M'(q,\omega)}{[\omega^2 - f_{un}(q)^2 - \omega M''(q,\omega)]^2 + [\omega M'(q,\omega)]^2}. \tag{2.54}$$

Here, $M(q,\omega)$ is the complex memory function $M(q,\omega) = M'(q,\omega) + iM''(q,\omega)$. In Table E.1 we have rewritten the expression for $S_{nn}^{\mathrm{sym}}(q,\omega)$ for various decay trees in such a way that the memory function can easily be identified. Since eqn 2.54 is nothing more than a rewriting of the effective eigenmode formalism, the memory function holds the same amount of information as the matrix $G(q)$. Also note that the memory function is to be applied to the correlation function function $S_{nn}^{\mathrm{sym}}(q,\omega)$ instead of to the dynamic structure factor $S_{nn}(q,\omega)$, and that the frequency that enters the formalism is $f_{un}(q)^2 = q^2/[\beta m S_{sym}(q)]$ instead of $q^2/[\beta m S(q)]$ (the classical limit of $f_{un}(q)^2$); these two facts combined ensure that the memory function formalism is valid both in the classical as well as in the quantum limit.

Whether to use the effective eigenmode formalism, or the memory function formalism is largely a matter of choice. In both formalisms decay processes are characterized by their decay times and by their amplitudes in $S_{nn}(q,\omega)$. In both formalisms the relevant number of decay channels can be adjusted: in the effective eigenmode formalism this is done by adjusting the size of the matrix $G(q)$ through the inclusion of additional microscopic variables, in the memory function formalism this is achieved by simply adding a decay channel to $M(q,t)$, the Fourier transform of $M(q,\omega)$. In order to make the choice easier, we review the advantages and disadvantages of both formalisms.

The effective eigenmode formalism automatically includes the frequency-sum rules on the amplitudes of the modes since it is set up by taking successive time derivatives of the basic microscopic variables and by incorporating the time evolution of the correlations between these derived variables.[11] The formalism also describes not just the density–density correlation function, but also the density–temperature and temperature–temperature correlation function by one and the same matrix. This is

---

[11] The frequency-sum rules are given by the short-time behavior of the intermediate scattering function; taking into account correlations between the time derivatives of the microscopic density automatically ensures that the short-time behavior of the density–density correlation function is properly reproduced.

particularly useful in analyzing computer simulations. The eigenvalues of the modes also automatically satisfy the sum rule on the decay rates and propagation frequencies of the modes: the sum of the eigenvalues adds up to the sum of the decay constants z (such as z_q and z_σ in Fig. 2.12) that show up on the diagonal of the matrix. A downside of the formalism is that the basic variables are expressed as frequencies and damping rates, and it is not immediately clear how these basic variables combine to yield the eigenvalues of $G(q)$ and the corresponding amplitudes of the modes in $S_{nn}(q,\omega)$. That is, one cannot directly associate a particular decay rate (such as $z_q(q)$ and $z_\sigma(q)$) with the damping rate of a particular mode. In general, the interpretation of combinations of matrix elements always seems to be a little more awkward than a direct fitting to decay times, as is done when using the memory function formalism. Of course, one can always simply fit to a sum of Lorentzians by using the intuitively more appealing eqn 2.37, however, the sum rules are more cumbersome to put in.

The memory function formalism identifies the decay channels by their decay times and amplitudes in the decay mechanism from the outset (Balucani and Zoppi, 1994), and these numbers are the ones that are fitted directly to the experimental data. As such, the interpretation of the fit results is very straightforward, with the amplitudes representing the strength of a particular decay mechanism (note that this is not necessarily the strength of the mode in $S_{nn}(q,\omega)$). In fact, the full expression of the resulting amplitudes associated with the various decay mechanisms in $S_{nn}(q,\omega)$ can get somewhat cluttered (Bafile *et al.*, 2006), but this is only a minor inconvenience. The drawback of the memory function formalism is that all the frequency-sum rules have to be put in by hand, and there is no straightforward way to incorporate the sum rules on the decay times. As a result, one often encounters that a memory function is being used that has the correct structure, but that has too many independent parameters. We will give an example of these potential problems in the following, where we make a direct comparison between the two formalisms.

The remainder of this chapter involves details regarding the various choices that one has in generalizing the basic hydrodynamics decay tree, and how these choices are linked to the interpretation of the scattering results on the atomic level. The reader may wish to skip this part until having seen the results of applying the formalisms to actual fluids (Chapters 4 and beyond). The equivalent memory function corresponding to the (ordinary) hydrodynamics decay tree in the effective eigenmode formalism (Fig. 2.11) is given by (Bafile *et al.*, 2006)

$$
\begin{aligned}
M(q,t) &= f_{un}^2(\gamma - 1)e^{-\gamma a q^2 t} + 2\phi q^2 \delta(t) \\
&= f_{uT}^2 e^{-\gamma a q^2 t} + 2\phi q^2 \delta(t).
\end{aligned}
\tag{2.55}
$$

Here, a is the thermal diffusivity and ϕ is the longitudinal kinematic viscosity (see also eqn 4.2 and Table 2.3). The delta function in time corresponds to the decay mechanism depicted by the curved arrow at the variable 'u' in Fig. 2.11. The exponential decay function corresponds to the decay mechanism shown by the curved arrow at the variable 'T' in Fig. 2.11. In general, a decay that directly affects 'u' will show up as an $\sim \delta(t)$ term in $M(q,t)$. For instance, the memory function corresponding to the damped harmonic oscillator (Fig. 2.16) is $M(q,t) = 2z_u\delta(t)$. The factor 2 in front of this delta function has to do with the Laplace transform from $t \to i\omega$ that only integrates over half the time domain of the delta-function.

The pre-factor in front of the exponential decay in eqn 2.55 is given by f_{uT}^2, reflecting the strength of the coupling constant shown as a spring in Fig. 2.11. This correspondence between the two formalisms can also be extended to more complicated models, such as the extended viscoelastic model shown in Fig. 2.15. Following the same 'rules' (or looking up the relevant functions in Tables E.1 and E.2) we can directly write down the equivalent memory function expression for the decay tree shown in Fig. 2.15:

$$M(q,t) = f_{u\sigma}(q)^2 e^{-z_\sigma t} + f_{uT}(q)^2 e^{-z_T t}. \tag{2.56}$$

In these examples, there is no daylight between the two formalisms, they are fully equivalent in both their functionality and ease of use. We list the comparisons for the other models discussed in the preceding section in Appendix E. The situation becomes more interesting when we look at fluids that have two very different timescales governing the relaxation of density fluctuations. These two timescales are of course related to the particle being locked up in the cage of its neighbors, and the slow structural relaxation of that cage (the particle escaping). Thus, the two timescales together model the entire process that is referred to as cage diffusion. Equation 2.56 (or the decay tree shown in Fig. 2.15) cannot properly describe both these mechanisms since it does not have enough decay mechanisms incorporated into it.

In order to include both structural relaxation as well as the cage rattling, the memory function is often generalized in an *ad hoc* manner as follows (Scopigno and Ruocco, 2005):

$$M(q,t) = f_{u\sigma}(q)^2[(1-\alpha)e^{-z_1 t} + \alpha e^{-z_2 t}] + f_{uT}(q)^2 e^{-z_T t}, \tag{2.57}$$

or sometimes as

$$M(q,t) = f_{u\sigma}(q)^2[(1-\alpha)e^{-z_1} + 2\alpha z_2\delta(t)] + f_{uT}(q)^2 e^{-z_T t}. \tag{2.58}$$

Here, z_1 is assigned the same meaning as z_σ, while z_2 is a new decay time. For ease of notation, we have not explicitly included the q dependence of the decay rates. The structure of this equation with the amplitude α ensures that the relevant frequency-sum rules are still kept in place, while the two timescales related to the relaxation of stress can now account for both structural relaxation as well as for the rattling motion (the particle bouncing around in its cage as depicted in Fig. 2.9). The relative amplitude α is a measure of the importance of the former. This generalization turns out to work particularly well in the case of liquid metals (see Chapter 7). The equivalent generalization of the effective eigenmode constitutes replacing the decay tree of Fig. 2.15 by the ones shown in Fig. 2.22. However, there is a subtle difference between these new decay trees, and the generalization given in eqns. 2.57 and 2.58.

First, the new decay tree shown in Fig. 2.22(b) corresponding to the new instantaneous decay channel of eqn 2.58 cannot be correct. The two coupling constants $f_{u\sigma}(q)$

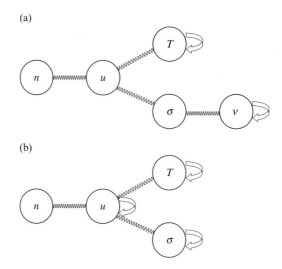

Fig. 2.22 The generalization of the decay tree shown in Fig. 2.15. Part (a) corresponds to eqn 2.57, part (b) corresponds to eqn 2.58. The symbol 'ν' represents the flux of 'σ'. In part (a) the additional decay mechanism shows up as a new coupling constant $f_{\sigma\nu}(q)$ and a replacing of the damping rate z_σ by z_ν. In part (b) the additional mechanism shows up as a new looping arrow at the microscopic quantity 'u'.

and $f_{uT}(q)$ exhaust all decay mechanisms; there simply is no additional instantaneous damping mechanism possible. This is not the end of the world, however, it just reflects that we made an additional assumption. Mathematically, there is no problem in adjusting the matrix $G(q)$, and from a philosophical viewpoint what is the difference between putting something into a model at an approximate level that represents a convenient way of capturing the essence of part of the decay tree, and between cutting off a decay tree at some point and ignore processes that take place on a timescale faster than our probing timescale? We will leave the judgement on this issue up to the reader. The fact is that eqn 2.58 (Fig. 2.22(b)) does a reasonable job at modelling the experimental data (Scopigno and Ruocco, 2005).

The *ad hoc* generalization of eqn 2.57 presents another potential problem. While the amplitudes (determined by α) are properly controlled such that $M(q, t = 0)$ yields the correct frequency-sum rules, there is no restriction on the two decay times z_1 and z_2 that describe the relaxation of the stress field 'σ' generated by the velocity disturbance through the coupling $f_{u\sigma}(q)$. For the case of two purely damped (non-propagating) modes, the decay tree shown in Fig. 2.22(a) would yield two decay rates based on: $z_{1,2} = z_\nu/2 \pm \sqrt{z_\nu^2/4 - f_{\sigma\nu}^2}$. Thus, while in eqn 2.57 the data are fitted to two independent decay rates, the decay tree shows that in this generalization there is only one underlying decay constant (z_ν). The coupling constant $f_{\sigma\nu}(q)$ in the decay tree is in practice an adjustable parameter, but its value also determines the relative amplitudes (α in eqn 2.57) of the two decay paths so it cannot be adjusted as a fully free parameter. Thus, while eqn 2.57 has three adjustable parameters (α, z_1 and z_2), the generalization of Fig. 2.22(a) only has two ($f_{\sigma\nu}$ and z_{ν}). This implies that eqn 2.57 by itself is not restrictive enough, potentially resulting in some loss of accuracy in the fitted parameters.

In practice, one lets the parameter α in eqns. 2.57 and 2.58 only vary between 0 and 1, ensuring that both contributions that constitute the $f_{u\sigma}$ decay channel produce positive amplitudes in $M(q,t)$. This restriction is not contained in the decay tree of Fig. 2.22(a); in fact, the opposite is true. To see that we explicitly write out the memory function corresponding to the decay tree of Fig. 2.22(a) (dropping the q dependence):

$$M(q,t) = f_{uT}^2 e^{-z_T t} + \frac{f_{u\sigma}^2}{z_1 - z_2}[z_1 e^{-z_2 t} - z_2 e^{-z_1 t}], \qquad (2.59)$$

with (as before) $z_{1,2} = z_{\nu}/2 \pm \sqrt{z_{\nu}^2/4 - f_{\sigma\nu}^2}$. For real $z_{1,2}$, corresponding to the case where $z_{\nu}/2 > f_{\sigma\nu}$, which is tacitly assumed when fitting scattering data to the extended memory functions, we would have that one of the exponentials in eqn 2.59 has a negative amplitude.[12] Note that this does not violate any known sum-rule, however, it does suggest that keeping α restricted to the range $0 < \alpha < 1$ is not only too restrictive, it might even stand in the way of obtaining the best possible fit to the data. The assumption that $z_{1,2}$ should always be real could well be correct, depending to what part of the ionic or electronic motion z_{ν} pertains. Should z_{ν} describe a relaxation channel associated with the electronic degrees of freedom, then it would be very surprising if $z_{1,2}$ were not real. Whether all these imposed restrictions present a problem in practice remains to be seen, but we believe it is good practice to not be overly reliant on restrictions on fit parameters when such restrictions do not have a sound theoretical justification. Overall, the differences between the effective eigenmode formalism decay tree and the memory function expression are fairly minor in the case of this extension of the viscoelastic model.

The identification within the memory function formalism of a slow and a fast decay channel for the momentum flux gives us the opportunity to say something more in general about the interpretation of the modes of decay, and about how best to generalize smaller decay trees. The main point to make about the interpretation of the fit results pertains to the parameter α. From scattering data on liquid metals – such as lithium (Scopigo *et al.*, 2000*a*) – it has been found that the fast part of the decay identified as a particle bouncing around in its cage accounts for upwards of 80–95% of the decay of the momentum correlations. However, this by no means implies that this rattling motion also accounts for the decay of density fluctuations at anywhere the same level. This is discussed in somewhat more detail in Section 4.5, but the crux of the matter is the following. Since the rattling motion describes a particle being locked up inside its cage, it cannot wander that far away from its initial position, and hence, the amount of correlation loss is limited. The most it can wander away is a distance Δ, where Δ is the average distance between the 'surfaces' of neighboring particles. The actual amount of correlation loss in the density–density correlation function due

---

[12] This equation also makes it clear how the amplitudes α and $1 - \alpha$ are in fact determined by z_1 and z_2.

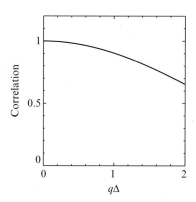

Fig. 2.23 Depicted is the amount of correlation loss that can be ascribed to the initial, rattling part of the cage-diffusion mechanism. Because of this rattling motion of this particle inside the cage, the particle can wander a distance Δ away from this position. The data are plotted as a function of $q\Delta$, with Δ the distance the particle has to cover before it is in contact with a particle that is part of the wall of the cage. Typical values for Δ are in the range of 0.2 to 0.6 Å for dense liquids. The function shown is $\int_V e^{i\vec{q}\cdot\vec{r}}d\vec{r}/V$, with $V = 4\pi\Delta^3/3$, and is evaluated to be $3[\sin(\Delta q) - (\Delta q)\cos(\Delta q)]/(\Delta q)^3$. A value of 1 implies that there is no correlation loss due to the rattling part of cage diffusion and therefore, this process would have zero intensity in the dynamic structure factor. Even at $q\Delta = 0.5$, roughly corresponding to position of the main peak in the static structure factor $S(q)$, the correlation loss is only 2.5%.

to the rattling motion depends on the probing wavelength $\lambda_{probe} = 2\pi/q$, and it is to good approximation given by $(q\Delta)^2/10$. One can view this expression as an effective form factor, not unlike the Debye–Waller factor in solids.[13] We plot this function in Fig. 2.23. From this figure it can be seen that the amplitude in $S_{nn}(q, \omega)$ of this decay mechanism resulting from the amplitude α of the fast decay mode in eqn 2.57 cannot exceed 0.01 for small q values (typically for $q < 0.5$ Å^{-1}).

At first sight, this limited loss of correlation due to the rattling motion is a rather curious finding as the generalization of the extended viscoelastic model through the inclusion of this one additional decay time was introduced (Scopigno and Ruocco, 2005) because of the problems in obtaining a satisfactory fit to the data at the smaller q values (typically for $q < q_{max}/2$ where q_{max} is the position of the main peak of the static structure factor). This apparent contradiction reflects the fact that a large value of the amplitude of the decay channel that describes the rattling motion does not necessarily imply a large amplitude in the decay of density fluctuations, rather it implies a large and fairly rapid decay in the momentum–momentum correlation function as we explain in the following paragraphs. However, note that if one models the relatively slow decay of density fluctuations by the behavior of the momentum–momentum correlation function (as is done in eqn 2.57), then it is important to include

[13] The form factor in this case would read $\int_0^\Delta d\vec{r}e^{i\vec{q}\cdot\vec{r}}/V$, with $V = 4\pi\Delta^3/3$.

both the rapid and the slow decay channels even when it does not directly translate in a correspondingly large effect in $S_{nn}(q, \omega)$.

Scrutinizing how the decay of momentum correlations shows up in the decay of density fluctuations also brings up the question of how to generalize decay trees. The preceding generalization of the decay tree shown in Fig. 2.15 is not the only possible one: one could indeed opt for the generalization shown in Fig. 2.22(a), or instead for the one discussed in the previous sections (Fig. 2.12). What is interesting about deciding on these choices is that it involves figuring out the physics behind cage diffusion, and it also touches upon the fact that in general it is not possible to assess the character of the eigenmodes from a study of the density–density correlation function alone. When a particle is locked up in the cage, it will collide with the atoms that form the cage. During this collision, there will be transfer of energy and momentum. Transfer of momentum shows up in the momentum flux ('σ'), transfer of energy in the heat flux ('q'). Thus, this rattling part of cage diffusion might well show up in both parts of the decay tree shown in Fig. 2.12.

We first look at the decay of the momentum correlation function during the initial part of the cage-diffusion process. The motion of the particle in between collisions shows up in the 'σ'–'ν' branch of the decay tree (Fig. 2.22(a)) as this motion represents the translational transport of momentum (that is, momentum flux). Regarding the collisional transfer of momentum, if a particle bounces back after a collision, there will have been the maximum amount of momentum transfer; if the particles escapes out of the cage, there will have been a minimum amount of momentum transfer during the collision (since the escaping particle presumably stayed more or less on its course). After a couple of collisions during which the particle remains inside the cage, there will be very little correlation left between the direction the particle is moving in initially and the direction it is moving in after a couple of collisions. Thus, these collisions have brought about a very rapid decay in the longitudinal momentum correlation function, yet they have done essentially nothing in the relaxation back to equilibrium of the original density fluctuation: the particle is still inside its cage. This illustrates that the process that is at the heart of the decay of momentum correlations might be very hard to observe through the decay of the density fluctuations; its relative amplitude in a scattering experiment will be given as a function such as the one depicted in Fig. 2.23.

The decay of the heat flux is more complex than that of the momentum flux. The density and temperature play a large role in determining how important cage diffusion is to the heat-flux decay mechanism. The temperature determines how much of a punch a particle can pack in trying to escape the cage, and the density determines to a large extent whether the particle will stay in the cage and whether the collisions will be elastic or inelastic. In the extreme case of purely elastic collisions, a particle will bounce right back without loss of kinetic energy, similar to a ball bouncing off a wall. When this happens, there is no decay of the energy self-correlation function of the particle during the collision. Thus, the spectra for the energy–energy and for the density–energy correlation functions at high density should display a very small characteristic width for probing wavelengths corresponding to the interparticle spacings (see Section 4.4).

On the other hand, if collisions are inelastic to some degree, then this will result in a large decline of the energy self-correlation function during a collision: after a few collisions that follow each other in quick succession the deviation from the average energy should have disappeared. In this case, we would expect the decay rates of the heat flux and of the momentum flux to be very similar, if not identical since both decays involve the same binary collisions. It could be that this is the reason behind the close similarity of z_q and z_σ shown in Fig. 2.17.

So where has this discussion left us, other than wanting to throw this book into the fireplace? On a length scale corresponding to the interparticle separations, we can expect to see cage diffusion in both the decay of the energy–energy correlation function, as well as the decay of the momentum–momentum correlation function. When a particle stays inside a cage for a few collisions, then the momentum–momentum correlation function should have decayed during this time. This process should not depend very sensitively on the density, as long as the density is high enough for the particle to remain inside the cage for a few collisions. Whether the energy–energy correlation function also decays on this timescale or not should depend much more strongly on the density; at very high density, collisions are bound to become more elastic since the wall of the cage has become more solid, resulting in a slow decay of this correlation function. Conversely, at somewhat lower densities (but not too low) or for soft (metallic) potentials the wall is more flexible, resulting in a rapid decay of this correlation function. While it is clear that the amplitude of the rattling motion to the overall decay of a density fluctuation will be small (as shown in Fig. 2.23), the process will play a major role in the decay of the momentum flux and of the heat flux, yielding decay rates for the two processes that should be very similar in magnitude. Therefore, when studying a system with non-negligible coupling between 'u' and 'T', one should probably generalize the hydrodynamics decay tree of Fig. 2.11 to the decay tree of Fig. 2.12 as opposed to the decay scheme shown in Fig. 2.15. By the same token, if the memory function of eqn 2.56 needs to be generalized to a form like the one shown in eqn 2.57, then the part of the memory function corresponding to f_{uT}^2 should be generalized at the same time.

By and large, it is difficult to decide a priori how to expand the decay tree. In a Lennard-Jones computer simulation representing argon at high densities it was found, through modelling of all three basic correlation functions, that the 'T'–'q' branch of the decay tree was essential in understanding the ultimate fate of density fluctuations. In liquid metals (see Chapter 7), it was found that the 'σ'–'ν' generalization (eqn. 2.57 and Fig. 2.22) provides a very satisfactory fit (with the caveat that this assertion is based on the density–density correlation functions alone). It is probably safe to say that computer simulations will play an essential role in elucidating how to exactly generalize the basic decay trees as the important processes that govern the decay of, for instance, the momentum flux might only have a very small amplitude in the density–density correlation function. Only a direct evaluation of all correlation functions will tell us exactly how to expand the decay tree. Whether one wishes to employ the effective eigenmode formalism or the memory function formalism is a matter of choice.

We are now ready to apply the effective eigenmode formalism to real systems. After a brief discussion of experimental techniques in Chapter 3, we will apply the formalism to simple liquids in Chapter 4, where we will encounter the experimental signature of both extended heat modes and extended sound modes. In Chapter 5 we will discuss systems that are dominated by the extended heat mode, after which we will make our way through systems in which the extended sound modes become increasingly more important, culminating in the chapter on superfluids where the extended sound modes are the only game in town.

3

Experiments and computer simulations

In this chapter we discuss the basic techniques that allow for the determination of the dynamic structure factor $S_{nn}(q, \omega)$ from experiment. This chapter does not purport to be a derivation of the theory of scattering for which there already exist many excellent texts (Squires, 1994; Lovesey, 1971; Berne and Pecora, 2000; Xu, 2000), rather it provides a graphical representation of what can be measured, and it briefly discusses some common problems in going from the experimentally measured cross-section to the desired $S_{nn}(q, \omega)$. We end with a discussion on various computer simulation techniques that are commonly used in liquids to determine $S(q, \omega)$, including those correlation functions other than the density–density correlation function, from knowledge of the two-particle interaction potential.

3.1 Graphical representation of scattering techniques

The most commonly used experimental techniques to determine $S_{nn}(q, \omega)$ are neutron scattering and light scattering. Both techniques can be used to measure the full dynamic structure factor $S_{nn}(q, \omega)$, while neutron scattering can also be used to measure the self-part of the dynamic structure factor $S_s(q, \omega)$. In addition, magnetic excitations in materials can be probed by both techniques, albeit that the results from neutron scattering are the most easy to interpret. Since magnetic scattering is not terribly important in liquids with the exception of the study of some metals (Bermejo *et al.*, 2007), we restrict our discussion on this part to the magnetic dynamic structure factor $S_{\mathrm{mag}}(q, \omega)$ as measured in neutron-scattering experiments.

Scattering is an interference technique, independent of whether one uses a neutron or a photon. In the following, we treat them on equal footing where possible, and we will refer to both the neutron and the photon as the particle. What is observed in scattering experiments is the interference pattern between the incident particle, and the scattered particle. Thus, we need to make use of the wave properties of a particle to understand how it can 'see' multiple atoms at the same time, thereby revealing their spatial arrangement and their relative motions.

A particle incident upon the sample can be represented by a plane wave, with its wavelength λ given by its momentum p: $\lambda = h/p$. This incident plane wave will scatter from the atomic nuclei (in the case of non-magnetic neutron scattering), or from the atomic electron clouds (X-ray scattering and magnetic neutron scattering), as shown in Figs. 3.1 and 3.2. The outgoing particle, consisting of a superposition of scattered

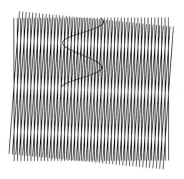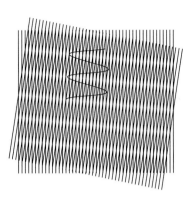

Fig. 3.1 In scattering experiments, we deal with an incoming plane wave, and an outgoing plane wave. The wavelength associated with the plane waves is given by the distance between the vertical lines, the wave crests. The incoming (moving from left to right) and outgoing (moving at an angle) plane waves give rise to an interference pattern that shows up as the white, almost horizontal bands. The distance between successive white bands corresponds to the probing wavelength, which has been plotted in the figure as the sinusoidal shape. The probing wavelength λ_{probe} depends both on the wavelength of the plane waves, and on the scattering angle. The latter is illustrated in the figure by comparing a scattering angle of 5° (left half of figure) and 10° (right half). The probing wavelength is related to the momentum transferred through $\lambda_{\mathrm{probe}} = 2\pi/q$.

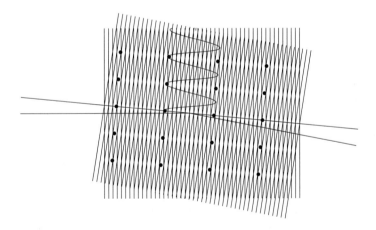

Fig. 3.2 When the interference pattern produced by the incoming wave and outgoing wave corresponds to a natural distance in the sample, then this will result in constructive interference. Here, if the probing wavelength λ_{probe} exactly matches the lattice spacing d of the solid, and if the sample is oriented in just the right way as shown in the figure, then the scattering process depicted might actually occur and a detector placed in the direction of the outgoing wave can detect a scattered neutron (or a photon in X-ray scattering). The solid lines depict that under these scattering conditions the radiation appears to bounce off a mirror-like surface formed by the array of atoms.

waves, will be detected far away from the sample. A particle has a good chance of being detected in a direction where these scattered waves interfere constructively, and a very low chance in the direction where those scattered waves interfere destructively. By the time the outgoing particle is detected, it is so far from the scattering center that it can, once again, be represented by a plane wave. This outgoing plane wave is shown in Fig. 3.1.

From Fig. 3.1 we can determine what spatial arrangement of the atoms would give rise to constructive interference as a function of the direction of the scattered particle. In this figure the areas of the sample that correspond to equal path differences, or integer multiples thereof, show up as bands. Atoms located within these bands, such as the ones shown in Fig. 3.2, all give rise to scattered waves with identical phases, leading to strong constructive interference. Conversely, if there is no one-to-one correspondence between the atomic positions and the bands, then there would be destructive interference. The distance between the white bands corresponds to the probing wavelength λ_{probe}, which depends on both the incident wavelength of the particle λ_i, and the scattering angle 2θ.[1]

In fact, Fig. 3.2 corresponds to Bragg scattering by a single crystal, and it gives rise to Bragg peaks in the diffraction pattern (Squires, 1994). That is, one will only observe scattered particles at some very well defined scattering angles. In the case of liquids, we can still get constructive interference, however, the condition for constructive interference will become fuzzier, resulting in smeared out peaks. We will now likely find some scattered particles at most angles, but there will still be angular ranges where we will find more scattered particles than average. This represents the short-range structure in a liquid that arises from the fact that liquid particles do not sit right on top of each other.

In liquids, at very small scattering angles we will find very few scattered particles. The reason for this can be seen by inspecting Fig. 3.1. Small scattering angles correspond to large probing wavelengths. In this case, the distances between the bands would correspond to many times the average atomic separation in liquids, and hence, there will not be any correlation in position between them. Therefore, scattering originating from these different regions in the liquid will not be in phase; for every scattered wave originating from one particular atom we can find another scattered wave originating from another atom that is exactly out of phase. As a result, we do not get constructive interference at low scattering angles, and we detect very few outgoing particles.

At the other extreme, if λ_{probe} corresponds to the average distance between the atoms in the liquid, then there will be a fair bit of constructive interference, and we can expect quite a few particles to be scattered at the corresponding angle. Finally, if the probing wavelength becomes much smaller than the interatomic spacings, then we will only see interference stemming from the scattered waves of one single atom while it is moving through its surroundings during the scattering event. The particles detected at the corresponding scattering angles will now give us information about the behavior of individual atoms.

---

[1] $\lambda_{\mathrm{probe}} = 2\pi/q = \lambda_i/(2sin\theta)$.

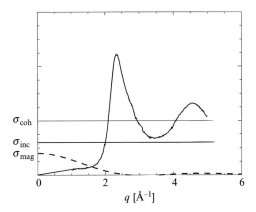

q [Å^{-1}]

Fig. 3.3 Figure adapted with permission from Patty *et al.* (2006). Schematic representation of the components that make up the scattering by a liquid as a function of (reduced) momentum transfer $q = 2\pi/\lambda_{\mathrm{probe}}$: the incoherent, the magnetic, and the coherent contributions. The relative strength of the three contributions depends on the momentum transfer and the element dependent cross-sections σ_{inc}, σ_{mag} and σ_{coh}, respectively. The coherent static structure factor, shown as $\sigma_{\mathrm{coh}}S(q)$ peaks at $q \approx 2\pi/d_{\mathrm{avg}}$, with $d_{\mathrm{avg}} = 2.8$ Å for this particular example representing liquid mercury.

The above is summarized in Fig. 3.3. The solid, oscillating line is the static structure factor $S(q)$ of liquid mercury (multiplied by σ_{coh} in this figure). $\hbar q$ is a measure of how much momentum is transferred to the liquid by the scattered particle, and it is directly related to the probing wavelength: $\hbar q = h/\lambda_{\mathrm{probe}}$. At large λ_{probe} (small q) there is very little possibility of constructive interference, and $S(q)$ is close to zero. When $\lambda_{\mathrm{probe}} \approx d_{\mathrm{average}}$, with d_{average} the average separation between the atoms, we see more constructive interference than average, and we observe a peak in $S(q)$. For dense liquids such as mercury, we can also see peaks corresponding to the next nearest-neighbor distances. For very small λ_{probe} (large q), we observe a single atom, and $S(q)$ will reach a constant value. In order to measure the structure in the liquid, the particle needs to be scattered by multiple atoms and therefore this type of scattering is referred to as coherent scattering. The measure for the overall likelihood of such a scattering event is the coherent cross-section σ_{coh}, an element- and isotope-dependent number in the case of neutron scattering, and an element dependent number in the case of X-ray scattering.

The neutron, being a fermion, also has a magnetic moment; this gives rise to two additional scattering mechanisms Squires (1994). Most, but not all nuclei possess an intrinsic angular momentum, giving rise to scattering through the electromagnetic force. The phase of the scattered neutron depends on the orientation of the nuclear moment. Since the nuclear moments of neighboring atoms do not line up in any pattern at temperatures of interest in liquids, we do not get any constructive interference out of this scattering mechanism. Instead, we only get information about a single atom, independent of the actual probing wavelength. Therefore, this scattering does not display any angular dependence. This process goes by the name of incoherent scattering, and the cross-section for this process (σ_{inc}) is both element and isotope dependent. This scattering mechanism is shown in Fig. 3.3 by a horizontal line at height σ_{inc}. It is through this scattering mechanism that we can measure the self-part of the dynamic structure factor $S_{\mathrm{s}}(q, \omega)$. Note that this scattering mechanism is entirely absent in light scattering since all isotopes have the same number of electrons and therefore the same scattering power. This absence can be a strong advantage since it allows for an easier identification of the coherent contribution to the observed signal.

A mixture of different isotopes will also give rise to incoherent scattering in liquids when probed with neutrons, but not when probed with X-rays. The neutron-scattering cross-section is isotope dependent (since it depends on the number of neutrons already present in the nucleus), hence a mixture of isotopes represents a mixture of cross-sections. Since X-rays scatter off the electron clouds such a mixture does not lead to incoherent scattering in X-ray experiments. We refer the reader to basic neutron-scattering texts for the details on the coherent and incoherent cross-sections when scattering from a mixture of isotopes. The difference between this process and the incoherent scattering process described in the previous paragraph is that the former can flip the intrinsic angular momentum (spin) of the neutron during the scattering process, whereas the latter cannot. This would only be relevant when the neutron-scattering experiment is carried out in full polarization mode, something that is hardly ever done in the study of liquids (for an exception see Section 7.4.2 and the work on liquid potassium by Bermejo *et al.* (2007)). Therefore, we will not go into any more detail about the incoherent scattering mechanisms in this book, it suffices to know that it represents scattering by individual atoms.

The second scattering process arising from the intrinsic angular momentum of the neutron is magnetic scattering. If an ion has unpaired electrons, then the total angular momentum of such unpaired electrons will interact with the neutron. The strength of this scattering mechanism for such electrons is comparable to the nuclear cross-section. Even for magnetic ions that are not lined up in any way with their neighbors, we do observe a strong dependence on λ_{probe}. The reason for this is that we are now scattering off of the electron cloud, which is of course much larger than the size of the nucleus. Therefore, once λ_{probe} becomes comparable to the size of this cloud, we see a diminished scattering cross-section because neutrons scattered by one side of the cloud are slightly out of phase with neutrons scattered by the opposite side, and the constructive interference condition is not fully met. Once λ_{probe} is much smaller than the size of the cloud, constructive interference is no longer possible and the cross-section becomes zero. All this is shown in Fig. 3.3: the magnetic cross-section at very large λ_{probe} is given by σ_{mag}, which in turn depends on the size of the electronic angular momentum; when λ_{probe} decreases, so does the scattering cross-section. The overall decrease of the cross-section as a function of q is known as the magnetic form factor, which turns out to be related to the Fourier transform of the shape and size of the electron cloud (Lovesey, 1971).

3.1.1 Inelastic scattering

A particle can also transfer some of its energy to the sample in the scattering process, or even gain some energy from its interaction with the sample. This process, in which the particle transfers an amount of energy $E = \hbar\omega$ is referred to as inelastic scattering, and the information gained about the liquid in this process is contained in the dynamic structure factor $S_{nn}(q, \omega)$. In the graphical depiction of Fig. 3.1 this would correspond to the scattered particle having a different wavelength from the incoming particle, leading to a slight change in the interference pattern. For instance, a neutron can create a sound wave such as the one depicted in Fig. 1.1 provided the probing wavelength matches the wavelength of the sound mode, and provided the energy transferred to the liquid corresponds to the energy of the sound wave. Thus, when one measures the likelihood that a particle scatters while transferring a certain amount of momentum and energy, one finds strong resonances when these amounts correspond to those of sound waves. This is shown in Fig. 3.4.

The process corresponding to collective diffusion (see Chapter 2) shows up as resonance centered around $\omega = 0$. Such lines are referred to as quasi-elastic lines. Since diffusive processes do not have any particular direction and since they are not propagating, they must be centered around $\omega = 0$. The width Γ of the resonances gives information about the timescale τ that these processes persist for according to the relation: $\Gamma = 1/\tau$. With the aid of the effective eigenmode formalism, the various decay processes that make up $S_{nn}(q, \omega)$ can be distilled from experiment, and the decay rates Γ and the propagation frequencies of particular excitations can be identified, even in those cases where there is a strong overlap between the quasi-elastic features and the true inelastic contributions to the scattering.

The incoherent nuclear scattering mechanism and the magnetic scattering mechanism can also be probed as a function of energy transferred. Since these processes are caused by the scattering by a single atom in liquids, we do not find collective behavior such as propagating waves. Rather, they show up as lines centered around $\omega = 0$, similar to the diffusive collective processes. As before, the width in energy of these somewhat blurry resonances gives information about the timescales involved. Both these processes would only show up in neutron-scattering experiments, they are absent in light-scattering experiments.

Since the cross-sections of all three basic scattering mechanisms are independent of each other, we measure the weighted sum of all three in neutron-scattering experiments. In X-ray scattering we only measure the coherent cross-section. We can distinguish the individual contributions by their angular dependence, as shown in Fig. 3.1. In addition, we can distinguish the coherent scattering from the incoherent (and magnetic) scattering if we also measure whether the neutron's angular momentum is flipped during the scattering process, or not. The strong nuclear interaction does not affect the neutron's spin, but the incoherent (and magnetic) scattering processes can affect the spin. The method of measuring where one also keeps track of the neutron's spin orientation is called polarized neutron scattering, and it is only very rarely used in experiments on liquids. However, it is frequently used in experiments on spin-liquids.

In summary, we find that the likelihood of a neutron being scattered while transferring a certain amount of energy and momentum to the sample – as quantized in the partial differential cross-section $\mathrm{d}^2\sigma/(\mathrm{d}\Omega\mathrm{d}E)$ – is directly proportional to the dynamical processes that take place in the liquid, that is, it is a direct measure of

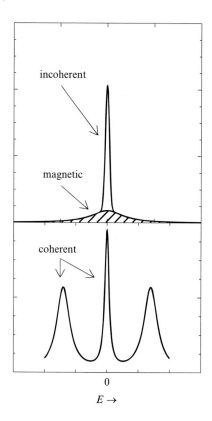

Fig. 3.4 Figure adapted with permission from Patty *et al.* (2006). Schematic representation of the components that make up the scattering by a liquid at low q, as a function of energy E transferred by the neutron: the incoherent, the magnetic, and the coherent contributions. The incoherent contribution consists of a sharp central line representing the self-diffusion and a very weak broad central line representing cage diffusion (the cage-diffusion mode is too small to be seen in this picture: $< 1\%$ of the total intensity). The magnetic contribution, if present, manifests itself as a broad central line. The intensity is element dependent, but it is typically of the same order of the coherent contribution. The coherent contribution consists of the Rayleigh–Brillouin triplet at low q. The central line is due to collective diffusion, the two side lines are the two propagating sound modes. Similar to the incoherent contribution, these three peaks sit on top of a broad cage-diffusion mode whose intensity is too small to be visible in the picture.

what we want to learn about the liquid. In the special case that a liquid is made up of identical atoms, we have (Squires, 1994)

$$\frac{\mathrm{d}^2\sigma}{\mathrm{d}\Omega\mathrm{d}E} = \frac{k_\mathrm{f}}{k_\mathrm{i}}\frac{\sigma_\mathrm{coh}}{4\pi}S_{nn}(q,\omega) + \frac{k_\mathrm{f}}{k_\mathrm{i}}\frac{\sigma_\mathrm{inc}}{4\pi}S_\mathrm{s}(q,\omega) + \frac{k_\mathrm{f}}{k_\mathrm{i}}\frac{\sigma_\mathrm{mag}}{4\pi}S_\mathrm{mag}(q,\omega). \tag{3.1}$$

Here, k_i and k_f are the wave numbers for the incident and scattered particle, respectively. The incident and scattered energy of the particle, E_i and E_f that yield the amount of energy transferred from the particle to the sample $E = E_i - E_f$, are related to k_i and k_f by the conversions listed in Appendix A. The magnetic cross-section σ_{mag} depends both on the direction of momentum transfer and the angular momentum direction of the atom that scatters the neutron, and in the notation of eqn 3.1 it includes the magnetic form factor. In liquids where neighboring magnetic moments, if present, are not likely to line up with each other, this directional dependence simply results in a numerical factor (2/3) reducing the magnetic cross-section (Squires, 1994).

3.2 Corrections to the data

The relationship between the partial differential cross-section and the dynamic structure factor only holds for particles that have been scattered exactly once, and by the sample only. In practice, one has to carry out various corrections before the data represents the relationship given in eqn 3.1. For instance, some particles that were detected at a particular scattering angle have been scattered more than once. Some particles were scattered, but got absorbed by the sample, or by the container. Some particles might be scattered by the container. Perhaps not all particles that are scattered are counted by the detector. Perhaps the number of incident particles is miscounted because the incident beam might have contained particles that have wavelengths given by λ/n that were also selected by the monochromating device. Needless to say, the incoming beam is not perfectly monochromatic, the sample is not point sized, the detectors have finite dimensions, all contributing to a finite experimental resolution.

3.2.1 Standard corrections

Somewhat remarkably, most of these corrections do not present a problem, and all of them can be overcome through careful planning of the experiment, and with some number crunching during the analysis stages. The instrumental resolution function that causes a broadening of the signal in both q and E is measured using a reference sample and folded into the model function when fitting the data. In general, the q resolution is not very important in scattering experiments on liquids (with the exception of scattering by superfluids), and steps taken to ensure a good enough E resolution tend to also yield an acceptable q resolution.

Sample containers are chosen so as to give a minimal amount of background scattering, and empty sample container runs are a standard part of any experiment. In some cases, when the sample is highly absorbent or very strongly scattering, additional precautions are required to take into account the fact that an empty sample chamber scatters more than a filled chamber since the liquid might absorb some of the neutrons (photons) that otherwise would have yielded detectable events. One can use a computer program to calculate the severity of this attenuation problem as a function of energy and momentum transfer, even for a sample (container) of irregular shape. One can also determine the attenuation correction experimentally in the case of neutron scattering: instead of running an empty sample-cell experiment, one can fill the sample cell with an amount of ^{3}He gas that absorbs just as much as the liquid of interest scatters (van Well and de Graaf, 1985). Or one can simply solidify the liquid and use such a dataset as the background dataset. By and large, even under the worst circumstances the attenuation problem has been solved, and there is no good reason not to correct the measured scattering intensities for attenuation effects.

The overall efficiencies of neutron and X-ray detectors are very straightforward to measure, and they can even be calculated for some detectors. The only potential problem that one might encounter is that of detector drift. Especially in neutron-scattering spectrometers where upwards of 1000 detectors are being used simultaneously, it is more likely than not that some detectors will have to be thrown out during the experimental analysis stages. As long as one makes sure that a reference sample such as vanadium is measured both before and after the experiment, then anything to do with malfunctioning detectors does not pose a problem in extracting $S_{nn}(q,\omega)$ from the experimental data.

3.2.2   Monitor-contamination correction

One correction that for some reason or another is frequently overlooked is associated with the counting of the incident number of particles. This is a problem that does not occur on time-of-flight spectrometers such as those that are found at neutron spallation sources, but it is a potential problem at reactor sources and even at synchrotron facilities since it changes the observed lineshape of $S_{nn}(q,\omega)$.

When a single crystal is used to select one particular wavelength λ from a beam of particles of all wavelengths, particles with wavelengths λ/n are also selected. These unwanted particles will also make it to the sample and scatter. However, by using filters such as graphite in the case of neutrons, these particles can still be prevented from making it to the detector. The problem here is that those particles will still have been detected by the incident beam monitor that is used to normalize the scattered intensity to the incident number of particles for each incident energy E_i. The result is that the measured lineshape as a function of energy transfer will be distorted. Note that this will only happen if the spectrometer is operated such that the incident energy of the particle is allowed to vary, while only scattered particles of a certain energy are allowed to make it to the detector. However, this is the standard mode of operation for most spectrometers since by keeping the final energy fixed one does not have to worry about the changing efficiency of the detector as a function of particle energy, and in addition for neutron-scattering experiments, the monitor efficiency is inversely proportional to the incident wave number of the particle, thereby effectively taking care of the k_f/k_i term in the expression for the cross-section (eqn 3.1). Thus, the measured signal will be directly proportional to $S_{nn}(q,\omega)$, rendering the experiment much easier to perform and analyze.

The severity of this monitor contamination problem depends on the incident energy, and on the filters that are present in the incident beam. For instance, for a neutron incident energy of 5 meV ($\lambda = 4$ Å) at a thermal reactor source, neutrons with wavelengths as short as $\lambda/5$ present a problem. As an example, for the Si(1,1,1) reflection at the N5 beam line at the NRU reactor at Chalk River, only 35% of the monitor counts at 5 meV correspond to the desired neutrons, the remainder is all due to higher-order neutrons. For higher incident energies, the problem rapidly diminishes. We plot the measured contamination in Fig. 3.5. There are three ways around this

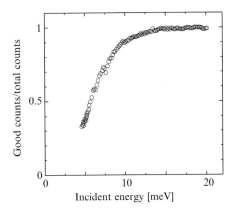

Fig. 3.5 The measured monitor contamination as a function of incident energy for the N5 triple-axis neutron spectrometer at the NRU reactor of the NRC Canadian Neutron Beam Centre. The spectrometer was operated using a Si monochromator with a forbidden second-order reflection. The incident beam had a cooled sapphire filter in it to reduce the fast neutron flux. The data points were obtained by using a set of 5 tin foils, and analyzed using the method described in the text. The data points were measured as part of an experiment on superfluid helium that used a final fixed energy of 5 meV. As can be seen in this figure, the monitor contamination factor changed from about 0.35 in the elastic channel, to about 0.6 for energy transfers of 1.5 meV. Failure to include this monitor correction would therefore have significantly skewed the lineshapes, and the interpretation of this experiment (presented in Figs. 8.3 and 8.4) would have been entirely different. Also note that the fact that the monitor correction is essentially absent for experiments using fixed final energies of 14 meV (a common choice in scattering experiments) is largely because of the use of a Si monochromator. Other monochromators would not eliminate second-order contamination, resulting in a noticeable effect even at these energies.

problem. First, if only $\lambda/2$ contamination would present a problem, then one can use a Si monochromator and use a reflection that does not allow for second-order scattering such as the Si(1,1,1) reflection frequently used in neutron-scattering experiments.

The second way around it only works in neutron-scattering experiments. One employs a velocity selector, which is a spinning collimator that only allows neutrons of a certain range of velocities to pass through; neutrons that go too slow or too fast will be absorbed by the velocity selector. The third solution is simply to measure the contamination. By placing well-characterized absorbers of varying thickness in front of the monitor, one can determine the contamination. We detail this procedure for a neutron-scattering experiment.

In a neutron-scattering setup for a given monochromator reflection, one measures the monitor count rate $R(\lambda_i)$ as a function of incident energy E_i, corresponding to an incident wavelength λ_i (see eqn A.1 in Appendix A). Because of higher-order neutrons also reaching the monitor, this count rate $R(\lambda_i)$ is given by

$$R(\lambda_i) = \sum_{j=1}^{N} \phi(\lambda_i/j)(1 - e^{-C\lambda_i/j}) = C \sum_{j=1}^{N} \phi(\lambda_i/j)(\lambda_i/j). \tag{3.2}$$

Here, C is a measure of the efficiency of the monitor, which is a small number by design. $\phi(\lambda)$ is the neutron flux for neutrons of wavelength λ that will make it to the monitor. The constant C will be eliminated since only the ratio $\phi(\lambda_i)/\sum_{j=1}^{N}\phi(\lambda_i/j)$ is needed to perform the monitor correction.

Next, one places a thin piece of neutron-absorbing material before the monitor, such as a tin foil of known thickness, and one repeats the measurement of the monitor count rate. Subsequently, one doubles the thickness of the foil, and so on, until one has at least as many foil measurements (m) as there are higher orders of contamination present in the incident beam $(m = N)$. For m layers of foil of thickness d, the monitor count rate is given by

$$R(\lambda_i, m) = C \sum_{j=1}^{N} \phi(\lambda_i/j)(\lambda_i/j)\mathrm{e}^{-mb/j}, \qquad (3.3)$$

with the foil parameter b is given by the number density of the foil n_f, the absorbtion cross-section of an atom in the foil σ_abs at neutron energy E_i, and its thickness: $b = dn_\mathrm{f}\sigma_\mathrm{abs}(E_i)$.[2] All measurements can be written in matrix form as $M\phi(\lambda_i/j) = R(\lambda_i, m)/C$, with the matrix elements given by

$$M_{j,m} = (\lambda_i/j)\mathrm{e}^{-mb/j}. \qquad (3.4)$$

Next, one finds the sought-after ratios for each λ_i by applying the least-squares formalism:

$$\phi(\lambda_i/j)/\phi(\lambda_i) = (M^T.M)^{-1}M^T[R(\lambda_i, m)/R(\lambda_i, 0)], \qquad (3.5)$$

with M^T the transpose of the matrix M.

We illustrate the importance of the monitor correction in Figs. 3.5 and 3.6. Figure 3.5 shows the measured contamination on the N5 triple-axis spectrometer at the NRU reactor as a function of incident energy. The N5 spectrometer was operated using a Si monochromator, hence second-order contamination was virtually absent. Nonetheless, it is clear from this figure that the monitor contamination results in a significant skewing of the spectra, especially in experiments where the fixed final energy would be a relatively low number (such setups are employed to obtain a better energy resolution). Figure 3.6 shows the lineshape distortion for data on liquid Ga resulting from the monitor contamination measured for the triple-axis spectrometer TRIAX (Miceli *et al.*, 2005) at the Missouri Research Reactor MURR. This thermal source spectrometer was operated at a fixed final energy of 13.7 meV, and the incident neutron energies were selected using the PG(0,0,2) reflection. As can be seen in this figure, even with sapphire and PG/Si filters in the incident beam, the lineshape distortion due to monitor contamination is rather substantial.

---

[2] The absorbtion cross-section depends inversely on the wavelength of the neutron, or equivalently, the cross-section is proportional to $1/v$. An almost correct way to think about this is that the longer a neutron spends in the nucleus of the scattering atom, the more chance it has of being absorbed. Thus, when we halve the wavelength of a neutron, we also halve the absorption cross-section. This process is behind the $\sim 1/j$ dependence in the exponents of eqns 3.2 and 3.3.

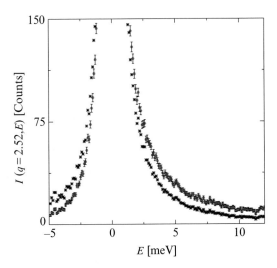

Fig. 3.6 The scattered intensity (circles plus error bars) for supercooled Ga at 293 K for $q = 2.52$ Å^{-1} as measured at a thermal-source triple-axis spectrometer. This particular q value corresponds to the peak in the static structure factor. Applying the measured monitor correction leads to a substantial change in lineshape (stars), illustrating the need to account for this correction before modelling the lineshapes.

3.2.3   Multiple-scattering correction

The direct correspondence between the partial differential cross-section and the dynamic structure factor as given in eqn 3.1 is only valid for particles that have been scattered exactly once. In an actual scattering experiment, this single-scattering condition can never be met. For instance, if a sample scatters 10% of the incident particles, then roughly 10% of the scattered particles will be scattered once more. If instead one opts for an experimental setup where only 1% of the particles are scattered, then the multiple-scattering problem will be of far lesser importance; however, one has to measure for 10 times longer, and the single-scattering signal might even disappear in the background noise. Whatever the setup, multiple scattering will present the largest headache at the low momentum transfers since the cross-section for collective single-scattering events is quite small here (see Fig. 3.3). Hence, at small momentum transfers, multiple scattering might constitute most of the detected signal, and one has to correct the data in order to re-establish the connection (eqn 3.1) between the cross-section and the dynamic structure factor.

Multiple-scattering events should not be confused with multiphonon scattering. We clarify the difference in Fig. 3.7 (Sears, 1975). In a multiphonon process, two excitations are created simultaneously in the scattering event. The multiphonon contribution might be an unwanted contribution since it could interfere with determining the density of states in a solid, or with determining the weight of the single-phonon scattering. Nonetheless, it is a part of the dynamic structure factor $S_{nn}(q, \omega)$, unlike

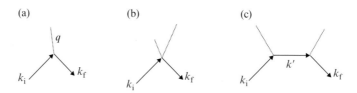

Fig. 3.7 When a particle with incident wavevector $\vec{k}_i$ (arrow) gets scattered by a sample it can excite a density disturbance of momentum $\vec{q}$ (dotted line, part a), or it can excite multiple disturbances (part b). The latter is referred to as the multiphonon component. A particle can also undergo multiple scattering events before exiting the sample (part c), a process referred to as multiple scattering.

multiple-scattering events. In liquids, only superfluid ^{4}He has been shown to exhibit multiphonon scattering (see Chapter 9).

In order to correct for multiple-scattering effects, one needs to know both the geometry of the sample, as well as the dynamic structure factor $S_{nn}(q, \omega)$. Provided the sample does not scatter more than $\sim 10\text{--}20\%$ of the incoming particles, one can calculate the multiple-scattering correction in a straightforward manner. For instance, as a first step one can model $S_{nn}(q, \omega)$ by $S_{nn}(q, \omega) = S(q)\Gamma_q/\pi(\Gamma_q^2 + E^2)$. Here, the linewidth Γ_q is given by the coefficient of self-diffusion D_s as $\Gamma_q = D_s q^2$. The reason why this initial guess for $S_{nn}(q, \omega)$ works remarkably well is because most of the multiple-scattering events originate from scattering processes corresponding to large values of $S(q)$ where the decay time of the excitations is dominated by the properties of individual atoms. In general, the multiple-scattering signal shows very little dependence on scattering angle. Once the multiple-scattering correction has been determined using this initial, somewhat crude model, a new iteration can now be performed where one can use the actual data minus the multiple-scattering correction as a better estimate for $S_{nn}(q, \omega)$.

The multiple-scattering correction, given an initial estimate of $S_{nn}(q, \omega)$, is calculated using a computer. On the one hand, one can use a Monte Carlo routine to generate random scattering events for a particle that enters the sample from one side with a certain initial energy, and that leaves the sample in a fixed direction with a given change in energy. The weight of the scattering events is determined by $S_{nn}(q, \omega)$. On the other hand, one can follow the procedure developed by Sears (1975) and use the computer to integrate over all multiple-scattering possibilities. We describe the latter procedure in some detail since it is very easy to carry out on liquids, yielding corrections to data that not only include corrections for when a particle scatters twice, but also for when it scatters more than two times. The procedure also generates the corrections for the attenuation by the sample as a function of scattering angle and energy transfer. The procedure is the same for both neutron- and X-ray-scattering experiments.

To calculate the relative strength of the multiple-scattering correction for a particle with incident energy E_i and wavevector $\vec{k}_i$ exiting the sample with final energy E_f and

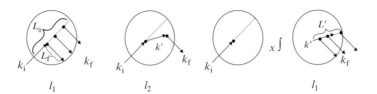

Fig. 3.8 Some of the potential paths that a particle can take when it is scattered by the sample. At every scattering event we have conservation of momentum and energy. The left panel depicts a particle with incoming momentum $\hbar \vec{k}_i$ that is scattered exactly once by the sample, somewhere along its path L_0. The length of the path L_f that the scattered particle has to travel before exiting the sample with momentum $\hbar \vec{k}_f$ depends on where it scattered initially. The three panels on the right sketch how we can decompose events where the particle is scattered twice by the sample into integrals over all possible intermediate scattering events.

wavevector $\vec{k}_f$ (see left panel of Fig. 3.8) the procedure is as follows. First, one comes up with a reasonable model for the dynamic structure factor $S_{nn}(q, \omega)$ as described in the preceding paragraphs. Next, one first calculates the strength of the single scattering contribution as shown in Fig. 3.8. In this figure, the length of the path in the sample that the particle would take if it were not scattered at all is designated L_0. The particle is then forced to scatter somewhere along this path at point x (with x ranging from 0 to L_0), transferring a fixed amount of energy $E = E_i - E_f$ and momentum $\hbar q = |\hbar \vec{k}_i - \hbar \vec{k}_f|$. The length of the path that the scattered particle has to cover before exiting the sample is calculated and designated $L_f(x)$. The weight $W(x)$ of this particular scattering event is

$$W(x) = \mathrm{e}^{-x\Sigma(E_i)} S(q = |\vec{k}_i - \vec{k}_f|, E)\mathrm{e}^{-L_f(x)\Sigma(E_f)}, \tag{3.6}$$

with $\Sigma(E_i)$ given by $\Sigma(E_i) = n(\sigma_{\mathrm{abs}}(E_i) + \sigma_{\mathrm{coh}} + \sigma_{\mathrm{inc}})$. The scattered intensity I_1 (left panel of Fig. 3.8) corresponding to all possible paths for single scattering in a liquid of number density n is then given by

$$I_1(\vec{k}_i, \vec{k}_f, E_i, E_f, L_0) = n\sigma_{\mathrm{coh}} \int_{x=0}^{x=L_0} \mathrm{d}x W(x). \tag{3.7}$$

Note that by replacing $S_{nn}(q, \omega)$ by 1 in eqn 3.6, one would get the formalism for calculating the attenuation correction.

Second-order multiple scattering is calculated in an analogous manner. The initial path L_0 is subdivided as before, but now the particle is allowed to scatter in any direction and emerge with an intermediate energy E' and wavevector $\vec{k}'$ (see Fig. 3.8). After the first scattering event the same procedure is followed as in single scattering by calculating the length of the path L' between the point of scattering and the exit point, and forcing the particle to scatter once more along this path to end up with a final energy of E_f and wavevector $\vec{k}_f$. One then integrates over all intermediate scattering states to get the scattered intensity I_2 corresponding to particles that have been scattered exactly twice:

$$I_2(\vec{k}_i, \vec{k}_f, E_i, E_f) = n\sigma_{coh} \int \frac{d\hat{\mathbf{k}}'}{4\pi} dE' dx S(|\vec{k}_i - \vec{k}'|, E_i - E') e^{-x\Sigma(E_i)} I_1(\vec{k}', \vec{k}_f, E', E_f, L').$$

(3.8)

In this equation, I_1 is calculated with the starting point of the particle's trajectory inside the sample, as shown in Fig. 3.8. Comparing I_1 (eqn 3.7) to I_2 yields the ratio of double scattering to single scattering for a particular momentum and energy transfer. Thus, one can carry out a multiple-scattering correction even when the data have not been absolutely normalized. The above procedure is easy to implement on a computer, and calculation of I_3 can be done in a similar recursive way, where I_3 is expressed as an integral over I_2. Modern computers have enough power to calculate second-order multiple scattering in a minute, and up to third-order multiple scattering corrections in a couple of hours. Hence, there is no longer a need to use approximate methods. Calculations up to third order are sometimes necessary when trying to obtain information on the single-scattering cross-section at low momentum transfers since single scattering can be so weak that it is comparable in strength to the third-order multiple-scattering contribution. Finally, for samples that do not have a constant thickness (such as the one shown in Fig. 3.8) one needs to integrate over all possible L_0.

3.2.4 Potential pitfalls in the analysis of scattering experiments

When all the data have been corrected for all unwanted instrumental effects, then one has obtained the dynamic structure factor $S_{nn}(q,\omega)$ for the liquid under study. Next comes the task of distilling the behavior of the various modes from the data. In the case of well-defined modes, with all characteristic frequencies well separated from each other, this task is rather unproblematic. However, in situations where the sound modes are not well defined, or when they show a strong overlap with, for instance, the heat mode, there are quite a few pitfalls that can easily be avoided. When the fluid is measured at low temperature, life becomes harder still. In this section we illustrate some of the most common errors in data analysis.

For the sake of argument, suppose we have the situation where we anticipate that $S_{nn}(q,\omega)$ is described by the sum of three modes: one diffusive mode (heat mode) and two propagating extended sound modes. We also assume that the decay rates of the modes are such that no clear peaks are visible in $S_{nn}(q,\omega)$; perhaps some weak shoulders at finite frequency ω at best, or perhaps not even that. One way frequently employed to tease the sound propagation frequencies out of this fairly featureless spectrum without too much data analysis is to convert $S_{nn}(q,\omega)$ to the longitudinal velocity–velocity correlation function $C_L(q,\omega)$, and to identify the peak positions in $C_L(q,\omega)$ with the sound propagation frequencies $\pm\omega_s$. We show in Fig. 3.9 that there is very little merit to this method as it only works well in the cases where the sound modes are well defined in the first place, that is, for sound modes whose decay rates are much smaller than their propagation frequencies. This problem has already been discussed in detail in the literature (Bafile *et al.*, 2008).

In order to make the case clear, we look at a very simple model, in which the entire spectra are dominated by the sound modes. Thus, our example is free from the obfuscating influence of the heat mode. In this example, the poles of the dynamic structure factor are given by $\sqrt{f_{un}^2 - \Gamma^2}$, with Γ the decay rate ($\Gamma = z_u/2$). This situation corresponds to $S_{nn}(q, \omega)$ being given by

$$S_{nn}^{\text{sym}}(q, \nu) = \frac{1 - e^{-\beta\hbar\omega}}{\beta\hbar\omega} S_{nn}(q, \omega) = \frac{S_{\text{sym}}(q)}{\pi} \frac{f_{un}^2 z_u}{(f_{un}^2 - \omega^2)^2 + (\omega z_u)^2}. \tag{3.9}$$

We plot the maxima of $S_{nn}(q, \omega)$ and $C_L(q, \omega)$ in Fig. 3.9, both for the classical case $\beta \to 0$ and for the quantum-mechanical case $\beta \to \infty$. As can be seen, the peak positions in $C_L(q, \omega)$ for the classical case are given by f_{un}, the undamped frequency, and these positions just reproduce the f-sum rule (see Appendix B). This latter fact can easily be verified from eqn 3.9. When we turn on quantum effects, the disagreements get even worse (Fig. 3.9). Note that the peak positions of $S_{nn}(q, \omega)$ are not of much help either, especially not in the high-temperature case. In summary, it is best to avoid taking the $C_L(q, \omega)$ shortcut altogether (Bafile *et al.*, 2008).

In order to fully describe the two propagating extended sound modes, only two parameters need to be given, yet this still sometimes leads to a rather confusing nomenclature. The two parameters $f_{un}(q)$ and $z_u(q)$ give a full description. Frequently, the results of the data analysis are presented by listing these two parameters, which most of the time go by the names of $\Omega(q)$ and $\Gamma(q) = z_u(q)/2$. Some groups opt to present their results in terms of two related parameters: $\omega_s(q) = \sqrt{\Omega(q)^2 - \Gamma(q)^2}$ and $\Gamma(q)$. This latter set carries the same information provided the two sound modes

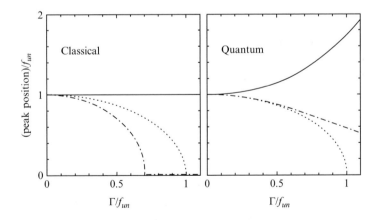

Fig. 3.9 The peak positions of $S_{nn}(q, \omega)$ (dashed-dotted line), those of the velocity–velocity correlation function $C_L(q, \omega)$ (solid line) and the poles of the dynamic susceptibility (dotted line, corresponding to the sound propagation frequency $\omega_s = \sqrt{f_{un}^2 - \Gamma^2}$). This figure was calculated for the case where the excitations are described by a damped harmonic oscillator model. The figure on the left corresponds to very high temperature, the figure on the right corresponds to $T = 0$ K. Note that the position of the peak in neither $S_{nn}(q, \omega)$ nor $C_L(q, \omega)$ is a good measure of the frequency of the underlying excitation when $\Gamma > 0.2 f_{un}$.

are not overdamped. Unfortunately, what is sometimes encountered is that $\Omega(q)$ is presented and interpreted as being $\omega_s(q)$, the propagation frequency of the extended sound modes. This is just plain wrong, but it is a practice that has taken hold such as in the description of the sound modes of water (see Section 6.5). The problem is the following. $\Omega(q)$ is the propagation frequency in the absence of any damping, but it fails to capture the interesting physics in the presence of damping. As an example, take a pendulum clock in the presence of damping (perhaps we hold the clock under water). When we switch on the damping, $\Omega(q)$ will not change, but the clock might well stop running. Therefore, while it is of course perfectly acceptable, and perhaps even recommendable to present the data in terms of $\Omega(q)$ and $\Gamma(q)$, one should not make the mistake of looking upon $\Omega(q)$ as an excitation frequency, or upon anything approximating the poles of the dynamic susceptibility.

While we are discussing the damped harmonic oscillator, another fairly common mistake is to impose propagating modes on the system, even when those two modes might well be overdamped. This mistake stems from the fact that for the case of propagating modes the propagation frequency is given by $\omega_s = \sqrt{f_{un}^2 - \Gamma^2}$ with $\Gamma = z_u/2$ (see eqn 3.9). What is sometimes done in practice is that the converse, namely $f_{un}^2 = \omega_s^2 + \Gamma^2$ is substituted in eqn 3.9 to read

$$S_{nn}^{\mathrm{sym}}(q,\nu) = \frac{S_{\mathrm{sym}}(q)}{\pi} \frac{f_{un}^2 z_u}{(\omega_s^2 + \Gamma^2 - \omega^2)^2 + (2\omega\Gamma)^2}. \tag{3.10}$$

Unfortunately, this is not correct for $f_{un} < z_u/2$, the case corresponding to over-damped modes, which would lead to imaginary values for $\sqrt{f_{un}^2 - \Gamma^2}$. When one tries to fit the experimental data to eqn 3.10 when $f_{un} < z_u/2$, the fit will not yield the desired result of $\omega_s = 0$. Rather, the fit will converge to some finite value for ω_s, and look rather good doing so in the process. This is just an artifact of using the wrong equation to describe the data, and mathematically it corresponds to trying to fit a function that is described by two parameters (f_{un} and z_u), by just one parameter Γ. Instead, when the modes are overdamped ($f_{un} < z_u/2$), the spectrum is described by the sum of *two* lines with different damping rates Γ_1 and Γ_2, and with different signs of amplitude. The proper fit function in this overdamped case is given by

$$S_{nn}^{\mathrm{sym}}(q,\nu) = \frac{S_{\mathrm{sym}}(q)}{\pi} \frac{\Gamma_{s,1}\Gamma_{s,2}(\Gamma_{s,1} + \Gamma_{s,2})}{(\Gamma_{s,1}\Gamma_{s,2} - \omega^2)^2 + (\omega(\Gamma_{s,1} + \Gamma_{s,2})^2)}. \tag{3.11}$$

In this case, the best option is to avoid picking either eqn 3.10 or eqn 3.11 before the fact, but instead to simply stick with eqn 3.9. The temptation to pick eqn 3.10 might be particularly hard to avoid in a very cold liquid such as liquid helium, since here we could have the case shown in the right panel of Fig. 3.9 where $S_{nn}(q,\omega)$ actually shows a nice peak well away from $\omega = 0$ because of the detailed balance pre-factor.

Another mistake that is easily made is leaving too many free parameters in the fit. Every mode, such as the heat mode and the two sound modes, corresponds to a Lorentzian line in $S_{nn}^{\mathrm{sym}}(q,\omega)$. It is tempting to analyze the observed spectra in terms of three Lorentzian lines with free amplitudes for the sound modes and the heat mode. However, the amplitudes of the modes are linked by the eigenvalues of the matrix $G(q)$

just as is the case in hydrodynamics where the amplitudes are linked by γ. In order to avoid introducing too many free parameters, it is best to use the matrix $G(q)$ as a part of the fit function from the outset, or to use one of the expressions written out in Appendix E for the particular model as these expressions contain all the linkages between the modes.

Finally, one oversight that one can make is the following. Let us look again at the damped harmonic oscillator. Equation 3.9 can be written as the sum of two Lorentzians, but these Lorentzians are asymmetric. The reason for this is that the amplitudes A_i of the Lorentzians are, in general, complex numbers (see eqn 2.44):

$$S_{nn}^{\text{sym}}(q,\omega) = \frac{S_{\text{sym}}(q)}{\pi} \text{Re} \sum_{i=1}^{2} \frac{A_i(q)}{i\omega + z_i(q)}. \tag{3.12}$$

When we take the real part of the right-hand side of this equation, the imaginary parts of A_i will combine with the imaginary part of the denominator to yield asymmetric Lorentzian lines (for the case of propagating modes $z_i(q) = \pm i\omega_s(q) + \Gamma_s(q)$):

$$S_{nn}^{\text{sym}}(q,\omega) = \frac{S_{\text{sym}}(q)}{2\pi} \left(\frac{2\Gamma_{\text{s}}(q) + \omega\Gamma_{\text{s}}(q)/\omega_{\text{s}}(q)}{\Gamma_{\text{s}}(q)^2 + [\omega + \omega_{\text{s}}(q)]^2} + \frac{2\Gamma_{\text{s}}(q) - \omega\Gamma_{\text{s}}(q)/\omega_{\text{s}}(q)}{\Gamma_{\text{s}}(q)^2 + [\omega - \omega_{\text{s}}(q)]^2} \right). \tag{3.13}$$

Presuming beforehand that the amplitudes $A_i(q)$ are real would remove the asymmetric parts of the lines. One consequence of this would be that the f-sum rule for $S_{nn}(q,\omega)$ would no longer be satisfied. Of course, the same holds true for the case where any pair of propagating modes combines with non-propagating modes, such as the combination of the heat mode with the two sound modes. Only the amplitudes of the non-propagating modes are real.

3.3 Computer simulations

Computer simulations are very successful in both the interpretation of scattering data, as well as in guiding new experiments. For instance, the existence of a sound mode unique to binary fluid mixtures was first discovered in a computer simulation of a Li–Pb mixture (Bosse *et al.*, 1986). In addition, computer simulations also yield the two other independent correlation functions that are not directly accessible to scattering experiments: the density–temperature and the temperature–temperature correlation function. Ever since the initial simulations of a hard-sphere fluid by Alley and Alder (1983) rapid progress has been made to the point that computer simulations now yield almost perfect agreement with experiments in classical liquids, ranging from simple liquids to colloidal suspensions. A good example of this agreement are the computer simulations that have been carried out on liquid metals (Bove *et al.*, 2005; Scopigno and Ruocco, 2005; Calderin *et al.*, 2008).

One thing computer simulations are not necessarily very good at is simulating quantum fluids. For cold liquids it is very difficult to put in the symmetry of the wavefunction, and one has to resort to the path-integral method. Even in liquids at higher temperature made up of light atoms, one can run into trouble since it is not possible to simulate the effects of quantum diffraction (the Ramsauer–Townsend effect, see Chapter 10) in modelling the collision between two atoms. Notwithstanding these limitations, computer simulations are an essential step in improving our understanding of liquids since they provide the connection between models that can be solved analytically and the complexity that is found in real liquids as measured in experiments.

3.3.1   Molecular dynamics computer simulations

In molecular dynamics computer simulations, the classical equations of motion are solved given an interaction potential. The outcome of such simulations are the intermediate scattering functions $F_{AB}(q, t)$. Starting with some configuration that has the correct number of particles per volume, the particles are assigned random velocities in such a way that their kinetic energy coincides with the desired temperature of the simulation. Next, the net force on a particle is calculated based on its distance to all the other particles, and the velocity of particle i is updated according to $\Delta \vec{v}_i = \vec{F}_i \Delta t' / m$. Here, $\Delta t'$ is a small time step, typically of the order of 0.01 ps (de Schepper *et al.*, 1984*b*). The position of the particle is updated in the same manner: $\Delta \vec{r}_i = \vec{v}_i \Delta t'$. This process is repeated until the simulation reaches equilibrium, which is when the total kinetic energy reaches a constant average value with small fluctuations around it.

Once equilibrium has been reached, the above process of updating the positions and velocities of the particles proceeds, while keeping track of quantities such as $\sum_i e^{i\vec{q}\cdot\vec{r}_i(t')}$ and the other ones listed in Table 2.1. Here, t' is the time of the simulation, it is not yet the time that enters the correlation functions. Once the simulation has generated the desired amount of corresponding real time (having run $N_{t'}$ time steps since reaching equilibrium), the correlation functions are calculated, using a boxcar averaging method to determine the average value $1/N_{t'} \sum_{t'} \sum_i e^{i\vec{q}\cdot\vec{r}_i(t')} \sum_j e^{-i\vec{q}\cdot\vec{r}_j(t'+t)}$. This is shown in

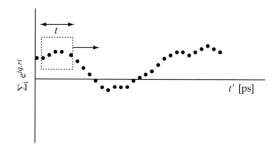

Fig. 3.10 In molecular dynamics computer simulations correlation functions are calculated in real time t by applying a boxcar average to the quantities of interest as they have been calculated in simulation time t'. The figure shows an example for the calculation of the real part of the density–density correlation function for t given by the width of the box: the product of the data points at the box's edges is calculated, after which the box is moved one time step t' and the process is repeated. In this way, one replaces the ensemble average by a time average.

Fig. 3.10. Thus, one uses the Ergoden postulate to replace the ensemble average by a time average, allowing one to calculate the density–density correlation function.

Similarly, one can keep track of the relevant microscopic quantities to calculate the density–temperature and the temperature–temperature correlation function. The entire simulation process, starting from an initial configuration, is then repeated multiple times, and the error bars on the correlation functions are estimated based upon the variance between the various runs.

These simulations are also an excellent way of calculating the self-correlation functions that are probed in neutron-scattering experiments. The density–density auto-correlation function is determined from $1/N_{t'} \sum_{t'} \sum_i e^{i\vec{q}\cdot(\vec{r}_i(t') - \vec{r}_i(t'+t))}$. From this, one can calculate the self-diffusion coefficient as well as the relative strength of the cage-diffusion mechanism as a function of q. This is shown in Fig. 3.11 for a simulation of liquid Hg (Bove *et al.*, 2002; González *et al.*, 2008). In addition, it is straightforward to simulate binary mixtures, and to determine not only the standard correlation functions, but also the partial correlation functions such as the helium–helium density–density correlation function in a helium–neon mixture (see Chapter 6). Of course, simulations are not restricted to classical liquids *per se*, one can also simulate spin-liquids.

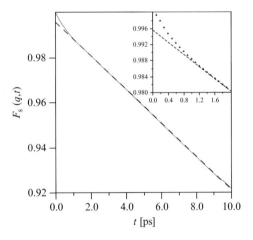

Fig. 3.11 Figure reproduced with permission from Bove *et al.* (2002). The simulated intermediate self-scattering function $F_s(q, t)$ for liquid Hg at 300 K for $q = 0.3 \text{Å}^{-1}$ (dotted curve). The data points deviate from the predictions of the law of self-diffusion (dashed curve) at short times, reflecting the cage-diffusion contribution. During the first 1 ps (see inset), there is a fairly rapid loss of correlation between the atom's initial position and its later position; this is due to the motion of the Hg ion inside the cage formed by its neighbors. The overall loss of correlation (0.004) is very small compared to 1, implying that scattering experiments should be largely insensitive to observing this cage-diffusion mode–as shown in Fig. 3.4 – at this momentum transfer since the $t = 0$ value of an intermediate scattering function yields the intensities in a scattering experiment (a 0.4 % contribution in this case).

The smallest q value that can be probed in simulations (q_{min}) is determined by the size of the simulation as $q_{\mathrm{min}} = 2\pi/L$, where L is the length of the box (if it is cubic). In order to reach lower q values without increasing the size of the simulation, one can employ a rectangular box. However, this limits the maximum correlation time that can be simulated: since one uses periodic boundary conditions, the maximum time is limited by the time it takes a sound wave to cover the shortest length of the box. The easiest way around this limitation is to buy a bigger computer. Overall, molecular dynamics computer simulations are very powerful at interpreting scattering experiments, and at guiding them.

The final note in this section pertains to how to determine the dynamic matrix $G(q)$ from a fit to the simulated $F_{AB}(q,t)$, the exact procedure for which can be found in de Schepper *et al.* (1988). Assume that the correlation functions $F_{AB}(q,t)$ have been determined for a set of unnormalized microscopic quantities $\{a_j(q)\}$ such as the ones given in Table 2.1. Associated with this set is a non-symmetric matrix $H^a(q)$ (eqn B.19) from which a symmetric matrix $G(q)$ can be constructed through $G(q) = U(q)H^a(q)U^{-1}(q)$. Here, $U(q)$ is the matrix that turns the set $\{a_j(q)\}$ (with m members) into the orthonormal set $\{b_j(q)\}$:

$$b_j(q) = \sum_{l=1}^{m} U_{jl}(q)a_l(q). \tag{3.14}$$

The matrix $U(q)$ is given by the equal-time correlation functions $V_{AB}(q) = F_{AB}(q,0)$ through (in matrix notation)

$$U^T(q)U(q) = V^{-1}(q). \tag{3.15}$$

There are two ways to proceed. Either one determines the elements of the non-symmetric matrix $H^a(q)$ through the fit to the simulated correlation functions by using the following function

$$F_{AB}(q,t) = [\mathrm{e}^{-tH^a(q)}V(q)]_{AB}, \tag{3.16}$$

where the matrix element $'AB'$ in the right-hand side of the equation refers to the numbering that has been employed. For instance, if $A = B = n$ has been assigned the number 1 in the set $\{a_j(q)\}$, then the density–density intermediate scattering function $F_{nn}(q,t)$ is given by the $'11'$-element of the matrix on the right-hand side of the equation. Once $H^a(q)$ has been determined, one can transform this matrix to $G(q)$, or not since both matrices carry the same amount of information.

Alternatively, one can directly fit to the symmetric matrix $G(q)$ by employing the following fit function:

$$F_{AB}(q,t) = [U^{-1}(q)\mathrm{e}^{-tG(q)}(U^T)^{-1}(q)]_{AB}. \tag{3.17}$$

3.3.2   Monte Carlo simulations

Monte Carlo simulations for liquids are mostly used to calculate the static, time-independent properties. Starting from some initial configuration and an interaction potential, random moves are generated; these moves will be accepted or rejected based upon whether a reference parameter decreases or increases in value. Typically, the total energy of the system serves as such a reference parameter. After many trial moves, the simulation will reach a state where the reference parameter reaches a stable value, with small fluctuations around this value. The system is then considered to be in equilibrium, and static properties such as $S(q)$ can be calculated. Such a simulation provides a good test bed for proposed interaction potentials, as it allows for a way to determine whether the proposed potential is consistent with measured thermodynamic quantities.

Another way of inferring information about the interaction potential is by doing a reverse Monte Carlo simulation (RMC). The reverse Monte Carlo method has been very successful in analyzing liquids for which only partial datasets are available (McGreevy, 2001). As in standard Monte Carlo, one starts with an initial configuration that represents the correct density. However, one does not use an interaction potential, rather one bases acceptance of a move upon better or worse agreement with experimental datasets. For instance, in the case where the static structure factor of a liquid is available over a limited range of momentum transfers, one first calculates $S(q)_{\mathrm{sim}}$ of the simulated configuration. Then, a random move is generated, and the new $S(q)_{\mathrm{sim}}$ is determined. If the new $S(q)_{\mathrm{sim}}$ agrees better with the measured $S(q)$ then the move is accepted, if the agreement is worse then the move is rejected with an exponential probability depending on how much worse the agreement got. This process is repeated until the level of agreement with the experimental data reaches a stable value, which would represent that equilibrium has been reached in this RMC simulation.

The advantage of RMC is that it allows for the calculation of the pair correlation function $g(r)$ even in cases where the measured $S(q)$ does not span a large enough range of q values to do a direct Fourier transform. In addition, it is easy to combine multiple experimental data sets, such as neutron- and X-ray-scattering measurements, in the acceptance criteria. The main disadvantage of RMC is that it is overdetermined. In essence, for a configuration of N particles, one has $3N$ degrees of freedom, ensuring that a perfect fit with the experimental data can be found provided N is large enough. To ensure that one does not end up with unphysical configurations, or with configurations that violate known parameters such as chemical bond angles, one puts in an additional set of restrictions. For example, one can put an excluded volume around each particle, and reject a move if it were to violate this exclusion zone. The success of a RMC simulation depends strongly on the careful choosing of such additional constraints. If the RMC modelling is carried out according to these principles, then one ends up with a configuration that is the most disordered configuration that is still consistent with all the experimental datasets and imposed chemical and physical constraints.

Once a satisfactory configuration has been obtained that does not violate any known constraints, one has access to additional static properties that cannot be measured directly in a scattering experiment. For instance, it is now possible to calculate the triplet correlation function $g(r, s)$ (Balucani and Zoppi, 1994; Hansen and McDonald, 2006) based upon the simulated configuration. Next, combining the pair and triplet correlation function, one can then solve the hierarchy equations to determine the interaction potential. Thus, by doing RMC modelling, one can (in principle) determine the two-particle interaction potential from experiment, even in the case of dense liquids.

A final Monte Carlo technique to mention briefly is the path-integral Monte Carlo simulation method (PIMC) as applied to quantum fluids. The method dates back to Feynman (Feynman, 1972), and has been applied with great success to superfluid helium by Ceperley (Ceperley and Pollock, 1986; Ceperley, 1995). Feynman showed that the static properties of a Bose-fluid can be mapped onto those of a classical system of interacting polymers, and that therefore a quantum fluid could be simulated using standard Monte Carlo techniques. Using the density matrix formulation of statistical mechanics, PIMC calculates the trajectories of these polymers in imaginary time, and as a result, PIMC produces correlation function in imaginary time. Thus, propagating modes show up as diffusive features in PIMC.

PIMC has produced remarkably accurate results for the pair correlation function in helium (see Fig. 3.12), as well as for the condensate fraction, the superfluid density and the zero-point energy. Perhaps even more impressive is that it has produced accurate values for the phonon–roton excitation energies in superfluid ^{4}He (see Chapter 9). In principle, one can get real-time correlation functions out of a PIMC simulation through a Laplace transform of the imaginary-time correlation functions; in practice, statistical noise makes this rather difficult. Nonetheless, good progress has been made

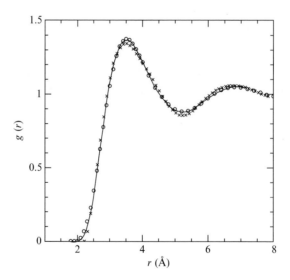

Fig. 3.12 This figure has been reproduced with permission from Ceperley (1995). The pair correlation function $g(r)$ of superfluid ^{4}He at low temperature for a number density of $n = 0.0218$ Å^3. The open circles are the results based on neutron-scattering experiments by Sears and Svensson (1979) for $T = 1.38$ K, the crosses are based on the X-ray-scattering data by Robkoff and Hallock (1981) at 1.38 K, and the solid line is the result of the PIMC simulation at 1.21 K (Ceperley, 1995).

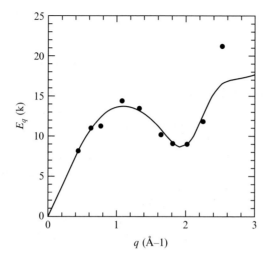

Fig. 3.13 This figure has been reproduced with permission from Ceperley (1995). The phonon–roton dispersion curve representing the excitation energies in superfluid helium (solid line, see Chapter 9) is reasonably well reproduced by PIMC calculations (symbols). In order to calculate these excitation energies from PIMC simulations a Laplace transform had to be carried out in order to convert the imaginary time density–density correlation function to $S_{nn}(q,\omega)$. Such a transformation is very difficult to do because of the presence of statistical noise; however, the results shown here (dots) obtained using maximum-entropy techniques (Silver *et al.*, 1990; Ceperley, 1995) demonstrate that it is not impossible.

(Silver *et al.*, 1990) and accurate values for the propagating modes in superfluid helium have been obtained by applying a maximum-entropy method (Ceperley, 1995). In Fig. 3.13 we show the PIMC results for the excitation energies. These figures (Figs. 3.12 and 3.13) convincingly demonstrate that computer simulations do not need to be restricted to classical systems.

4
Simple liquids

In this book, the term simple liquids refers to those classical monoatomic liquids where the interaction between the particles can be written as the sum of pairwise interactions. Or at least, corrections to the pairwise additive interaction should be fairly negligible. Also, the outer electron shell of the atoms should be completely filled, as it is in the alkali metals. The noble gases are good examples of manifestations of the idealized systems, and of course, so are the systems modelled by computer simulations. It is through the study of such simple liquids that we have learned about the decay mechanisms of fluctuations. In this chapter, we go over the most salient features of the behavior of simple liquids, and we use these features to establish the concept of the extended hydrodynamic modes and their relationship to the microscopic quantities listed in Table 2.1.

4.1   Density fluctuations: general behavior of the extended modes

We know that the spectra of density fluctuations in dense simple liquids can be decomposed as consisting of only three modes. This knowledge has come from molecular dynamics computer simulations, starting with the early work by Alley and Alder (1983) on a system of hard spheres, to the present-day work using very realistic interaction potentials. It has come from the early neutron-scattering experiments on inert gases and their fluids to the present-day X-ray-scattering experiments on every imaginable liquid. And it has come from the theoretical work on the dynamics of hard-sphere systems and on more realistic fluids. In all, it has been a process that has spanned almost half a century. These three modes go by the name of extended heat mode and extended sound modes.

By itself it is most remarkable that the complexity of 10^{23} degrees of freedom can be captured by such a small number of variables. It is as if the liquid acts like one enormous molecule with only very few degrees of freedom available to it to respond to an external perturbation (Cohen and de Schepper, 1990). We call those degrees of freedom the effective eigenmodes of the fluid, and since at very long probing wavelengths these modes are identical to what one observes in hydrodynamics, they are also called the extended effective hydrodynamic eigenmodes, or extended modes for short.

First, the fact that perturbations decay according to exponential functions in time, or equivalently, that their decay gives rise to Lorentzian lines in the dynamic structure factor, is entirely non-surprising. It simply tells us that how much of the disturbance is left after some time depends on how much there was of it in the first place. The fact that we see more than one exponential tells us that some disturbances give rise to others, with the result that we see a coupled decay. What is surprising is that the number of such couplings relevant to the decay of density fluctuations is so limited. The decay of density fluctuations that last for over a nanosecond down to the decay of fluctuations that persist for less than a picosecond, covering length scales comprising thousands of atoms down to just a couple, can all be described by just three exponential functions.

The fact that only three exponentials, or three effective eigenmodes are required does not imply that only three microscopic quantities are relevant to the decay of density fluctuations. This has to do with the character of the modes, by which we mean their amplitudes in scattering experiments. The amplitudes of the three modes relevant to the hydrodynamic region are listed in Table 2.3. The three eigenmodes describing the decay of collective excitations in hydrodynamics, the heat or Rayleigh mode and the two sound or Brillouin modes, have relative amplitudes determined by the thermodynamic state of the liquid (Hansen and McDonald, 2006; Balucani and Zoppi, 1994). When we probe the liquid on shorter and shorter length scales, we find that the relative amplitudes of the three modes change as a function of wave number q. In order to describe these relative changes, we need to include more than three microscopic variables. In fact, it turns out that we need (at least) five. These five variables are listed in Table 2.1 and they are sketched in Fig. 2.12.

We do not fully understand why it is that if we require five microscopic variables, we only observe three exponential functions in the decay mechanism of density fluctuations rather than five. It is almost as if density fluctuations of a certain wavelength decay most effectively by setting up a stress field that disappears quickly, while fluctuations of a different wavelength tend to decay mostly by giving rise to a temperature fluctuation that decays by setting up a heat current. But there does not appear to be a mixing between the two paths of decay of a density fluctuation of a given wavelength. Why this should be the case has not yet been resolved and we will revisit this issue in multiple places in this book.

Alley and Alder (1983) calculated all three independent correlation functions in their molecular dynamics simulations of hard-sphere fluids. They found that the time evolution of all three correlation functions (density–density $F_{nn}(q,t)$, density–temperature $F_{nT}(q,t)$, and temperature–temperature $F_{TT}(q,t)$) could be expressed in terms of three exponentials. It is important to note that the arguments of the three exponentials are the same for all three basic correlation functions, only their amplitudes differ.[1] This implies that all three correlation functions are determined by the same dynamic matrix $G(q)$, such as the ones described in Chapter 2. Subsequent molecular dynamics simulations on fluids that interact through a more realistic Lennard-Jones interparticle potential (e.g., de Schepper *et al.* 1984*b*; Bruin *et al.* 1986) revealed that in fact five exponential functions were required to fully describe all three

---

[1] For example, the decay rate and propagation speed of the extended sound mode would also show up in the density–temperature correlation function. However, the strength (that is, the amplitude) of this mode would be different, implying that this mode would be less or more effective in the decay of the density–temperature correlation function than it is in the decay of the density–density correlation function.

basic correlation functions, albeit that the density–density correlation function still only required three exponentials. Moreover, these later simulations showed that while the density–density correlation function only required three exponentials to describe the decay of fluctuations, in order to understand the relative amplitudes of these three modes over the entire q range one required in fact five microscopic variables (de Schepper *et al.*, 1988).

This three versus five discussion might be a little confusing, so here is a quick summary. When the fluid is probed on a wavelength scale outside of the hydrodynamic region, fluctuations persist for shorter and shorter times. In order to describe the decay of such short-wavelength fluctuations, one requires five microscopic variables, as listed in Table 2.1. The five exponentials generated by these five microscopic variables are sufficient to describe the decay of all basic correlation functions over the entire q range of interest, from large-scale hydrodynamic fluctuations to fluctuations that happen on lengths scales less than the separation between the particles. The relative amplitudes of the various modes of decay are determined by the strengths of the couplings between these five variables, and by the decay rates (see Fig. 2.12). The fact that the number of visible modes in $S_{nn}(q,\omega)$ is reduced from five to three in the description of the decay of the density fluctuations implies that the fluid has preferred channels of decay, potentially those that restore the state of equilibrium most quickly.[2] Finally, there actually might be an exception to this 'three'-rule in very dense supercritical helium. This is discussed in Section 4.6.

In Table 4.1 we summarize the various loosely delineated regions in q space as pertinent to the decay of the correlation functions. These regions are distinguished based on the wavelength of the density fluctuations compared to the three basic length scales of the liquid: the hydrodynamic length scale $l_{\mathrm{H}} = 2\pi/q_{\mathrm{H}}$, the interparticle distance $d = 2\pi/q_{\max}$, and the mean-free path between collisions l_{free}. In dense liquids we have that $l_{\mathrm{free}} \ll d$. The mean-free path between collisions, introduced by Clausius is 1858 (Brush, 1976), serves as a cut-off length scale below which the effective eigenmode formalism loses its usefulness. At length scales shorter than l_{free} we are following individual particles in between collisions, and the decay of correlation functions will be better and better (with decreasing length scales) described by a Gaussian decay function. Of course, a sum of Lorentzian lines is quite a good approximation of a Gaussian lineshape, but it is just not very useful. At this point the approach to ideal-gas behavior is much better described by the approximate theories outlined in Section 2.2.3 rather than by the formalism set up to deal with the decay of collective excitations.

The hydrodynamic length scale l_{H} above which fluctuations decay according to the predictions of hydrodynamics is determined by the criterion that the time it takes for a density fluctuation to decay should greatly exceed the time between collisions, which in the frequency domain corresponds to $\omega\tau_{\mathrm{decay}} \gg 1$. Since the frequencies of interest go up to the frequency of sound propagation $\omega_{\mathrm{s}} = cq$, with c the velocity of

---

[2] We put this statement in not because we necessarily think that it has a good chance of being correct, but rather to draw attention to the fact that we do not have a good explanation for this remarkable property of liquids.

Table 4.1 The change in decay mechanism of density fluctuations as a function of their wavelength $\lambda = 2\pi/q$. The decay trees given in this table are the ones most successfully applied in that particular q region, however, they are not necessarily unique. In particular, for $q \approx 2\pi/d$, more than one decay scheme has been found to yield a satisfactory description of $S_{nn}(q,\omega)$.

length scale	decay tree	main features
$q < q_{\mathrm{H}}$		Propagating sound modes + one diffusive heat mode
$q_{\mathrm{H}} < q < \dfrac{2\pi}{d}$		Propagating extended sound modes + one diffusive extended heat mode. Strong damping of the modes
$q \approx \dfrac{2\pi}{d}$		Strongly to overdamped sound modes + one diffusive extended heat mode
$\dfrac{2\pi}{d} < q < \dfrac{2\pi}{l_{\mathrm{free}}}$		Propagating extended sound modes + one diffusive extended heat mode
$q > \dfrac{2\pi}{l_{\mathrm{free}}}$		Free-streaming + binary collisions

sound, we find that l_{H} is in fact determined by the condition that sound waves should propagate over many wavelengths before they decay. Clearly, the decay rate depends on the fluid in question and on its thermodynamic state, so that the hydrodynamic limit q_{H} differs from fluid to fluid.

In practice, it has been found that $q_{\mathrm{H}} < 0.1$ Å^{-1}. This value is an annoyingly small number since it is very difficult to do scattering experiments – in particular neutron-scattering experiments – at such small momentum transfers. However, it is not impossible to carry out neutron-scattering experiments in the hydrodynamic range, as has been demonstrated in experiments at the Institut Laue-Langevin (ILL), some of which are shown in the next section. Conversely, in Brillouin light-scattering experiments on dense fluids one always measures in the hydrodynamic limit, given the small amount of momentum transferred to the system. The gap between these two techniques can be bridged by X-ray-scattering experiments, however, the resolution in these experiments is not (yet) sufficiently good to be able to resolve all the features of $S_{nn}(q,\omega)$. What is frequently done in practice is that one extrapolates the findings from beyond the hydrodynamic region to $q \to 0$. This procedure seems to work reasonable well for simple liquids, but it does not work as well in the case of more complex liquids, such as binary gas mixtures (Chapter 6) and liquid metals (Chapter 7).

In the following sections we review the behavior of the correlation functions on the various length scales, and we discuss some (very) approximate models for density fluctuations with wavelengths corresponding to the interparticle spacings. The approximate models for these latter short-wavelength fluctuations only serve as a very coarse reference frame for making a modicum of sense out of propagating modes that decay before they have travelled much more than a quarter of a wavelength.

4.2   The hydrodynamic limit and the approach thereof

We center our discussion in this section around the 5×5 matrix $G(q)$ (eqn 2.36) since it is straightforward to illustrate the transition between the hydrodynamic region and the non-hydrodynamic region using this matrix. For small wave numbers q with $q < q_H$ the five eigenmodes of the 5×5 matrix $G(q)$ (see Fig. 2.12) reduce to the usual three hydrodynamic modes plus two kinetic modes for which there is no hydrodynamic equivalent. In particular, the four coupling constants and the three transport coefficients relevant to the decay of density fluctuations show the following limiting behavior for $q \to 0$ (de Schepper *et al.*, 1988), with c the hydrodynamic (adiabatic) speed of sound:

$$
\begin{aligned}
f_{un}(q) &= (1/\gamma)^{1/2}cq + O(q^3) \\
f_{uT}(q) &= (1 - 1/\gamma)^{1/2}cq + O(q^3) \\
f_{u\sigma}(q) &= v_{u\sigma}q + O(q^3) \\
f_{Tq}(q) &= v_{Tq}q + O(q^3) \\
z_{q\sigma}(q) &= O(q) \\
z_\sigma(q) &= z_\sigma(0) + O(q^2) \\
z_q(q) &= z_q(0) + O(q^2).
\end{aligned}
\tag{4.1}
$$

The four coupling constants approach 0 linearly in q, with the constants of proportionality such as $v_{u\sigma}$ given by the thermodynamic quantities describing the liquid state. Explicit expressions for all these quantities are given in de Schepper *et al.* (1988) and in Appendix B.2; here we do not reproduce the low-q expressions for z_σ and z_q as in practice they are determined from experiment.

Using perturbation theory on $G(q)$ in the limit given by eqn 4.1 reproduces the standard hydrodynamic modes listed in Table 2.3. The transport coefficients a and ϕ that enter the decay rates of the hydrodynamic modes are given by:

$$
\begin{aligned}
\gamma a &= \frac{v_{Tq}^2}{z_q(0)} \\
\phi &= \frac{v_{u\sigma}^2}{z_\sigma(0)}.
\end{aligned}
\tag{4.2}
$$

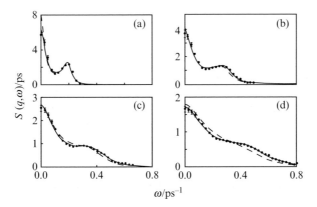

Fig. 4.1 Neutron-scattering data (dots) for $S_{nn}(q, \omega)$ for Ar at $T = 301$ K and $n = 0.0504$ Å^{-3}. The figure has been reproduced with permission from Bafile *et al.* (1990). The dashed curves are the predictions of the usual hydrodynamics equations, the full curves are the extended hydrodynamics equations with fitted parameters. The q values are 0.050 Å^{-1}, 0.075 Å^{-1}, 0.10 Å^{-1} and 0.125 Å^{-1} for panels (a), (b), (c) and (d), respectively. Note that the transition from standard to extended hydrodynamics takes place for $q < 0.1$ Å^{-1}.

In practice, one determines the matrix elements of $G(q)$ from a fitting procedure to the experimental data for all q values,[3] and then one tries to extrapolate these matrix elements to $q \to 0$ according to eqn 4.1. From the limiting values one can then determine the macroscopic quantities of the fluid such as the speed of sound, the specific heat ratio, the thermal diffusivity and the longitudinal viscosity, and these quantities can then be tested against the published values of the same. This procedure clarifies whether the hydrodynamic limit can be reproduced from the scattering experiments. In molecular dynamics computer simulations this limit serves as a reality check on how well the simulation is able to reproduce the modelled state of the liquid. By and large, in simple liquids this extrapolation procedure seems to work well, whereas in binary mixtures and in liquid metals its merits are still in question (see Chapters 6 and 7).

As mentioned, advances in scattering techniques (in particular at the ILL) have made it possible to actually observe the transition from the standard hydrodynamic behavior to the realm of microscopic fluctuations. At least, this has been done for simple liquids such as argon. We show the data obtained by Bafile *et al.* (1990) in Fig. 4.1 for argon at a density of $n = 0.05$ Å^{-3} and temperature $T = 300$ K. These data nicely demonstrate the continuous transition from $q < q_H$ to $q > q_H$, thereby validating the extended eigenmode approach for such simple liquids.

Another way to visualize the hydrodynamic limit of our extended modes is to roll-up the 5×5 matrix $G(q)$ into a 3×3 matrix $G^H(q, z)$, where the inclusion of the z

---

[3] In molecular dynamics computer simulations the coupling constants are directly available as a combination of equal-time correlation functions, see Appendix B.2.

dependence ensures that no information has been lost. Specifically, $G^{\mathrm{H}}(q, z)$ reads (de Schepper *et al.*, 1988):

$$G^{\mathrm{H}}(q, z) = \begin{pmatrix} 0 & \mathrm{i}f_{un}(q) & 0 \\ \mathrm{i}f_{un}(q) & z_\phi(q, z) & \mathrm{i}f_{uT}(q) + \mathrm{i}\Delta(q, z) \\ 0 & \mathrm{i}f_{uT}(q) + \mathrm{i}\Delta(q, z) & z_T(q, z) \end{pmatrix}. \qquad (4.3)$$

Here, the generalized q and z-dependent transport quantities $z_\phi(q, z)$, $z_T(q, z)$ and $\Delta(q, z)$ are given by

$$z_\phi(q, z) = \frac{f_{u\sigma}(q)^2}{z + z_\sigma(q) + z_{q\sigma}(q)^2/[z + z_q(q)]},$$

$$z_T(q, z) = \frac{f_{Tq}(q)^2}{z + z_q(q) + z_{q\sigma}(q)^2/[z + z_\sigma(q)]}, \qquad (4.4)$$

$$\Delta(q, z) = \frac{-f_{u\sigma}(q)f_{Tq}(q)z_{q\sigma}(q)}{[z + z_\sigma(q)][z + z_q(q)] + z_{q\sigma}(q)^2}.$$

When taking the $z = 0$ limit and upon taking into account the hydrodynamic limit as given by eqn 4.1, it is straightforward to see that this matrix reduces to the standard hydrodynamic matrix with the transport coefficients given by eqn 4.2. In addition, eqn 4.5 also provides a working definition of the generalized transport coefficients as encountered in the memory function approach. Based upon these expressions it is possible to generalize the memory function approach from the viscoelastic models (Figs. 2.14 and 2.15) to more extended models that include more decay channels. We return to this in Section 7.1.1.

The above generalized hydrodynamic matrix $G^{\mathrm{H}}(q, z)$ also illustrates an important experimental difficulty. The 2×2 submatrix involving the z-dependent quantities is fully symmetric in $z_\sigma(q)$ and $z_q(q)$. Therefore, if the experimental fitting procedure reveals that only one of the decay rates is required at finite q in order to produce a satisfactory description of the data, then it is not possible from the study of the density–density correlation function alone to assess which one of the decay channels is irrelevant. In order to make the latter assessment, one needs to have access to the two other independent correlation functions, namely the density–temperature and the temperature–temperature correlation function. At present, these latter two can only be evaluated by doing molecular dynamics computer simulations. One might argue that it should be possible to assess the character of the eigenmodes by simply following the change of character from the hydrodynamic region to the region of interest. In practice, this method is far from reliable. First, the hydrodynamic region is often not accessible to experiments, in particular in the study of mixtures (Chapter 6) and liquid metals (Chapter 7). Secondly, the changes in character of the modes can occur over a very narrow window in q range, as shown in the figures in Appendix D. Thirdly, when modes cross, or when parameters such as the two damping rates z_σ and z_q reach similar values, it is simply not possible to disentangle the resulting character of the modes based on the study of the density–density correlation function alone.

4.3   Beyond hydrodynamics

The region with $q > q_H$ is the traditional realm of neutron- and X-ray-scattering experiments. The early development of the effective eigenmode formalism (de Schepper and Cohen, 1982; de Schepper *et al.*, 1984*c*; Zuilhof *et al.*, 1984; de Schepper *et al.*, 1984*b*; McGreevy and Mitchell, 1985; Cohen *et al.*, 1986) is largely based on the results for the inert gases krypton, argon and neon, and on the knowledge gleaned from molecular dynamics computer simulations. We show an example of neutron-scattering data on high-density neon (van Well and de Graaf, 1985) in Fig. 4.2. While these experiments can nowadays be performed with higher statistical accuracy, all the essential information is already visible in these spectra.

The data on neon were fitted to the sum of an extended heat mode and two extended sound modes (eqn 2.37). The first thing to notice in Fig. 4.2 is the very good description that the three extended eigenmodes provide. For all q values shown in this figure, from just above the hydrodynamic range to the position of the main peak in $S(q)$ at $q = 2.2$ Å^{-1}, the fitted modes follow the data extremely well. At the lowest q values the extended sound modes show up as distinct shoulders in the spectra. With increasing q, these shoulders become part of the broad central ($\omega = 0$) feature. Nonetheless, the extended eigenmode analysis reveals that this central feature is not well represented by a single Lorentzian contribution, not even for a q value corresponding to the main

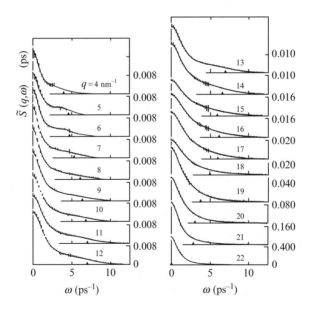

Fig. 4.2 Neutron-scattering data (vertical bars) for $S_{nn}(q,\omega)$ for Ne at $T = 35.1$ K and $n = 0.0346$ Å^{-3}. The figure has been reproduced with permission from van Well and de Graaf (1985). The full curves show the result of a fitting procedure with three extended eigenmodes and the black triangles denote the propagation frequencies of the extended sound modes. The q values of the spectra are shown in the figure in nm^{-1}.

peak in $S(q)$. One note on the fitting procedure employed in the data analysis of these neon scattering experiments is that not one particular decay tree was chosen a priori. What was done instead (van Well and de Graaf, 1985) is that the data were fitted to a sum of three Lorentzian lines – one diffusive one and two propagating ones that were allowed to become overdamped – in such a way that the amplitudes were not used as free parameters, rather they were determined from the restrictions on the frequency-sum rules (see Appendix B and eqn 2.38).

Starting out at the lowest q values, the results for the q-dependent propagation frequencies of the extended sound modes (denoted by the triangles in Fig. 4.2) can be seen to increase with increasing q, after which the propagation frequency reaches a maximum around $q \simeq 1.2$ Å^{-1}, followed by a marked softening, and a so-called propagation gap at $q = 2.2$ Å^{-1}. This type of behavior of the sound mode is fairly typical for a fluid of comparable density, and a sound-propagation gap has now been observed in a range of fluids. In Section 4.4.1 we discuss a crude physical picture to aid with the interpretation of microscopic sound modes with a wavelength comparable to the interatomic spacings d.

The difference between propagating and non-propagating short-wavelength modes is more than mere semantics. For instance, the superfluid to normal fluid transition in pressurized ^{4}He is marked by a complete softening of the extended sound modes, and this softening takes place in a very narrow temperature region just below the superfluid transition temperature (see Chapter 9). The extent (in q) of the propagation gap is strongly density dependent. In Fig. 4.3 we show that this propagation gap in Ar (de Schepper *et al.*, 1983) extends over a range of q values, implying that sound modes of a range of wavelengths λ with $\lambda \approx d$ cannot propagate; instead we find two diffusive modes, one of which has a negative amplitude.

The density dependence of the extent of the propagation gap has been determined by molecular dynamics computer simulations (Bruin *et al.*, 1986), and by the Enskog

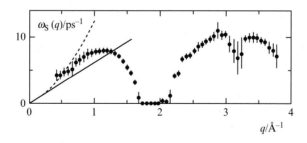

Fig. 4.3 The propagation frequency (dots) of the extended modes as obtained from an extended eigenmode analysis of neutron-scattering data for $S_{nn}(q,\omega)$ for Ar at $T = 120$ K and $n = 0.0176$ Å^{-3}. The figure has been reproduced with permission from de Schepper *et al.* (1983). The propagation gap around $q = 1.9$ Å^{-1} – the position of the main maximum in $S(q)$ – is clearly visible. The solid line is the hydrodynamic dispersion curve given by cq, and the dotted line is the prediction of mode-coupling theory $cq + aq^{5/2}$ (Hansen and McDonald, 2006).

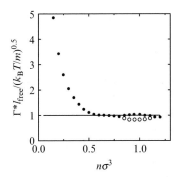

Fig. 4.4 The calculated damping Γ of sound waves of wavelength $\lambda = \sigma$ based on the Enskog theory for hard spheres (circles) as a function of reduced density $n\sigma^3$. The damping has been scaled using the mean-free path between collisions l_{free} and the thermal velocity $(\beta m)^{-1}$. The region where Γ is double valued is the region where the sound waves are overdamped (as in an harmonic oscillator), resulting in two values for the damping. Note that this propagation gap where the sound waves are overdamped disappears again at the very highest densities. The horizontal line is the prediction of eqn 4.6.

theory (Kamgar-Parsi *et al.*, 1987) for hard spheres (see Fig. 4.4). The main findings indicate that the appearance of a gap is a manifestation of a packing problem, albeit with a small twist. At low densities, the extended sound modes show a propagating, yet strongly damped behavior for $\lambda \approx d$. Then, with increased density, a gap appears, and the extent (in q) of the gap also increases upon a further increase in density. Then, upon an even further increase, the extent of the propagation gap decreases, and at even higher densities yet the propagation gap disappears (Verkerk *et al.*, 1987). At these solid-like densities it is found that short-wavelength sound modes can once again propagate. The reason behind all this is not entirely clear yet; for instance, there is no model that predicts at what density a propagation gap should appear, and when it should disappear again. Also, there does not appear to be any direct connection between the macroscopic transport coefficients in simple classical liquids, and the appearance of a propagation gap.

The above analysis of $S_{nn}(q,\omega)$ in terms of three effective eigenmodes does not tell us whether the velocity disturbance that arose from the density fluctuation decays by setting up a stress field, or whether it mainly decays through the coupling with the temperature (which in turn generates a heat flux). The only way that this information can be inferred is by analyzing other, independent correlation functions. This has been done in the effective eigenmode analysis of molecular dynamics computer simulations for a fluid interacting though a Lennard-Jones potential. As in the earlier simulations on hard-sphere fluids (Alley and Alder, 1983), in this simulation all three independent correlation functions were determined (de Schepper *et al.*, 1988), and fitted to the 5×5 dynamic matrix $G(q)$ (eqn 2.36). In Fig. 4.5 we show the full set of matrix elements that was determined in this study for a liquid state that is representative of argon at its triple point.

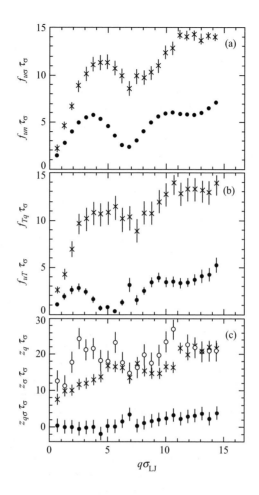

Fig. 4.5 Figure reproduced with permission from de Schepper *et al.* (1988). Shown are the seven matrix elements of $G(q)$ (see Fig. 2.12) for a Lennard-Jones fluid corresponding to liquid argon at its triple point. The matrix elements have been determined from a fit to all three basic correlation functions. Note the linear in q behavior for small q values of the four coupling constants, and that $z_{q\sigma} \approx 0$ over the entire q range. The data have been plotted in reduced units, with $\sigma_{\rm LJ} = 3.36$ Å^{-1} and $\tau_\sigma = 0.8$ ps.

The matrix elements of $G(q)$ shown in Fig. 4.5 display the expected limiting low-q behavior as given by eqn 4.1. The four coupling constants are linear functions of q for small q values, whereas the two decay rates $z_q(q)$ and $z_\sigma(q)$ do indeed approach finite values when $q \to 0$. The predicted low-q behavior of $z_{q\sigma}(q)$ (the coupling parameter given by the dotted line in Fig. 2.12) could not be ascertained as this parameter appears to be very close to zero for all q values probed.

Once all the matrix elements of $G(q)$ have been determined, the eigenvalues and eigenfunctions of the matrix can be calculated, and the individual modes can be scrutinized. For this Lennard-Jones computer simulation, $S_{nn}(q, \omega)$ is well described by the sum of three effective modes, which as usual are referred to as the extended heat and sound modes. These modes are plotted in Fig. 4.6. As can be seen in this figure, these modes tend continuously to their hydrodynamic equivalent in the limit $q \to 0$. The other two modes following from the 5×5 matrix description are the extensions of the two kinetic modes at $q = 0$. The eigenvectors of these two modes do not have a component on the density at the lowest q values, that is, $\psi_1(q)$ is zero or very close to zero for these two modes (see eqn 2.44). Therefore, these two modes are not visible in $S_{nn}(q, \omega)$, but they could well be visible in other correlation functions such as $S_{TT}(q, \omega)$. Note that the two 'invisible' modes are not necessarily representative of the microscopic quantities 'σ' and 'q'; this is only strictly the case near $q = 0$. When we scrutinize the modes at finite q, we find that these two modes can acquire components on the density and that they will start to change the character of the other eigenmodes, or even become an eigenmode themselves such as the heat mode. Examples of this changing character are shown in Figs. 2.13, 2.20, 2.21 and in the figures of Appendix D.

Thus, the character of the three extended modes changes as a function of q. At $q = 0$, the eigenvalues of these three modes only have components on 'n', 'u' and 'T'. However, with increasing q components on 'σ' and 'q' start to mix in, representing that in order to describe the decay of density fluctuations we have to expand the decay tree from the hydrodynamics decay tree; it now becomes important how, for instance, the temperature disturbance relaxes through the heat current (also referred to as heat flux). The same also holds for the other correlation functions such as $S_{TT}(q, \omega)$. We could have the scenario that the heat current set up by the temperature disturbance does not act as a total power sink. Instead, some power could come back through the coupling constant f_{Tq}, which in turn could be partly dissipated through the $f_{uT} \to f_{u\sigma} \to z_\sigma$ route. This is just a confusing way of saying that intermixing of the microscopic variables occurs. Thus, despite the names of the extended heat mode and sound modes, we should not picture these modes as scaled-down versions of the modes in hydrodynamics, accomplishing the same thing by the same means except on a smaller length scale. Once again, this also illustrates why we cannot fully determine the character of the extended heat and sound modes by simply measuring $S_{nn}(q, \omega)$: we can certainly determine the eigenvalues of these modes, but we cannot ascertain whether their eigenvectors have component on 'σ', or on 'q'.

Another interesting aspect of Fig. 4.6 is that the Enskog theory for an equivalent hard-sphere fluid (van Beijeren and Ernst, 1973) does a very good job at predicting the behavior of the extended heat mode, but a pretty poor job when it comes to the sound modes. On the one hand, this nicely represents that we are still severely lacking in a microscopical understanding of what happens in liquids given that the Enskog theory is the best theory that we have to date in terms of a full microscopic description of dense fluids. On the other hand, it tells us that in dense classical fluids the heat mode is by far the most dominant contribution to the relaxation of density fluctuations. This latter statement is even more correct for the region $q \simeq 2\pi/d$. In the next section we will look more closely at this region, and we discuss that the heat mode ends up dominating the scattering spectra in this region.

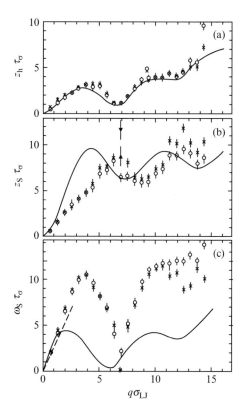

Fig. 4.6 Figure reproduced with permission from de Schepper *et al.* (1988). Shown are the three extended eigenmodes of the Lennard-Jones fluid described in Fig. 4.5. The top panel shows the extended heat mode, the middle panel the decay rate of the extended sound modes, and the bottom panel its propagation frequency. The solid lines are the predictions for an equivalent hard-sphere fluid, as calculated using the Enskog theory. While this prediction is accurate for the extended heat mode, it leaves quite a bit to be desired when it comes to the extended sound modes. The reduced units of this figure are the same as in the preceding figure.

4.4 The response for $q\sigma \simeq 2\pi$

When the probing wavelength λ_{probe} corresponds to a natural length scale in a system, we can expect a much stronger response than when there is no match. In liquids, the natural length scale is the average distance d_{avg} between the particles. In very dense liquids, this average distance is almost equal to the diameter of the particles σ. Indeed, when $\lambda_{\text{probe}} \simeq d_{\text{avg}}$ we find that the static structure factor $S(q)$ peaks (Fig. 3.3) and that the characteristic width (in energy) of the spectra reaches a minimum. Sometimes, we find that the sound modes cease to propagate. This is the region in q space where we see the dynamics of a liquid in its purest form, that of individual particles moving between collisions, and colliding with their neighbors.

If we ultimately are to make the connection between the microscopic dynamics and the macroscopic transport coefficients, it is imperative that we fully understand the region where $\lambda_{\text{probe}} = 2\pi/q \simeq d_{\text{avg}}$, or $qd_{\text{avg}} \simeq q\sigma \simeq 2\pi$. At the moment we do not fully understand this region, but we are not entirely in the dark either.

4.4.1 Visualizations of the excitations for $q\sigma \simeq 2\pi$

The average distance d_{avg} between the atoms can be estimated from the density of the liquid. Actually, it can only be guessed since there is no watertight procedure for determining the density dependence of d_{avg}. Of course, the separation d_{avg} depends on the number density n of the liquid as $d_{\text{avg}} \sim n^{-1/3}$. Estimating the proportionality factor is a somewhat nebulous undertaking because unlike in a solid, we do not have a nice periodic arrangement. For our estimate we use the fact that a dense liquid fairly accurately resembles a system of closely packed spheres. We clarify what we mean by this in Fig. 4.7. For a very dense fluid, such as a hard-sphere fluid close to melting, we have that the arrangement of the atoms is starting to resemble a close-packed structure. The hard-sphere atoms do not necessarily all touch each other as they do in a close-packed structure, since the freezing point is actually determined by the point when the particles can no longer stream past each other. The volume per particle V_{melt} at this point is given by $V_{\text{melt}} \approx 1.52 V_{\text{cp}}$ (see Appendix C). Here, V_{cp} is the volume per particle at closest packing. We have sketched this situation in the left panel of Fig. 4.7. As can be seen in this panel, we would get a very good estimate of the average distance between the atoms if we assume a close-packed arrangement for the *volume cut out* by each individual particle, which is essentially the volume of the cage that the particle finds itself locked up in. Given that the fraction of the volume occupied by closely packed atoms in a liquid is $\pi/3\sqrt{2} = 0.741$, we estimate the atomic separation for this case resembling close packing to be

$$d_{\text{avg}} = [\pi/(3n\sqrt{2})]^{1/3} = 0.905/n^{1/3}. \tag{4.5}$$

Fig. 4.7 The average separation between atoms is, by and large, given by a close-packed structure. The atoms, depicted by the gray disks, cut out a volume for themselves (shown by the larger, transparent disks) by colliding with their neighbors and thereby keeping them at a distance. Since all atoms have very similar kinetic energies, we can expect the cages that they create to be of very similar sizes. These cages are stacked in a pattern that closely resembles a close-packed structure. The three panels depict the situation at high density (left panel), intermediate density (middle panel), and low density (right panel). The low density is not so low that cages have no longer formed, but it is close to the borderline. We have plotted the change in density by plotting smaller atoms, rather than drawing bigger boxes.

This separation is slightly lower than that for a simple cubic structure ($d_{avg} = (1/n)^{1/3}$), which simply tells us that the atoms are closer together in a close-packed structure. At lower densities, the situation does not essentially change as long as the particles find themselves locked up in cages. This situation is depicted in the middle and right panels of Fig. 4.7. Thus, also for the case of lower densities we can expect eqn 4.5 to be an accurate predictor of the average distance between particles. The only times that we would expect this equation to break down is at very low densities, or if we are dealing with liquids that do not solidify in a close-packed structure, such as gallium and mercury.

Not only can we expect the static structure factor to display its main peak at $q \simeq 2\pi/d_{avg}$, we also expect the characteristic width of $S_{nn}(q, \omega)$ to exhibit a minimum here. The reason behind this is easy to see: when imposing a disturbance on a liquid, it costs a certain amount of energy. When the wavelength of this disturbance matches a natural length scale, then the amount of energy required should be minimal since the average structure of the liquid is conducive to supporting such a disturbance. By the same token, once such a disturbance has been created, it should persist for a relatively long time τ because the structure of the liquid prevents a quick relaxation, and the characteristic width $\omega_H = 1/\tau$ should reach a minimum. This process is known as structural slowing down, and it leads to the well-known de Gennes minimum (de Gennes, 1959; Hansen and McDonald, 2006). In Fig. 4.8 we plot the position $q_{min} = 2\pi/d_{avg}$ of this de Gennes minimum for a range of fluids, and compare it to our estimate (eqn 4.5) for d_{avg}. Clearly, the estimate is accurate enough to be useful in setting up experiments on liquids. We note that this relationship is already known for superfluid helium, where the position of the roton minimum (see Chapter 9) is accurately given by eqn 4.5 for all superfluid densities.

Next, we try to develop a visual image for the extended heat and sound modes for wavelengths $\lambda = d_{avg}$. In Fig. 4.9 we sketch a collision between two particles. It is clear from this figure that such a collision is probed when $\lambda_{probe} \approx d_{avg} \approx \sigma$. By the same token, the movement of a single particle between collisions is also probed on this wavelength, as shown in Fig. 4.10 This identification immediately tells us that a model that describes the dynamics on this length scale is bound to have the coefficient of self-diffusion D_s in it. After all, the movement between collisions as well as binary collisions are what determines this coefficient. Indeed, a very successful approximation to the half-width of $S_{nn}(q, \omega)$ for q values near $2\pi/\sigma$ does contain the coefficient for self-diffusion as the main parameter. This model is detailed in the next section, what we want to establish in the remainder of this section is an order of magnitude estimate for the heat and sound modes. Note that the following is a mere order of magnitude estimate intended to serve as an illustration of what ingredients go into the heat and sound modes at these short length scales, it does not contain enough physics to produce accurate results. In other words, do not quote us on this (unless it happens to work).

In order to get to our order of magnitude estimate, we need to make an additional assumption. The dynamic structure factor is a measure of the density–density correlations. While it is clear from Figs. 4.9 and 4.10 that the probing wavelength corresponds to the right length scale to follow collisions, it is not exactly clear what the density fluctuation is that we are probing. The additional assumption that we have to make

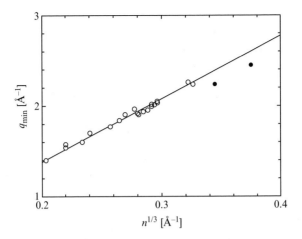

Fig. 4.8 The position q_{min} of the de Gennes minimum for a range of fluids (symbols) compared to the estimate $q_{\mathrm{min}} = 2\pi/d_{\mathrm{avg}} \simeq 2\pi/[\pi/(3n\sqrt{2})]^{1/3}$ (solid line, see eqn 4.5). The data in this figure have been compiled by Dylan Moore from published literature data. The simple fluids (inert gases and alkali metals – open symbols) follow the prediction quite well, whereas Hg and Ga (filled symbols) clearly deviate from this prediction. The latter is not entirely unexpected as these two elements do not condense into a close-packed structure. The data plotted are for Kr (Egelstaff *et al.*, 1983), K (Cabrillo *et al.*, 2002), Ne (van Well and de Graaf, 1985), Ar (van Well *et al.*, 1985), superfluid He (Dietrich *et al.*, 1972; Montfrooij and de Schepper, 1995) where the position of the roton minimum has been plotted, Cs (Bodensteiner *et al.*, 1992), Hg (Badyal *et al.*, 2003), and Ga (Scopigno *et al.*, 2005*a*). Typical error bars on q_{min} are ± 0.1 Å^{-1}.

is that there is no enormous difference between a density disturbance of wavelength $\lambda = \sigma$, and the atom itself, provided that the atoms are close together (as in a dense liquid). This is clarified in Fig. 4.10. The center of the atom represents a density higher than the average density, whereas the edge of the particle represents a density that is lower than average, provided we are dealing with a dense liquid. We use this identification – of an atom being very similar to a short-wavelength density fluctuation – for our estimates for the heat and sound modes.

We will start with the sound modes. For the ease of calculation, we assume that the two colliding particles in Fig. 4.9 are hard spheres of diameter σ. The average speed of a particle is given by the Maxwellian average $v_{\mathrm{rms}} = \sqrt{3k_{\mathrm{B}}T/m}$, and it will therefore have a velocity of $\sqrt{k_{\mathrm{B}}T/m}$ along any particular direction, such as along the line of collision. Thus, without too much loss of generality, we can assume that one particle in a collision has a speed along the line of collision of $\sqrt{k_{\mathrm{B}}T/m}$, while the other particle has on average a speed of zero along the line of collision. Viewed in the center-of-mass system, this collision corresponds to two particles each approaching the other one at a speed of $\sqrt{k_{\mathrm{B}}T/m/2}$ along the line of collision, after which they move away from each other at $\sqrt{k_{\mathrm{B}}T/m/2}$ (Fig. 4.9). Using our additional assumption, this process is equivalent to two sound waves of very short wavelength, propagating in opposite directions.

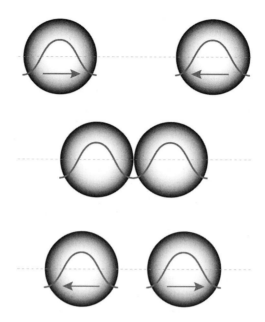

Fig. 4.9 The collision between two particles as seen in the center-of-mass system. There is a strong resemblance between this collision, and a pair of sound modes of wave length $\lambda \simeq \sigma$ propagating in opposite directions. Figure rendered by Alexander Schmets.

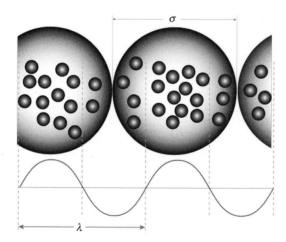

Fig. 4.10 Just as a disturbance of a wavelength spanning quite a few particles (smaller spheres) can be viewed as a departure from the average density (see also Fig. 1.1), on a much smaller scale an atom itself (larger spheres) represents a departure from the average density. Figure rendered by Alexander Schmets.

We now calculate the damping of such sound waves. Clearly, the sound waves as depicted in Fig. 4.9 will cease to exist if either one of the particles collides with a third particle, because then the motion of the two original particles will no longer be correlated. If we denote the mean-free path (the average distance between collisions) by l_{free} then we find for the damping rate Γ of the sound waves

$$\Gamma = \frac{\sqrt{k_{\mathrm{B}}T/m}/2 + \sqrt{k_{\mathrm{B}}T/m}/2}{l_{\text{free}}} = \frac{\sqrt{k_{\mathrm{B}}T/m}}{l_{\text{free}}}; \quad \frac{\Gamma l_{\text{free}}}{\sqrt{k_{\mathrm{B}}T/m}} = 1. \qquad (4.6)$$

Thus, the damping rate of the sound modes should be inversely proportional to the distance between collisions. We have calculated the left-hand side of eqn 4.6 for a hard-sphere fluid using the Enskog theory for hard spheres for density disturbances of wavelength $\lambda = \sigma$ as a function of density. The results are shown in Fig. 4.4. As can be seen, the agreement between hard-sphere results and our simple model (eqn 4.6) is very good for densities $n\sigma^3 \geq 0.55$. As expected, our simple picture breaks down for low densities since the average separation between the atoms then greatly exceeds the diameter of the particles. Perhaps the most surprising observation is how accurate the model is in the case of a classical hard-sphere fluid. This is probably fortuitous, but the fact that the damping rate is inversely proportional to the mean-free path should also be relevant in real dense fluids, and at the very least we have gained some insight into what these short-wavelength sound modes are, and what determines their damping rates.

The other mode in hydrodynamics, the heat mode associated with the decay of temperature fluctuations and whose characteristic time is determined by the thermal diffusivity of a liquid, also has an equivalent at short wavelengths. As mentioned in the introduction, when a density fluctuation is present in a liquid, one also encounters a temperature fluctuation. This excess heat will diffuse through collisions of atoms with their neighbors. The overall process is determined by a standard diffusion equation and as in any diffusive process its characteristic decay time τ is given by $\tau^{-1} = B(2\pi/\lambda)^2$, with B the unknown microscopic diffusion coefficient relevant to this process (see also eqn 2.15).

Using the same simple model that gained us some insight into the mechanism of short-wavelength sound modes, we can link the coefficient of diffusion for this process for wavelengths $\lambda \simeq \sigma$ to the single-particle diffusion coefficient D_{s}. In fact, since in our model the atom itself represents a density fluctuation of wavelength $\lambda \simeq \sigma$, the movement and collisions of this particle represents the decay of this short-wavelength density fluctuation. On these length scales, the density fluctuation is virtually identical to the temperature fluctuation since the particle represents both an excess in the density and a deviation in the average energy (temperature). The close similarity between the density and the temperature mode is nothing new; it was already noted by Kirkpatrick (1985) who stated that the heat mode has a large component on the density. Given the similarity between the density and temperature fluctuation (de Schepper *et al.*, 1984*a*), we have that B must be very closely related to D_s. In fact, if the fluid were as dense as possible (close packing with available volume per particle of $\sigma^3/\sqrt{2}$, or 4 particles per cubic unit cell of side $\sqrt{2}\sigma$), then there would not be a difference between the two. We can now try to extend this connection to lower densities by proposing that $B = D_s n\sigma^3/\sqrt{2}$. This seems to be a reasonable choice since it reproduces $B = D_s$ when the system is at maximum density, and it yields a value smaller than D_s at lower densities reflecting that in between collisions our particle is not pushing any other particles out of the way. Thus, we can view the factor $n\sigma^3/\sqrt{2}$ as the free-volume correction relevant to very short distances. However, other than that our correction factor should work well at the very highest densities, the above reasoning is not very tight, and the calculation for the characteristic decay time should not be taken too seriously.

It is straightforward to compare the model to experiment. In neutron-scattering experiments one can directly measure the characteristic time since the heat mode is the dominant process in determining how a disturbance created by a neutron decays. van Loef and Cohen (1989) have collected the experimental results for argon and krypton for a range of densities at equivalent temperatures. They deduced the following phenomenological relation (shown in Fig. 4.11) between the self-diffusion coefficient and the characteristic decay time for wavelengths corresponding to $\lambda \sim \sigma$:

$$\tau^{-1} = 27.7 D_s n\sigma. \tag{4.7}$$

Using our model we find

$$\tau^{-1} = B(2\pi/\sigma)^2 = D_s(n\sigma^3/\sqrt{2})(2\pi/\sigma)^2 = 27.9 D_s n\sigma. \tag{4.8}$$

As with the description of sound modes of short wavelength, our model (dotted line in Fig. 4.11) gives a satisfactory description of what determines the characteristic

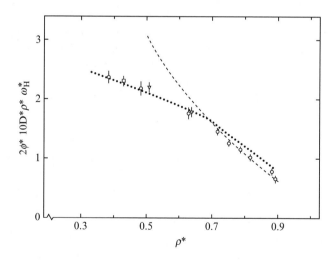

Fig. 4.11 Figure reproduced with permission from van Loef and Cohen (1989). Shown are the measured diffusive relaxation rates (symbols) of density fluctuations with wavelength $\lambda = \sigma$ for Ar and Kr, and the prediction using our simple model (dotted line) as a function of reduced density $\rho^* = n\sigma^3$. The reduced temperature $T^* = k_B T/\epsilon$, when measured with respect to the depth ϵ of the potential well, is the same (1.92) for both Ar and Kr.

decay of a diffusive mode of short wavelength $\lambda \sim \sigma$. Note that the only temperature dependence that enters the simple model is through D_s. Thus, if one wants to form a mental image of what a short-wavelength heat mode looks like, picturing an atom finding its way through a sea of others – thereby transporting excess energy from one place to another – is a good start.

4.4.2   Realistic model for $q\sigma \simeq 2\pi$

No theoretical expression has been determined for the half-width of $S_{nn}(q,\omega)$ for real fluids, or even for model fluids like a Lennard-Jones fluid. The only expression that has been calculated, albeit approximately, is for a dense hard-sphere fluid (Cohen *et al.*, 1984; Kirkpatrick, 1985):

$$\omega_{\mathrm{H}}(q) = \frac{D_{\mathrm{E}}q^2}{S(q)}d(q) = \frac{D_{\mathrm{E}}q^2}{S(q)[1 - j_0(q\sigma) + 2j_2(q\sigma)]}. \tag{4.9}$$

Here, D_{E} is the Enskog self-diffusion coefficient $D_{\mathrm{E}} = D_{\mathrm{B}}/\chi$ (see Appendix C), with D_{B} the Boltzmann diffusion coefficient that describes how a particle would diffuse at low density. χ is the value of the pair correlation function $g(r)$ at contact $\chi = g(r = \sigma)$ and it accounts for the increased collision frequency with increased density. j_0 and j_2 are the zeroth- and second-order spherical Bessel functions, respectively. Thus, the only input in this expression is the (equivalent) hard-sphere diameter provided $S(q)$ is calculated using hard-sphere theory; otherwise, the measured $S(q)$ is also an input parameter.

The various terms in eqn 4.9 can be understood as follows: The $\sim D_{\mathrm{E}}q^2$ dependence simply states that at the heart we are following a particle trying to diffuse away, as explained in the previous subsection. The term $\sim 1/S(q)$ takes the structural slowing down into account, representing how the surrounding neighbors make it increasingly more difficult for a particle to escape its cage. Note that the Enskog diffusion coefficient $D_{\mathrm{E}} = D_{\mathrm{B}}/\chi$ already takes into account the increased collision frequency resulting from the increased density (Cohen and de Schepper, 1990). The term $d(q)$ represents how effective a particle is at pushing its neighbor out of the way during a collision. We show this correction term in Fig. 4.12. Note that $d(q)$ equals 1.5 for $q\sigma = 2\pi$, somewhat negating the effects of a particle's confinement (given by $S(q)$). We compare this expression (eqn 4.9) to the measured half-widths in Fig. 4.13. The agreement is very impressive, not just at $q\sigma = 2\pi$, but also for a reasonable large q range near this minimum. This implies that the correction factor $d(q)$ correctly describes the change in decay time due to the effectiveness of momentum and energy transfer during a collision (see Fig. 4.12). Also shown in Fig. 4.13 is the expression (de Gennes, 1959) that only takes into account the free-streaming of the particles in conjunction with the average structure of the liquid: $\omega_{\mathrm{deGennes}} = q/\sqrt{\beta m S(q)}$.[4] The latter expression fails to provide a meaningful description of the experiments. Equation 4.9 also gives a good description of the characteristic width of colloidal suspensions (see Fig. 5.3), and even for the characteristic width in liquid metals (see Fig. 7.15). The case of liquid metals is particularly surprising as the core of the interatomic potential in this case is very soft. We will revisit this issue in Chapter 7.

---

[4] Note that this equals the classical limit for $f_{un}(q)$.

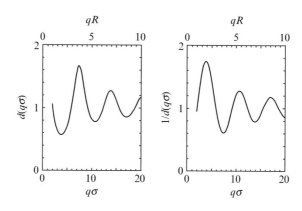

Fig. 4.12 The correction function $d(q\sigma)$ (eqn 4.9) that describes how effective a particle is at pushing the particle it is colliding with out of the way. The function shows an extremum at $q\sigma = 7.4$, or $qR = q\sigma/2 = 3.7$. The role of the correction function is such that it results in an increase in half-width for $q\sigma \approx 7.4$, implying that the residency time of the particle inside the cage is diminished. We display both $d(q\sigma)$ and $1/d(q\sigma)$ for ease of comparison with experiments, such as the ones shown in Chapter 5.

4.4.3   Dominance of self-diffusion for $q\sigma \simeq 2\pi$

The hard-sphere expression (eqn 4.9) that describes the characteristic width of the scattering spectra so well near $q\sigma \simeq 2\pi$ is in fact an expression for the extended heat mode, it ignores the extended sound modes. The fact that this expression works so well reflects the dominance of the heat mode at these wavevectors. It is as if the entire decay tree of the effective eigenmode formalism can be replaced with the one shown in Fig. 4.14, representing a simple diffusive process.

   The fact that the heat mode dominates the spectra in simple liquids also explains how multiple models, such as the viscoelastic model, yield a good description of $S_{nn}(q, \omega)$ in this q range. As long as the models include the density fluctuations – which they all do – then they should be adequate for usage near $q\sigma \simeq 2\pi$. To illustrate this, we will look at the viscoelastic model (eqn 2.48) and the extended hydrodynamics model (eqns. 4.3 and 4.5). The procedure is straightforward. We simply ignore all the time derivatives except that of the density–density correlation function. This amounts to replacing the z dependence by zero. Perhaps this procedure is best illustrated by using the continued fraction expression for the viscoelastic model (Fig. 2.14 and Appendix E) and replacing $i\omega$ by zero:

$$\frac{\pi S_{nn}^{\text{sym}}(q, \omega)}{S_{\text{sym}}(q)} = \text{Re} \left[i\omega + \frac{f_{un}(q)^2}{i\omega + \frac{f_{u\sigma}(q)^2}{i\omega + z_\sigma(q)}} \right]^{-1} \rightarrow \text{Re} \left[i\omega + \frac{f_{un}(q)^2 z_\sigma(q)}{f_{u\sigma}(q)^2} \right]^{-1}. \quad (4.10)$$

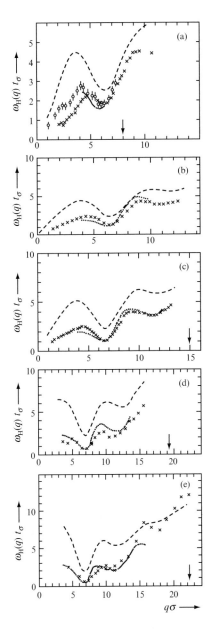

Fig. 4.13 Figure reproduced with permission from Cohen *et al.* (1987). The solid line is the prediction of eqn 4.9 compared to the half-widths of $S_{nn}(q,\omega)$ for a range of liquids ordered by increasing density. The vertical arrows represent the upper range of validity of eqn 4.9 given by $q = 1/l_{\text{free}}$, which equal the upper range of usefulness of the effective eigenmode formalism. The dashed curve is given by $\omega_{\text{deGennes}} = q/\sqrt{\beta m S(q)}$. Part (a) corresponds to a reduced density of $n\sigma^3 = 0.64$ (Kr, crosses) and 0.63 (Ar, circles); part (b) to $n\sigma^3 = 0.75$ (Ar); part (c) to $n\sigma^3 = 0.81$ (Ar); part (d) to $n\sigma^3 = 0.88$ (Ar); part (e) to $n\sigma^3 = 0.92$ (Rb).

Fig. 4.14 The decay tree for density fluctuations of wavelength $\lambda \approx d$. The extended heat mode dominates all decay processes, resulting in simple diffusive behavior.

This expression is equivalent to a simple diffusion process with diffusion constant

$$D_{\text{reduced}}^{\text{visco}}(q) = [f_{un}(q)/(q f_{u\sigma}(q))]^2 z_\sigma(q). \tag{4.11}$$

We can apply (de Schepper *et al.*, 1988) a similar 'roll-up' procedure to eqn 4.3 to arrive at the diffusion constant in that model:

$$D_{\text{reduced}}^{\text{extended}}(q) = (f_{un}(q)/q)^2 [z_\phi(q,0) + [f_{uT}(q) + \Delta(q,0)]^2 / z_T(q,0)]^{-1}, \tag{4.12}$$

where z_T, z_ϕ and Δ are defined in eqn 4.5. Thus, in both cases a diffusive process would be fitted to the data, and provided that the eigenfunctions corresponding to the extended heat modes in both models have their main component on the density, both models would provide an adequate fit for the spectra near $q\sigma \simeq 2\pi$. And conversely, if one finds experimentally that one particular model gives a good description in this q range, then only very limited conclusions can be drawn about the validity of that particular model for the rest of the q range.

4.5 Cage diffusion

The cage-diffusion process at intermediate densities is easy to visualize as an individual atom bouncing around in the cage formed by its nearest neighbors, as shown in Fig. 2.9, before it escapes its cage and the process repeats itself. At high densities the cage as a whole moves through the fluid (Pilgrim and Morkel, 2002). So how does cage diffusion show up in experiments and computer simulations? The timescale of the bouncing around part of cage diffusion is determined by the size of the cage, and by the average speed of the particle. An example of what can be expected for the intermediate scattering function describing the dynamics of a single particle is shown in Fig. 3.11. Even though the data plotted in this figure are for liquid mercury – admittedly not a simple liquid – the overall shape of the curve is very similar for both liquid metals and simple liquids.

Figure 3.11 demonstrates that the decay of the self-correlation function consists of two components. There is a rapid, but small loss of correlation between the initial position of the particle and its position a short time later. This part of the decay is associated with the multiple collisions of the particle within its cage, and the degree of correlation loss is determined by the size of the cage; while the particle is in its cage it can never wander far away from its initial position. We can capture the size of this correlation loss by $\int_0^\Delta d\vec{r} e^{i\vec{q}\cdot\vec{r}}/V$, with Δ the distance between the 'surfaces' of neighboring particles, and $V = 4\pi\Delta^3/3$. This function is plotted in Fig. 2.23 and for small q values this integral reduces to $1-(q\Delta)^2/10$. We write 'surfaces' since our particles are of course not hard spheres, but being simple liquids, our particles have fully filled outer electron shells and they resemble hard spheres fairly closely in this respect. Typical values for Δ in dense liquids are in the range of 0.2 to 0.6 Å, so that for small q values the initial loss of correlation due to this cage effect is limited to less than 1%. In Fig. 3.11 the loss corresponds to less than 0.5% (the difference between 1 and the extrapolation of the simple diffusion result to $t = 0$).

Experimentally, it is as good as impossible to observe this initial part of cage diffusion in simple liquids.[5] In order to observe it in the self-part of the dynamics, it of course requires that the self-part of the dynamics has been isolated from the experiment. This is something that is quite possible in neutron-scattering experiments, given that for dense liquids the coherent part of the scattering is very weak at low q values because of the smallness of $S(q)$. However, one then has to observe an additional component besides the simple diffusion component (given by eqn 2.15) whose total magnitude is less than 1% of that of the simple diffusion component. Alternatively, one can go to higher q values (shorter probing wavelengths) so that the initial loss of correlation becomes more prominent. However, at the same time the coherent contribution increases very rapidly so that once again, the small signal will be difficult to observe. Nonetheless, some systems should lend themselves to probing this part of the dynamics, such as hydrogen and vanadium. In order to see the contribution in the collective part of the dynamics one faces similar difficulties: one has to discern a signal that is simply too weak compared to the other scattering mechanisms, including the unwanted contributions from instrumental background.

A much easier way to observe cage diffusion is to look at momentum transfers beyond the main peak of the static structure factor. The probing wavelength in this region will be less than the distance between the particles, and therefore, the loss of correlation should mostly reflect single-particle dynamics (see the reasoning in Section 2.2.3) so that we can now observe how long a particle spends inside its cage. The characteristic width of the scattering spectra is inversely proportional to the residency time τ_{res} inside the cage $\omega_H = 1/\tau_{res}$. Thus, the smaller the characteristic width, the longer the residency time. We show this characteristic width in Fig. 4.15 for liquid mercury (Badyal *et al.*, 2003). As mentioned, liquid Hg is not a simple liquid since we have both the ionic fluids and the electron fluid, however, Hg does a remarkably good job at imitating a simple liquid.

It can be seen in this figure that the characteristic width oscillates around the characteristic self-diffusion width $D_s q^2$ showing that we are indeed mostly dealing with the dynamics of individual atoms. The oscillations are in anti-phase with the oscillations of the static structure factor, demonstrating the effects of confinement to a cage. Whenever the static structure factor reaches a local maximum, the characteristic width reaches a minimum, implying that the time it takes for the density fluctuation to decay reaches a maximum. In other words, at these probing wavelengths we see the structure of the cage. When a density fluctuation is created by the probing radiation that matches the natural structure of the system, we observe that such fluctuations

---

[5] In Chapter 5 it is shown that the initial part of cage diffusion is very straightforward to follow in colloidal systems where the dynamics take place on time-scales of the order of a second.

take much longer to decay than when there is a mismatch between the fluctuation and the liquid structure. Thus, the measured spectra are consistent with the anticipated behavior for a particle that is inside a cage, and that stays inside that cage for a relatively long while.

From Fig. 4.15 it is clear that cage diffusion is a very important mechanism in the relaxation of density fluctuations, and it is clear that the local structure plays a dominant role, but it is not the whole story as the particle also has to (and does) escape its cage. This was already demonstrated in Fig. 4.13, where it was shown that the characteristic half-width is not entirely given by $D_s q^2 / S(q)$; one also has to

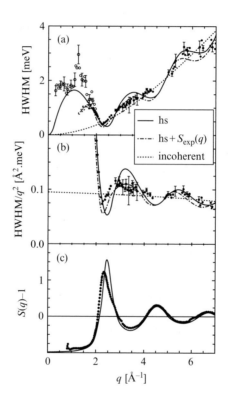

Fig. 4.15 Figure reproduced with permission from Badyal *et al.* (2003). Part (a): the characteristic width (circles) of $S_{nn}(q, \omega)$ (denoted HWHM in the figure) as a function of wave number q. The curves are calculated using the Enskog theory with $\sigma_{hs} = 2.81$ Å. The dotted line is the HWHM describing the self-part of the dynamics, the solid curve is the prediction for the coherent HWHM using the calculated $S(q)$ for the equivalent hard-sphere fluid shown in part (c) and denoted 'hs' in the legend, while the jittery dashed-dotted curve is the prediction for the collective HWHM using $S(q)$ from experiment ('hs+$S_{exp}(q)$' in the legend). Part (b) is the same as part (a), except everything has been divided by q^2. Part (c): the static structure factor $S(q)$ measured by neutron-diffraction experiments and compared to $S(q)$ calculated using the Percus–Yevick theory (Hansen and McDonald, 2006) with Henderson–Grundke correction factor (solid line).

include a term that incorporates how easy it is for particles to push the others out of the way, namely, a term like $d(q)$ in eqn 4.9 that takes collisional transfer of energy and momentum into account. Overall, the characteristic behavior of the half-width being in anti-phase with the oscillations in the structure factor has been observed is a very large number of fluids. As a final remark in this section, we note that the cage-diffusion mechanism can be followed in much greater detail in colloidal suspensions (see Chapter 5) than it can be in simple fluids.

4.6   Temperature fluctuations and oscillations

So far we have discussed how density fluctuations give rise to velocity disturbances, which in turn can give rise to temperature fluctuations. In the fluids discussed thus far, these temperature disturbances set up a heat current (or flux), which decays fairly rapidly. What we mean by this is that z_q is large compared to f_{Tq} (see Fig. 2.12) so the power that is put into this decay channel is lost right away. Since no power is coming back to 'T', oscillations cannot be set up between 'T' and 'q'. As a result, in experiments we only observe oscillations between the semiconserved microscopic quantities. However, there is no microscopic rule that states that $z_q/2 > f_{Tq}$, so in principle one could see temperature oscillations.

Temperature oscillations have been investigated extensively in superfluid helium, where they go by the name of second sound. While the temperature does not couple to the velocity in superfluid helium since $\gamma = 1$ (that is, $f_{uT}(q) = 0$), and therefore, density fluctuations cannot give rise to a temperature disturbance, propagating temperature fluctuations can nonetheless be studied, albeit on a macroscopic scale. In superfluid helium these temperature oscillations can be described on the basis of the two-fluid model (Griffin, 1993) where the oscillations occur between the superfluid and the normal-fluid component without affecting the total density. Since only the normal-fluid component carries entropy, oscillations between the two components represent a temperature oscillation. It is of course this ability of superfluid helium to sustain propagating temperature waves that makes it such an incredible conductor of heat, some two orders of magnitude better than copper. But what about normal fluids, can we have propagating temperature waves in those systems, even if these waves do not propagate very far, as they do in superfluid helium?

There are some indications that temperature waves do indeed exist in normal systems, but the evidence for this is not beyond contention. And we state the latter, being the authors of the study that proposed the existence of such temperature waves. We measured (Montfrooij *et al.*, 1992*a*) the dynamics of normal fluid helium at 13.3 K at a density of $n = 0.032$ Å^{-3} (which actually is higher than the liquid density). The dynamics of this state were analyzed with the decay tree shown in Fig. 2.12, and it was found that two pairs of propagating modes existed in the range $0.3 < q < 1.1$ Å^{-1}. One set of propagating modes was attributed to propagating temperature fluctuations, in agreement with the findings that $f_{Tq}(q) > z_q(q)/2$ in this q range. The other set of propagating modes were the 'standard' modes of the viscoelastic part of the decay tree.

The potential source of error in this study is that the effective eigenmode formalism was not applied to the correct correlation function, which in turn may, or may not have skewed the findings. Instead of using the relaxation function $S_{nn}^{\text{sym}}(q,\omega)$ to account for the difference between a classical and a quantum system, the symmetrized function $S_{\text{cl}}(q,\omega) \equiv e^{-\beta\hbar\omega/2} S_{nn}(q,\omega)$ was used instead. While this function can also be fitted by employing a (large enough) matrix $G(q)$, the eigenvalues that come out of this fitting procedure do not correspond to the poles of the dynamic susceptibility. Moreover, it is possible that the lineshape was distorted sufficiently by employing a slightly wrong temperature dependent pre-factor $[e^{-\beta\hbar\omega/2}$ versus $(1 - e^{-\beta\hbar\omega})/\beta\hbar\omega]$ that a function that could have been described by the sum of three Lorentzian lines now required five such lines. Since the original data are no longer available, we cannot be certain whether propagating temperature fluctuations have indeed been observed in a normal fluid, or not. If they do indeed exist, then this thermodynamic state of helium would be the exception to the rule that only three exponentials are required to describe the fate of a density fluctuation in a liquid; in this case we would require all five exponentials.

5
Colloidal suspensions

While the idea of effective eigenmodes can also be applied to colloidal suspensions, the effective eigenmode formalism does not add to the analysis of the dynamics in these systems. The only reasons why we (very briefly) cover this topic is to illustrate the ubiquity of cage diffusion, and because of the fact that some colloidal systems are fairly well described as a collection of interacting hard spheres. Other than that, the dynamics of colloidal suspensions are far too complex to cover in this book. It is because of this complexity that this chapter is not a subsection of Section 4.4 where we discuss the behavior of simple liquids near $q\sigma \approx 2\pi$.

Colloidal suspensions – or colloids for short – consist of a dispersing medium such as water or glycerol in which particles are suspended rather than dissolved (Hunter, 1989). The sizes of these particles can range from a few nanometers to a few hundred nanometers. A colloid in which the distribution of sizes of the suspended particles is very narrow is referred to as a monodisperse colloid, a suspension with a broad distribution is referred to as a polydisperse colloid. The density n of a colloid is typically expressed as the fraction of the occupied volume ϕ of the suspended particles: $\phi = n\pi\sigma^3/6$, where σ is the effective hard-sphere diameter of the particle.

The particles in a colloid are suspended in the dispersing medium rather than dissolved; in order to prevent them from aggregating or sinking to the bottom or floating to the top, the suspension needs to be stabilized. This stabilization is achieved either through electric repulsive forces of charged suspended particles (charged colloids) or through repulsive steric forces (neutral colloids) when a surface layer has been grafted onto the colloidal particles. The static properties of both charged and neutral colloidal suspensions resemble those of hard-sphere fluids. While the dynamic properties of charged suspensions are reasonable well explained by hard-sphere theory, neutral suspensions are not described nearly as well; the dynamics of these two types of colloidal systems also differ greatly from each other.

The dynamics of colloids are investigated by means of light scattering. Impressive progress has been achieved in the scattering techniques themselves, essentially overcoming the problem of multiple scattering. In particular, the emergence of the two-color dynamic light-scattering method (Drewel *et al.*, 1990; Segrè *et al.*, 1995*b*) and the extension of the dynamic light-scattering technique to the X-ray region (Dierker *et al.*, 1995) have extended the availability of experimental results to wave numbers q below the main peak of the static structure factor $S(q)$, a region previously inaccessible, given the dominance of multiple scattering over single scattering at low q values.

The development of better scattering techniques for the study of colloids coincided with the advent of molecular dynamics computer simulations that managed to overcome the difficulties of modelling the time-dependent hydrodynamic interactions between the suspended particles (Ladd, 1994a,b; Behrend, 1995). Experimental results for the dynamics of the particles and those obtained by means of computer simulations are in close agreement, an example of which is shown later in this chapter in Fig. 5.4. Of course, experimental results would still have been very hard to interpret had it not been for advances in the ability to manufacture colloids that are almost monodisperse with a typical range of sizes limited to 5% or less.

We refer the reader to recent and classical texts (Hunter, 1989) on colloids for the state of the theory and for detailed information on the advances in experimental and simulation techniques. As mentioned, here we restrict ourselves to some striking similarities between the behavior of these complex colloidal systems and the behavior of ordinary liquids. We do not discuss the theory behind the dynamics of the suspended particles as detailed in the seminal paper by Beenakker and Mazur (1984), and extended by various groups. As such, this chapter is more about the behavior of ordinary fluids than it is about the dynamics of colloidal suspensions.

In light-scattering experiments on colloids one probes the intermediate scattering function $F_{nn}(q,t)$, the same function that is being probed in molecular dynamics computer simulations on simple liquids. As an example, we show the results for a neutral colloid at a volume fraction of $\phi = 0.465$ as measured by Segrè and Pusey (1996) in Fig. 5.1. This figure can be compared to the results for the intermediate self-scattering function for liquid Hg as determined from computer simulations (Bove *et al.*, 2002) shown in Fig. 3.11. In both cases the intermediate scattering function displays a fairly rapid decrease, followed by a much slower decay. For colloids, both the initial and the long-time decrease result in linear curves when plotted as $\sim q^2$ as is done in Fig. 5.1, demonstrating that the relevant processes are diffusive in character. The limited loss of correlation during the initial part of the decay is caused by the particles moving around in the cage formed by their neighbors, the long-time decay is caused by the particles escaping their cages as well as by the slow diffusion of the cages over a much longer timescale.[1] This is very similar to what happens in ordinary liquids, as follows from a direct comparison of Figs. 3.11 and 5.1. The fact that we are comparing the intermediate scattering function of colloids to the intermediate self-scattering function in a liquid metal does not affect this interpretation since both the collective as well as the self-scattering function would show a rapid initial decay if particles can move around within the cages before the cage collapses, or before the particles escape.

The bottom panel of Fig. 5.1 demonstrates that the oscillations in both the short-time and the long-time diffusion coefficients are in phase with those of the static structure factor $S(q)$, at least for this particular suspension of neutral particles. This

---

[1] Note that both for the short-time as well as for the long-time behavior diffusion coefficients can be determined directly from the light-scattering experiments by measuring the slopes of the curves shown in Fig. 5.1.

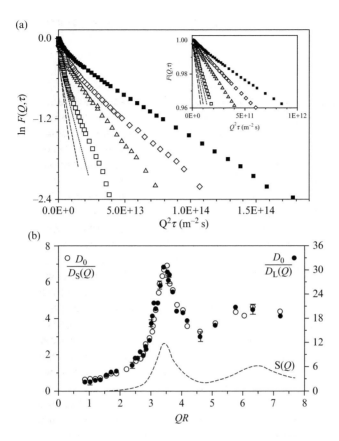

Fig. 5.1 Figure reproduced with permission from Segrè and Pusey (1996). The top panel shows the intermediate scattering function $F(q, t)$ for a neutral suspension of sterically stabilized particles of polymethyl-methacrylate (PPMA) of radius $R = \sigma/2 = 178$ nm suspended in *cis*-decalin at a volume fraction of $\phi = 0.465$. The data, obtained using the two-color dynamic light-scattering method, have been plotted on a single log-scale in order to emphasize the two distinct time domains that govern the decay of density fluctuations. The four topmost curves depicted by symbols are for different q values, corresponding from top to bottom to $qR = 3.5, 3.9, 3.0, 2.5$, respectively. The length and timescales are such that $q^2\tau = 10^{14}$ s/m^2 corresponds to a time of 0.25 s for the topmost curve. The inset shows the initial decay of $F(q, t)$. The sharp initial decrease, proportional to q^2, is related to particles moving around in the cages formed by their neighbors; the long-time decay, which is also linear in q^2, is related to the movement of the particles over a distance larger than their diameter. The bottom panel displays the two diffusion constants that can be determined from the slopes of the short-time and of the long-time decay. Both the short-time diffusion constant $D_S(q)$ (open symbols, left-hand scale) and the long-time diffusion constant $D_L(q)$ (closed symbols, right-hand scale) exhibit oscillations that are in phase with the static structure factor $S(q)$ (dashed curve, left-hand scale). D_0 is the Brownian diffusion coefficient of the suspended particles at infinite dilution.

observation lends credence to the identification of the main diffusive mechanisms with those of cage diffusion: obviously the local structure of the suspension plays an important role in its dynamics. Note that the short-time and the long-time diffusion constants are seen to differ only by a numerical factor of 4.3 over the entire q range and also note that for $qR < 2.5$ the static structure factor and the diffusion constants cannot be made to scale. These observations have not been fully explained yet, illustrating the complexity of colloid dynamics.

Notwithstanding these unexplained features, a quantitative comparison between the behavior of ordinary liquids and colloidal suspensions is possible, as we discuss in the following. However, the details of this comparison depend strongly on whether the colloidal particles are charged or neutral, and should therefore not be confused with a detailed theory of the dynamics of colloidal suspensions. Rather, they should be viewed as colloidal suspensions offering insight into the behavior of ordinary fluids, in particular when it comes to the characteristics of the heat mode.

5.1   Charged colloidal suspensions

Overall, it would appear that charged colloidal suspensions resemble ordinary liquids more closely than do neutral suspensions. We will first make a comparison between ordinary liquids and charged colloidal suspensions. Charged colloidal suspensions closely resemble a system of interacting hard spheres, both in their static properties as well as in their dynamics. It has been found that the static structure factor of a colloidal system is very well described by that of an equivalent hard-sphere system where the colloidal Debye spheres have been replaced by impenetrable particles of hard-sphere diameter σ. The forces between the charged colloidal particles are governed by the electrostatic repulsion between them; the forces between the neutral colloidal particles involve the van der Waals attraction between them, the excluded volume of the particles and the hydrodynamic interactions mediated by the dispersing medium. As with a system of hard spheres, packing the particles beyond the limit where they can slide pass each other results in crystallization.

The main problem in understanding the behavior of neutral colloidal suspensions arises from the hydrodynamic interactions between the particles as they are mediated by the dispersing medium. These interactions resulting from the fluid flowing over the surface of the particles originate at the solid/fluid interface and diffuse throughout the fluid. It is in these hydrodynamic interactions that neutral and charged colloids differ greatly from each other. Note that the particles in neutral colloids are actually very large, in the sense that their diameter is of the same order as the distance between the particles. The particles in charged colloids are very small, their actual radius is one or two orders of magnitude smaller than the distance between the particles (Hunter, 1989). As such, hydrodynamic interactions between the particles as mediated by the fluid play no role in charged colloids since the particles themselves are simply too far away from each other. On the other hand, the size of the Debye spheres describing the extent of their charge clouds is of the same order as the separation between the particles. When we choose the effective hard-sphere diameter to correspond to this Debye sphere, as opposed to the actual diameter of the particles, then we find the close correspondence between the dynamics of the charged colloids and those of the hard-sphere fluid.

On the shortest timescales $t < \tau_B = m/6\pi\eta_0 R$ the dynamics of the charged particle and of the fluid are not decoupled yet, the particle is still slowing down and its motion evolves from ballistic to diffusive. Here, η_0 is the shear viscosity of the dispersing fluid, m is the mass of the suspended particle of radius R and τ_B stands for the Brownian relaxation time. On the longest timescales $t > \tau_R = R^2/D_0$ the particles diffuse over distances comparable to their diameters, with D_0 the diffusion constant at infinite dilution $D_0 = k_B T/6\pi\eta_0 R$. For times in between these two regimes $\tau_B \ll t \ll \tau_R$ the particle is locked up in the cage of its neighbors (for sufficiently large volume fractions ϕ). The experimental diffusion constant $D_{\text{eff}}(q)$, as measured in experiments on charged colloidal suspensions, is well described by $D_{\text{eff}}(q) = D_0/S(q)$. This corresponds to the case where the suspended particles interact strongly with each other through (long-range) repulsive electric forces while the hydrodynamic interactions between the particles can be neglected. Thus, the diffusion rate of the particles is given by the bare diffusion constant modulated by the local structure. The equivalent expression for hard-sphere liquids would be $D_{\text{eff}}(q) = D_B/S(q)$, with D_B the Boltzmann diffusion coefficient $D_B = k_B T/6\pi\eta_0 R = (3/8n\sigma^2)\sqrt{k_B T/\pi m}$. Here, $R = \sigma/2$ and η_0 is the shear viscosity of the hard-sphere liquid itself. Examples of the diffusion constant associated with this part of the motion of the particles responsible for the fast initial decay of the intermediate scattering function $F(q,t)$ are shown in Fig. 5.2 for polystyrene spheres in water. It is clear from this figure that the relationship for the short-time diffusion coefficient $D_{\text{eff}}(q) = D_0/S(q)$ is indeed borne out by experiments on these charged colloidal suspensions.

As far as we are aware there exist no comparable experimental data on the initial short-time decay in ordinary fluids. First, the timescales governing the decay of the density disturbances are not well separated in ordinary fluids at $q\sigma \approx 2\pi$. The timescales are well separated at the smaller q values, however, the intensity of the signal associated with this short-time decay is vanishingly small in scattering experiments where one probes the dynamic structure factor, rather than the intermediate scattering function, as is done in experiments on colloids. This has already been discussed in great detail in the preceding chapters. The only avenue open to testing a relationship such as $D_{\text{eff}}(q) = D_B/S(q)$ would therefore appear to be through molecular dynamics computer simulations, such as the ones shown in Fig. 3.11.

The diffusion coefficient $D_L(q)$ associated with the long-time decay of the intermediate scattering function of charged colloids is also open to a direct comparison with the behavior of ordinary liquids. Given the good agreement between $D_0/D_{\text{eff}}(q)$ and $S(q)$ shown in Fig. 5.2, one can extend the comparison to the longer times incorporating the diffusion (out) of the cage itself. In ordinary liquids this leads to eqn 4.9, where one incorporates the factor $d(q)/\chi$ to approximate the effects of collisional transfer of momentum and energy during hard-sphere interactions when a particle collides with the particles that make up the wall of the cage. The factor $1/\chi$, with χ the value of the pair correlation function at contact $g(r = \sigma)$, accounts for the increased collision frequency due to the excluded volume of the particles. For charged colloids, the equivalent expression therefore reads:

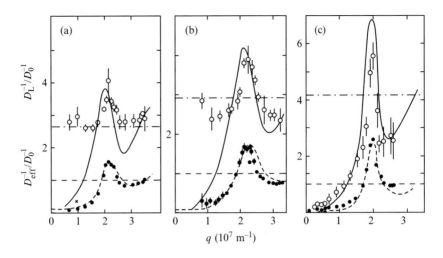

Fig. 5.2 Figure reproduced with permission from de Schepper *et al.* (1989). Shown are the short-time and long-time diffusion coefficients (denoted $D_{\text{eff}}(q)$ and $D_{\text{L}}(q)$, respectively) for three suspensions of polystyrene spheres in water. The light-scattering data in part (a) have been measured by Pusey (1978), those in part (b) by Dalberg *et al.* (1978) and those in part (c) by Grüner and Lehman (1982). The data for $D_{\text{L}}(q)$ in parts (a) and (b) have not been corrected for incoherent scattering except for one point in (a) (cross at $q = 10^7/\text{m}$). The lower dashed curves in all three panels are given by $D_0/D_{\text{eff}}(q) = S(q)$, where $S(q)$ has been calculated for a system of hard spheres of equivalent diameter σ of $\sigma = 310$ nm (a), 296 nm (b), and 343 nm (c). The hard-sphere densities are given by $n\sigma^3 = 0.59$ (a), 0.65 (b), and 0.82 (c). The solid lines are the prediction near $q\sigma \approx 2\pi$ for the long-time diffusion coefficient for charged colloids as given in eqn 5.1. The lower horizontal line at height 1 represents free Brownian motion, the upper line is at a height of χ representing the enhanced collision frequency resulting from excluded-volume effects.

$$D_{\text{L}}(q) = \frac{D_0}{S(q)} \frac{d(q)}{\chi}. \tag{5.1}$$

The only essential difference between this equation and eqn 4.9 is that the diffusion constant is now the diffusion constant of a single particle in the dispersing medium, relevant to times when the dynamics of the particle and the medium have decoupled. The other non-essential difference between the two expressions is the vast difference in time and length scales, with the suspended particles up to three orders of magnitude larger than the atoms of a liquid, and the timescales up to twelve orders of magnitude larger. Given that the overall, long-time decay will be governed by the long-time diffusion mechanism, we can also identify the characteristic decay time $\tau(q) = 1/\omega_{\text{H}}(q)$ and estimate the equivalent heat-mode damping rate $\Gamma_{\text{L}}(q)$ for colloids to be given by

$$\Gamma_{\text{L}}(q) = D_{\text{L}}(q)q^2 = \frac{D_0 q^2}{S(q)} \frac{d(q)}{\chi}. \tag{5.2}$$

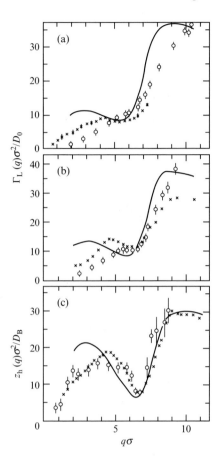

Fig. 5.3 Figure reproduced with kind permission of Societa Italiana di Fisica from Cohen and de Schepper (1990). This figure demonstrates the similarity in dynamics between charged colloidal suspensions and ordinary liquids. The characteristic widths of the scattering spectra (Γ_L for colloids and z_h for ordinary liquids) have been scaled using the equivalent hard-sphere diameter and the diffusion coefficients at infinite dilution. The open symbols represent the colloids, the crosses the ordinary liquids. The densities of the colloidal systems correspond to the densities listed in Fig. 5.2 ($n\sigma^3 = 0.59$, 0.65 and 0.82 for the top, middle and bottom panels, respectively). As can be seen, the scaling immediately brings out the strong resemblances in the dynamics, not only in the q dependence of the half-widths of the spectra of ordinary fluids and the long-time diffusion equivalent for colloids (eqn 5.2), but also in their absolute values. The similarities between the two systems get stronger with increased density. The solid lines in the figures are the prediction of eqn 5.2; these predictions are also seen to get increasingly better with increased density. The open symbols are the light-scattering data for charged polystyrene spheres in water (see caption Fig. 5.2), the crosses are the ordinary fluid data for Ar at 212 K (a), Kr at 297 K (b) and Ar at 120 K (c). The densities for the ordinary fluids are very close to those of the colloidal systems, namely $n\sigma^3 = 0.50$ (a), 0.64 (b) and 0.81 (c).

The comparison between $D_L(q)/D_0$ and the prediction of eqn 5.1 is given in Fig. 5.2 by the solid curves. The comparison (de Schepper *et al.*, 1989) is encouraging in the sense that the overall shape of the curves is very similar, the data and the prediction are in phase and there is fairly good quantitative agreement. Having said that, the data themselves have not been corrected for multiple-scattering effects except for the dataset shown in Fig. 5.2(c). A similar level of agreement is obtained when a direct comparison is made between various ordinary liquids and charged colloidal suspensions at comparable densities. This comparison (Cohen and de Schepper, 1990) is reproduced in Fig. 5.3, where the heat-mode eigenvalues for ordinary liquids are plotted and compared to the colloidal equivalent given by eqn 5.2 and based on the measured long-time diffusion coefficients. First, note that the mere scaling process employed in this figure has brought to the fore the strong similarities between the two systems, independent of any underlying theory or model. Secondly, the agreement between ordinary liquids, colloidal suspensions and the cage-diffusion expression of eqn 5.2 is encouraging at the lower densities, and the agreement appears to improve with increasing density, as one would expect should eqn 5.2 capture the cage-diffusion mechanism in colloids. Therefore, the one thing we can most definitely say for sure is that the cage-diffusion mechanism is just as relevant in charged colloidal suspensions as it is in ordinary liquids.

5.2   Neutral colloidal suspensions

The situation becomes much more murky when we compare the short-time and the long-time dynamics of neutral colloidal suspensions to the equivalent dynamics in ordinary fluids. The main difference appears to be the influence of the hydrodynamic interactions as mediated by the dispersing fluid. While the decay of the intermediate scattering function still shows two very distinct timescales (as already shown in Fig. 5.1), the short-time diffusion constant $D_S(q)$ is no longer simply given by $D_S(q) = D_{eff}(q) = D_0/S(q)$. This is clearly demonstrated in Fig. 5.4 where we show the results by Segrè *et al.* (1995a) for the short-time diffusion coefficient for the neutral colloidal suspension previously described in Fig. 5.1. In Fig. 5.4 Segrè *et al.* plot their results as $D_S(q)S(q)/D_0$, which would be identical to 1 should hydrodynamic interactions not be important. Instead, one observes a strong q dependence and packing fraction dependence. For the lower packing fractions $\phi < 0.4$ the theory of Beenakker and Mazur (1984) gives good agreement with the data, especially in the region $q\sigma = 2qR \approx 2\pi$. For the packing fraction at the crystallization density the theory overestimates the experimental findings, however, the shift in peak position for $q = q_{max}$ is still captured by the theory.

Verberg *et al.* (1999) proposed that the curves shown in Fig. 5.4 for the short-time diffusion coefficient can still be interpreted in terms of the familiar cage-diffusion mechanism. They proposed that eqn 5.2 could now be used to describe the short-time diffusion mechanism while also still forming the basis of the long-time diffusion mechanism. Presumably, the reasoning is that the hydrodynamic interactions mediated by the dispersing medium when neutral particles move through it already make the particles interact with each other even before the actual collisions – associated with the

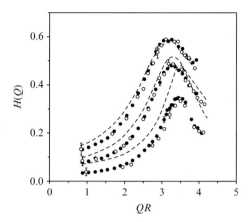

Fig. 5.4 Figure reproduced with permission from Segrè *et al.* (1995*a*). Shown are the enhancements of the short-time diffusion coefficient over simple structural effects in a neutral colloid. These enhancements are captured in the quantity $H(q) = S(q)D_S(q)/D_0$. For the charged colloids shown in Fig. 5.2 this quantity was identical to 1. For the neutral colloids shown here it is obvious that hydrodynamic interactions have modified this behavior extensively. The filled symbols have been measured for polymethyl-methacrylate spheres dispersed in *cis*-decalin at volume fractions ϕ of 0.311 (top curve), 0.382 (middle) and 0.491 (bottom). The open symbols were calculated by doing computer simulations (Segrè *et al.* 1995) demonstrating the power of computer simulations. The dashed curves are the predictions based on the theory of Beenakker and Mazur (1984). This theory works best for $\phi < 0.4$, however, it captures the shift in peak position for all ϕ. According to Verberg *et al.* (1999), $H(q)$ can be approximated by $H(q) = d(q)/\chi$. For these packing fractions this comparison would result in a peak height of 0.65 (top curve), 0.49 (middle) and 0.29 (bottom). The position of the maximum of all curves $d(q)/\chi$ would be at $qR = 3.7$. See Fig. 4.12 to aid with a direct comparison between the shape of the function $d(q)$ and the experimental data.

part of the cage-diffusion mechanism where a particle migrates to new surroundings – take place. As such, transfer of energy and momentum can already occur during the initial stages of cage diffusion when the particles are still locked up in their cages, and the short-time diffusion coefficient needs to be modified through the inclusion of the term $d(q)/\chi$. Should this reasoning be correct then the curves shown in Fig. 5.4 should be identical to $d(q)/\chi$. The agreement turns out to be reasonable at the lower densities, and quite good at the higher density. The heights of the peaks are well reproduced, but the shift in peak position observed in Fig. 5.4 cannot be reproduced since the function $d(q)$ (Fig. 4.12) is independent of packing fraction ϕ and always displays a peak at $qR = 3.7$. Therefore, eqn 5.2 would be a good starting point in describing the short-time behavior, especially at the higher densities, but there are remaining differences between the short-time behavior of neutral colloids and of ordinary fluids.

The long-time diffusion coefficient, as shown in the bottom panel of Fig. 5.1 has an identical q dependence as the short-time diffusion coefficient in this particular neutral colloidal suspension. The reason behind this is unclear, and it is not known whether many neutral colloids display such scaling behavior. In order to understand the behavior of the long-time diffusion coefficient one (most likely) needs to resort to mode-coupling theory. Various attempts have been made in this direction (e.g., Ngai 2007), and we refer the reader to the literature to judge the success of these attempts as such detailed theories are outside the scope of this book. It will be interesting to see in the future once these theories become more and more sophisticated whether similar mode-coupling effects, or remnants and perhaps even ghosts thereof, can also be observed in ordinary fluids at wave numbers near $q\sigma \approx 2\pi$.

As mentioned, the scaling for all q values between the short-time and long-time diffusion coefficient measured by Segrè and Pusey (1996) (Fig. 5.1) is not observed in charged colloids. This is clear from Fig. 5.2; subsequent experiments using light-scattering techniques in the X-ray regime on a charged colloid consisting of polystyrene latex spheres in glycerol (Lurio *et al.*, 2000) also failed to find any scaling relationship between $D_\mathrm{L}(q)$ and $D_\mathrm{S}(q)$. We reproduce their findings in Fig. 5.5. The most interesting part – at least from the point of view of ordinary liquids – about these new, highly accurate experiments is that the shape of the curves $D_\mathrm{L}(q)/D_\mathrm{S}(q)$ at the higher packing fractions once again appears to follow the oscillations contained in $d(q)$ (see eqn 5.2 and Fig. 4.12) implying that charged colloids and ordinary fluids could indeed be fairly similar.

In conclusion, cage diffusion is as important in colloids – both charged and neutral – as it is in ordinary liquids. The behavior of charged colloids and ordinary liquids is fairly similar, even at a quantitative level, whereas hydrodynamic interactions in

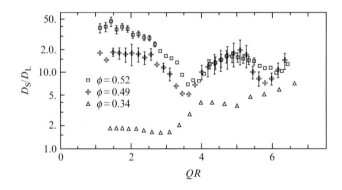

Fig. 5.5 Figure reproduced with permission from Lurio *et al.* (2000). The ratio of the short-time and long-time diffusion coefficients measured by means of X-ray scattering for polystyrene latex spheres suspended in water for the packing fractions ϕ given in the figure. Unlike the ratio for the neutral colloidal suspensions shown in the bottom panel of Fig. 5.4, in this charged colloid the two coefficients do not demonstrate a simple scaling relationship over the entire q range. For the highest packing fraction, the q dependence is similar in shape to the function $1/d(q)$, which would show a maximum at $qR = 2$ and a minimum at $qR = 3.7$. This demonstrates that the dynamics of charged colloidal suspensions are much more closely related to those of ordinary liquids than their neutral counterparts are. Note the vertical log scale.

neutral colloids create mode-coupling effects that are far more important to the overall dynamics than they are in ordinary liquids. Colloids provide us with a wonderful opportunity to study, through scattering experiments, the effects of cage diffusion in much more detail than we can in ordinary liquids. As such, we are much more likely to learn about the behavior of ordinary fluids by studying colloids than the other way around. In particular, by studying the behavior of charged colloids we will learn more about the behavior of the heat mode in ordinary fluids as colloids represent a system in which the heat mode exists in the absence of sound modes. This is the opposite of superfluid helium where we can study the behavior of the extended sound modes in the absence of the heat mode. Lastly, once the mode-coupling effects that strongly influence the dynamics in neutral colloids are better understood, it is likely that this knowledge will be directly relevant to a class of liquids not touched upon in this book, namely supercooled liquids. The slow dynamics in the latter will be strongly influenced by mode-coupling effects that are not visible in ordinary fluids.

6
Binary mixtures

The interest in the microscopic dynamics of binary mixtures received a boost in 1986 with the discovery by Bosse *et al.* (1986) of the existence of a new propagating mode in a lithium-lead mixture. In molecular dynamics computer simulations they discovered that short-wavelength density fluctuations were found to propagate at a speed considerable faster than the hydrodynamic sound velocity. In a far-sighted move that spurred the interest in this phenomenon, they dubbed this mode 'fast sound'. Fast sound appeared to be the equivalent of a sound wave that was propagating through the light particles only. The equivalent sound wave propagating through the heavier particles in the mixture, the so-called 'slow-sound' mode, has also been found to exist. In this chapter we look at the details of these modes, and how the dynamics of a binary mixture can be incorporated in the effective eigenmode formalism.

6.1   The experimental signature of fast and slow sound

Fast and slow sound are phenomena that occur outside of the hydrodynamic region, and they involve a separation of the dynamics of the light and heavy constituents of the mixture, at least according to the original interpretation of the phenomenon (Bosse *et al.*, 1986). In molecular dynamics computer simulations, one determines all possible partial dynamic structure factors, giving access to the behavior of the two constituent species. In scattering experiments one only has access to the total dynamic structure factor, which may be dominated by the scattering by one of the mixture's components. As such, fast and slow sound might both exist in a mixture, but it might not be possible to observe the existence of both in one experiment. In Fig. 6.1 we show the original simulation data by Bosse *et al.* (1986) for the partial dynamic structure factor $S_{\mathrm{Li-Li}}(q,\omega)$ for an 80–20% mixture of Li and Pb, respectively. The side peaks, corresponding to propagating oscillations in the Li atoms, are clearly visible over a large range of momentum transfers. In fact, they are much more visible than the shoulders that one typically encounters in a monoatomic fluid over the same q range. For instance, one can compare $S_{\mathrm{Li-Li}}(q,\omega)$ to the spectra for liquid neon shown in Fig. 4.2.

This enhanced visibility of the propagating modes turns out to be a common feature of mixtures, even in scattering experiments where one measures a combination of the various partial dynamic structure factors. This combination depends on the concentration of the two species x_1 and x_2, and on their respective scattering powers. The latter is of course technique dependent, and for mixtures it is most conveniently

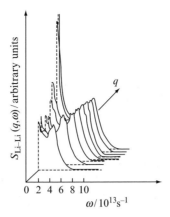

Fig. 6.1 Figure reproduced with permission from Bosse *et al.* (1986). Shown is the partial dynamic structure factor for the Li–Li correlations in a 80–20% Li–Pb mixture, as determined by molecular dynamics computer simulations. The q values range from 0.2 to 1.0 Å^{-1}. The side peaks in the dynamic structure factor correspond to the fast-sound mode, and its propagation frequency corresponds to a sound velocity that exceeds the hydrodynamic sound velocity of the mixture by a factor of three.

expressed in terms of scattering lengths b_1 and b_2 as follows (de Schepper and Montfrooij, 1989):

$$S_{\text{total}}(q,\omega) = x_1 b_1^{*2} S_{11}(q,\omega) + x_2 b_2^{*2} S_{22}(q,\omega) + 2\sqrt{x_1 x_2} b_1^* b_2^* S_{12}(q,\omega). \qquad (6.1)$$

Here, the normalized scattering lengths of the species j is given by

$$b_j^* = b_j / \sqrt{x_1 b_1^2 + x_2 b_2^2}. \qquad (6.2)$$

Thus, depending on the respective values of b_j and concentrations x_j, the scattering spectra might be more sensitive to one species than to the other. In practice, one can use this to one's advantage in order to try to deduce all partial scattering functions from experiment through, for instance, substitution of isotopes (in neutron-scattering experiments).

Theoretical calculations of mixtures of hard spheres revealed (Campa and Cohen, 1988) that fast sound should also occur in mixtures where the mass of the two components did not necessarily differ by a factor of 30. This was confirmed by molecular dynamics simulations on a He–Ne mixture (Montfrooij *et al.*, 1988) and subsequently, fast sound was observed experimentally in a neutron-scattering experiment on a He–Ne mixture (Montfrooij *et al.*, 1989) and slow sound was observed in light-scattering experiments on dilute H$_2$–Ar mixtures (Wegdam *et al.*, 1989) and H$_2$–SF$_6$ mixtures (Clouter *et al.*, 1990). We show the neutron-scattering results for a 65–35% He–Ne mixture in Fig. 6.2.

The fast-sound modes show up as clearly discernible side peaks in the total dynamic structure factor. The positions of these side peaks are virtually indistinguishable from

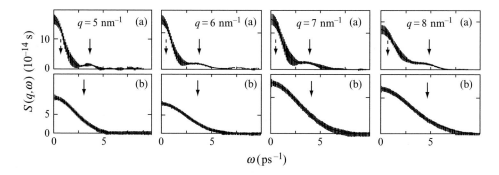

Fig. 6.2 The total dynamic structure factor for a 65–35% He–Ne mixture (top panels), compared to the spectra for pure He (bottom panels) under the same thermodynamic conditions ($T = 39.3$ K at a pressure of 114 bar). This figure has been reproduced with permission from Westerhuijs *et al.* (1992) The solid arrows in the top panels point to the fast-sound excitations, which occur at the same frequencies as the 'hidden' extended sound-mode excitations (solid arrows in bottom panels) in the pure helium fluid. This lends credence to the notion that fast sound involves an oscillation of the lighter atoms. The full dispersion curve for the mixture is shown in Fig. 6.3.

the extended sound-mode propagation frequencies inferred for a pure helium fluid under the same pressure and at the same temperature (see bottom panel of Fig. 6.2). This is exactly what would be expected if the fast-sound mode represents an oscillation sustained by the lighter particles only (Bosse *et al.*, 1986). The propagation speed of such an oscillation should only depend on the temperature and pressure of the fluid. We show the full dispersion of the fast sound mode in Fig. 6.3, where it can be seen that the propagation speed is identical to that of the pure helium fluid, and that this speed is considerably higher than the hydrodynamic speed of propagation. These findings, namely the observation of a mode whose speed of propagation greatly exceeds that of hydrodynamic sound, have since been reproduced with much higher accuracy on a range of mixtures, as well as by means of molecular dynamics simulations (Enciso *et al.*, 1995; Anento and Padró, 2001; Bryk and Mryglod, 2002).

The standard interpretation of the fast-sound mode is that outside the hydrodynamic region the heavier particles cannot follow the oscillations of the lighter particles (Campa and Cohen, 1990). The heavier particles can still sustain propagating waves, however, but at a much lower speed of propagation. These propagating slow-sound waves are not directly visible in the neutron-scattering experiments shown in Fig. 6.2; presumably they are hidden in the shoulders of the central feature, similar to the poor visibility of the extended sound modes in pure fluids. However, the slow-sound modes show up very clearly as separate excitations in light-scattering experiments on other mixtures. In the $q \to 0$ limit, both slow and fast sound should merge into the hydrodynamic sound mode, or one mode should transform itself into ordinary hydrodynamic sound, while the amplitude of the other mode decreases

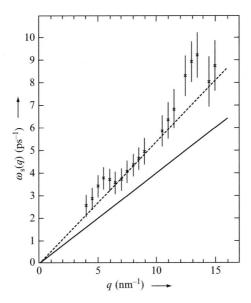

Fig. 6.3 The frequency of fast sound for a 65–35% He–Ne mixture at 39.3 K and 114 bar (crosses with error bars) greatly exceeds that given by the hydrodynamic speed of sound (solid line), but it is coincident with the extended sound mode propagation frequency of a pure helium fluid (dotted line) at the same temperature and pressure. This figure has been adapted with permission from Montfrooij *et al.* (1989).

with decreasing q until it vanishes at $q = 0$; how this might happen is discussed in Section 6.3.

Besides the observed fast- and slow-sound modes, there are a few more modes that might occur in binary mixtures, but that are absent in pure fluids. Since we now have two species in the fluid, we should be able to see two types of diffusive processes, as well as interdiffusion. These processes have indeed been inferred from experiment. In addition, the equivalent of optical sound modes could show up in mixtures, characterized by an out-of-phase oscillation of the light and heavy particles. This latter process has not (yet) been observed in an experiment, although they have been observed in computer simulations (Bryk and Mryglod, 2002), as we shall discuss later in this chapter. Next, we will adapt the effective eigenmode formalism in order to incorporate the increased number of modes in a binary mixture.

6.2   Effective eigenmode formalism for mixtures

In this section, we will generalize the basic hydrodynamics decay tree (see Fig. 2.11) to the case of binary mixtures. We will need to expand our set of three basic microscopic quantities ('n', 'u' and 'T') to five. This set is shown in Fig. 6.4. Of course, this set can be further generalized by including the equivalent of the microscopic quantities

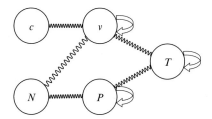

Fig. 6.4 The forces and decay rates pertinent to a binary mixture. This figure is comparable to Fig. 2.11 for simple fluids with the decay tree cut off after the velocity and temperature variables. However, it does include an additional variable ('v') that is not present in the hydrodynamic equations for a mixture; this new variable is necessary in order to be able to accommodate both fast and slow sound, while maintaining the proper $q \to 0$ limit.

'σ' and 'q' (see Fig. 2.12). We shall not do that here since the set shown in Fig. 6.4 is adequate in describing the (limited) scattering results available.

The five basic, orthonormal microscopic quantities of the mixture are constructed out of the number density of species 1 and of species 2, the longitudinal velocity of species 1 and species 2, and the total microscopic energy density of the mixture (Westerhuijs *et al.*, 1992). These five basic variables are listed in Appendix F. From these five quantities an orthonormal set $\{b_1, \ldots, b_5\}$ is crafted whose precise definitions are given in Appendix F. However, there is no need to resort to definitions in order to ascertain that these new microscopic variables make sense. The new set consists of the total number density ('N'), the total momentum ('P'), the total temperature ('T'), the mutual number concentration ('c') and the mutual relative velocity of the constituents ('v').

There is one issue, i.e. a problem in the interpretation of scattering results, that is unique to mixtures. In simple fluids the connection between the dynamic matrix $G(q)$ and the scattering results goes through the lowest-order frequency-sum rule: $S(q)$ in the classical case, and $\chi_{nn}(q)$ in the quantum case. In mixtures the equivalent procedure requires all three partial sum rules (Westerhuijs *et al.*, 1992). Thus, for a classical mixture we would require knowledge of $S_{11}(q)$, $S_{22}(q)$ and $S_{12}(q)$. Typically, in scattering experiments we only have access to the total static structure factor $S_{\text{total}}^{\text{exp}}(q)$ given by

$$S_{\text{total}}^{\text{exp}}(q) = x_1 b_1^{*2} S_{11}(q) + x_2 b_2^{*2} S_2(q) + 2\sqrt{x_1 x_2} b_1^* b_2^* S_{12}(q). \qquad (6.3)$$

In the absence of experimental knowledge of the partial structure factors they can either be left as adjustable parameters to vary within a sensible fitting range dictated by, for instance, the predictions of an equivalent hard-sphere mixture, or they can be fixed by assuming the following ideal mixing rules (Westerhuijs *et al.*, 1992):

$$S_{11}(q) = 1 + x_1[S_T(q) - 1]$$
$$S_{22}(q) = 1 + x_2[S_T(q) - 1] \qquad (6.4)$$
$$S_{12}(q) = \sqrt{x_1 x_2}[S_T(q) - 1].$$

Here, $S_T(q)$ is given by

$$S_T(q) = x_1 S_{11}(q) + x_2 S_2(q) + 2\sqrt{x_1 x_2} S_{12}(q) = 1 + \frac{S_{\text{total}}^{\text{exp}}(q) - x_1 b_1^{*2} - x_2 b_2^{*2}}{(x_1 b_1^* + x_2 b_2^*)^2}. \quad (6.5)$$

The dynamic matrix $G(q)$ that determines all 25 correlation functions between the five basic variables is given by (Westerhuijs *et al.*, 1992)

$$G(q) = \begin{pmatrix} 0 & 0 & 0 & 0 & if_{cv}(q) \\ 0 & 0 & if_{NP}(q) & 0 & if_{Nv}(q) \\ 0 & if_{NP}(q) & z_{PP}(q) & if_{PT}(q) & 0 \\ 0 & 0 & if_{PT}(q) & z_{TT}(q) & if_{Tv}(q) \\ if_{cv}(q) & if_{Nv}(q) & 0 & if_{Tv}(q) & z_{vv}(q) \end{pmatrix}. \quad (6.6)$$

While this matrix is straightforward, the determination of its elements from experiment is, as mentioned in the preceding, more complicated owing to the presence of the partial structure factors. In Appendix F we give the full procedure, as well as expressions for the coupling parameters $f_{cv}(q)$, $f_{Nv}(q)$ and $f_{NP}(q)$. With the aid of the equations in this appendix it is possible to determine the matrix elements of $G(q)$ from a fit to a scattering experiment.

The 5×5 matrix given in eqn 6.6 has five eigenvalues. For q values outside of the hydrodynamic region, but not as high as the positions of the main peaks of the static structure factors, two of those eigenvalues would correspond to the fast-sound modes, two would correspond to the slow-sound modes, and one would be the equivalent of the heat mode. One of the main questions in the study of mixtures is how these modes change into the hydrodynamic modes of a mixture upon $q \to 0$. We will discuss this limit in the next section.

6.3   The hydrodynamic limit

First, the hydrodynamic limit can be studied in light-scattering experiments on dilute mixtures, such as H_2–Ar. The results of such experiments performed by Wegdam *et al.* (1989) are shown in Fig. 6.5. These experiments quantitatively confirm the predictions of Campa and Cohen (1988) that fast and slow sound should also be present in dilute mixtures, and how the two modes should merge into a single hydrodynamic sound mode.

The situation for dense mixtures is a little more complicated. The hydrodynamic equations for a mixture predict four modes: two propagating sound modes, one heat mode and one interdiffusion mode. In order to be able to describe the emergence of both fast and slow sound using the effective eigenmode formalism, the non-hydrodynamic variable 'v' had to be introduced (Fig. 6.4). It is not known whether this minimalistic generalization is good enough to describe the results for scattering experiments on mixtures over a range of densities and concentrations; it is just known that it did the job for the particular He–Ne mixture most studied. We note that in more recent molecular dynamics computer simulations larger sets have been utilized (Bryk and Mryglod, 2002).

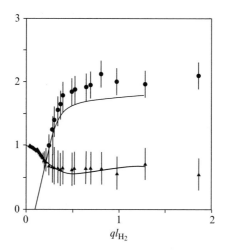

ql_{H_2}

Fig. 6.5 Figure reproduced with permission from Wegdam *et al.* (1989). The figure shows the reduced propagation frequencies $\omega_{prop}/(c_{hydro}q)$ of the fast- and slow-sound modes in a dilute mixture of H_2–Ar as a function of reduced wave number $ql_{free} = ql_{H_2}$. The propagation frequencies were determined from the peaks of the current–current correlation function $C_L(q,\omega)$. Here, c_{hydro} is the hydrodynamic sound velocity involving the in-phase motion of both species. In these experiments the wavevector transfer was varied by changing the density of the mixture, thereby changing the mean-free path $l_{free} = l_{H_2}$ between collisions. The solid lines are the theoretical predictions based upon the Enskog theory for mixtures (Campa and Cohen, 1989). With increasing q one can observe the change from a single hydrodynamic sound mode to a slow- and fast-sound mode with the fast-sound propagation frequency about three times faster than that of slow sound.

The hydrodynamic limit of the five eigenmodes for the 65–35% He–Ne mixture is shown in Fig. 6.6. The figure shown is based on an extrapolation of the fitted results in the range $q > 0.3\,\text{Å}^{-1}$ down to $q \to 0$. According to this extrapolation, which is based on the predicted[1] low-q behavior of the matrix elements of $G(q)$, the hydrodynamic sound mode gives way to the fast- and slow-sound mode; slow sound appears to be a continuation of ordinary hydrodynamic sound (see top panel in Fig. 6.6), while fast sound is seen to come into existence around $q = 0.07\,\text{Å}^{-1}$.

As it turns out, the above extrapolation does not agree with experimental results that were obtained subsequently. In neutron-scattering experiments designed to probe the behavior at the lowest q values (Bafile *et al.*, 2001) of the same He–Ne mixture, it was in fact found for the q range $0.07 < q < 0.18\,\text{Å}^{-1}$ that only one propagating mode was present in the mixture, and that the frequency of this mode was given by the hydrodynamic velocity of sound. These results are shown in Fig. 6.7. It took molecular dynamics computer simulation data to show exactly how fast sound emerges from ordinary hydrodynamic sound (Sampoli *et al.*, 2002). Interestingly enough, the computer simulation that was able to resolve the transition from the hydrodynamic

---

[1] The off-diagonal elements vary $\sim q$ for $q \to 0$.

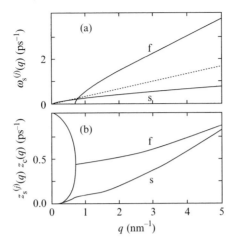

Fig. 6.6 Figure reproduced with permission from Westerhuijs *et al.* (1992). The eigenvalues of the matrix $G(q)$ (eqn 6.6) as determined from an extrapolation to $q \rightarrow 0$ of the fit results for a 65–35% He–Ne mixture (Fig. 6.2). The labels 'f' and 's' refer to fast and slow sound, respectively. The top panel shows the propagation frequencies, the bottom panel displays the decay rates. The dashed straight line in the top panel is given by the hydrodynamic sound velocity. According to this extrapolation, fast and slow sound should change to ordinary hydrodynamic sound around $q \approx 0.07$ Å^{-1}.

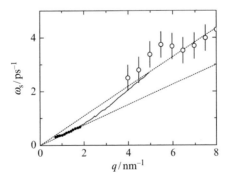

Fig. 6.7 Figure reproduced with permission from Bafile *et al.* (2008). The measured and simulated dispersion curve of the acoustic mode in a He–Ne gaseous mixture. Full dots are the neutron-scattering data from Bafile *et al.* (2001), the open symbols are from Montfrooij *et al.* (1989). The molecular dynamics calculations (Sampoli *et al.*, 2002) provide the connection (solid line) between the hydrodynamic dispersion and the fast-sound mode. The straight lines correspond to the hydrodynamic speed in the mixture ($c_s = 374$ m/s) and to the equivalent (Sychev *et al.*, 1987) for pure helium (544 m/s – top dashed curve) under the same pressure.

region to the fast-sound region, did not find any evidence for the slow-sound mode excitation in this mixture. However, when the density of the mixture was increased in the simulation, evidence for both fast- and slow-sound modes appeared for $q > 0.4$ Å^{-1}. We will revisit this issue in the next section when we discuss potential optical modes in a mixture.

There probably is a moral to the story of the results detailed above. Let us just say that extrapolations should sometimes be viewed with some healthy scepticism. We already saw an example of this in the case of determining mode-coupling effects in the velocity autocorrelation function, and we will find (Chapter 7) that the hydrodynamic transition in liquid metals is also quite difficult to get a handle on. What is very encouraging though is the excellent agreement between computer simulations and experimental results, and how computer simulations were able to provide a clearcut answer for this particular mixture. Perhaps this is only fitting since after all, fast and slow sound were first discovered by means of computer simulations (Bosse *et al.*, 1986). In fact, newer computer simulations (Bryk and Mryglod, 2002, 2005) on a host of mixtures have demonstrated that the approach to hydrodynamic behavior is very much dependent on the mixture under scrutiny.

6.4 Optical modes and missing modes in a mixture

At small wave numbers, both the heavy and light atoms oscillate in phase, producing ordinary hydrodynamic sound. This of course begs the question of whether we can also observe an out-of-phase oscillation, the equivalent of optical modes in solids. Experimentally, such an optical mode has not been observed, but molecular dynamics computer simulations (Bryk and Mryglod, 2002) not only leave open the possibility of their existence, they also show under what conditions these optical modes might be observed.

Fast and slow sound signal the decoupling of the dynamics of the light and heavy species in the mixture. At least that is what it looks like given the limited amount of scattering data available. If such a decoupling does indeed take place outside of the hydrodynamic region (at $q \approx 0.4$ Å^{-1} for the He–Ne mixture discussed above), then one would not expect to find the equivalent of an optical mode in this region, since this mode would involve the coordinated out-of-phase motion of the heavy and light particles. Nonetheless, the data shown in Fig. 6.2 do leave open the possibility of the presence of an optical mode (Westerhuijs *et al.*, 1992) at frequencies $\omega \approx 7$ ps^{-1}.

A repeat of the measurements on a different spectrometer with a higher energy resolution (Montfrooij and Svensson, 1996) did not produce any evidence for an optical mode at these momentum transfers. These later experiments are reproduced in Fig. 6.8. The fast-sound mode can be seen as a side shoulder in the spectra at $q = 0.5$ Å^{-1} (bottom panel of Fig. 6.8). However, these spectra do not show any features at higher frequencies that can be attributed to some optical excitation. On the positive side, its absence in the 65–35% He–Ne mixture is at least consistent with the interpretation of a decoupling of the dynamics on microscopic length scales.

In fact, these observations are reasonably consistent with molecular dynamics simulations by Bryk and Mryglod (2002, 2005) that predict the absence of optical

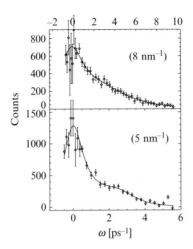

Fig. 6.8 Neutron-scattering data taken on a 65–35% He–Ne mixture show no evidence of any optical modes. The q values are given in brackets in the figure; note that the horizontal axes differ between the two panels. The data in this figure have not been deconvoluted with the spectrometer resolution function; this is in contrast to the data shown in Fig. 6.2. The bottom panel shows the fast-sound mode as a side peak. This figure has been adapted with permission from Montfrooij and Svensson (1996).

excitations in the He–Ne mixture. These simulations predict, as shown in Fig. 6.9, that the He and Ne dynamics in a 65–35% He–Ne decouple for $q > 0.6$ Å^{-1}; this would be in line with the assertion that fast sound propagates through the lighter particles. Moreover, these simulations predict that the only mode visible in $S(q, \omega)$ as measured in scattering experiments on this particular mixture in the range $q < 0.5$ Å^{-1} would be the mode that becomes the ordinary hydrodynamics mode (see Fig. 6.10). This is of course what was observed in the experiments (Bafile *et al.*, 2008) as depicted in Fig. 6.7. However, the finer details still need to be worked out as the fast-sound mode clearly shows up in scattering experiments at $q = 0.5$ Å^{-1} (see Figs. 6.2 and 6.8), while the simulations predict that it should not show up until higher q values (Figs. 6.9 and 6.10).

Bryk and Mryglod (2002) showed by means of computer simulations that optical modes should exist in a 50–50% Kr–Ar mixture over a wide q range ($q \leq 2$ Å^{-1}). We reproduce their results in Figs. 6.9 and 6.10. According to these computer simulations, the Kr–Ar mixture should not only be able to sustain a coupled out-of-phase motion of the Kr and Ar atoms, this mode should be visible (Fig. 6.10) in scattering experiments at fairly low q values. The reason why the He–Ne and the Kr–Ar mixtures show such a different response is largely determined by the mutual coefficient of diffusion D_{12} (Bryk and Mryglod, 2002). This coefficient determines the damping rate of collective out-of-phase oscillations and provided this damping rate is small enough compared to the strength of the microscopic forces this can result in optical-like excitations. Bryk

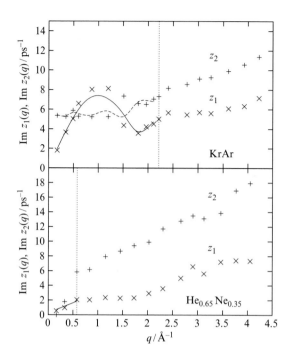

Fig. 6.9 Figure reproduced with permission from Bryk and Mryglod (2002). The top panel shows the results for a molecular dynamics simulation on a 50–50% mixture of Kr–Ar at 116 K and at a density $n = 0.0182$ Å^{-3}, the bottom panel shows the results for a 65–35% He–Ne mixture at 39.5 K and a density of $n = 0.0186$ Å^{-3}. The dotted vertical lines in the panels are a rough measure of the boundary between collective dynamics of the two constituents, and decoupled dynamics. The mode labelled z_1 (crosses) reduces to the ordinary acoustic hydrodynamic mode at the smallest q values. This figure shows that an optical-like mode (labeled z_2, pluses) is possible in the Kr–Ar mixture in the coupled dynamics region, but that such a mode should not exist in the He-Ne mixture. In the latter, the z_2 mode would only show up in the region where the dynamics are already decoupled, resulting in a fast-sound mode that propagates through the lighter atoms only.

and Mryglod (2002) captured this in the following criterion:

$$\frac{D_{12}^2 q^4 M_{cc}^{(4)}(q)}{[2M_{cc}^{(2)}(q)]^2} < 1. \tag{6.7}$$

Here, we have introduced the nth reduced frequency moments $M_{xy}^{(n)}$ of the partial (classical) dynamic structure factors $S_{xy}(q, \omega)$ (with $x, y =$ 'c' and 'N') by

$$M_{xy}^{(n)} = \frac{\int \omega^n S_{xy}(q, \omega) \mathrm{d}\omega}{\int \omega^{n-2} S_{xy}(q, \omega) \mathrm{d}\omega}. \tag{6.8}$$

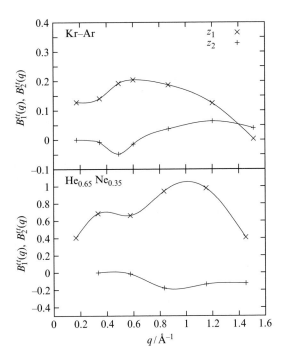

Fig. 6.10 Figure reproduced with permission from Bryk and Mryglod (2002). The corresponding amplitudes in the total dynamic structure factor $S(q,\omega)$ of the sound modes shown in Fig. 6.9. B_1 (crosses) is the amplitude of the mode that becomes the acoustic sound mode in the mixture in the hydrodynamic limit, B_2 (plus signs) is the mode that becomes the optical mode provided the criterion in eqn 6.7 is met. Note that the predicted amplitude of the optical mode in the Kr–Ar mixture is very close to zero at the smallest q values, but the predicted amplitude implies that optical-like excitations should be visible in this mixture for $q > 0.4$ Å^{-1}.

This criterion is satisfied for the Kr–Ar mixture for $q < 2$ Å^{-1}, but it is not satisfied for the He–Ne mixture down to the lowest q value ($q = 0.17$ Å^{-1}) of the simulation shown in Fig. 6.9. This difference largely stems from the fivefold difference in D_{12}. The only thing that is left to do now is to actually measure the dynamic structure factor of a 50–50% Kr–Ar mixture.

Whether, or how well a particular excitation is visible in the measured total dynamic structure factor depends also on the scattering lengths of the two species. In Appendix F we calculate the relative contributions of the partial dynamic structure factors $S_{cc}(q,\omega)$, $S_{NN}(q,\omega)$ and $S_{cN}(q,\omega)$ to the total measured dynamic structure factor given in eqn 6.1. For instance, the contribution of the partial dynamic structure factor $S_{cc}(q,\omega)$ relevant to concentration fluctuations (and hence also to optical modes) depends on the difference in scattering cross-sections $b_1^2 - b_2^2$. This difference is large in X-ray-scattering experiments, but relatively small in most neutron-scattering experiments. As a result, excitations consisting of concentration fluctuations are expected to be almost absent in neutron-scattering experiments. This was indeed observed in the He–Ne scattering experiments discussed in this chapter, where the quasi-elastic line characteristic of mutual diffusion appeared to be absent from the spectra.

In fact, neutron scattering has an advantage over X-ray and light scattering for the study of mixtures because in addition to the scattering lengths for all elements being of the same order of magnitude, the scattering lengths can also be negative. Thus, through a judicious choice of mixture elements and relative concentrations, it is possible to completely eliminate $S_{NN}(q, \omega)$ and $S_{cN}(q, \omega)$ from the scattering in favor of $S_{cc}(q, \omega)$, or vice versa (see eqn F.9). For instance, an 83–17% Li–Pb mixture should be dominated by the contributions arising from $S_{cc}(q, \omega)$, while a Ne–Xe mixture of any concentration will mostly display the dynamics associated with $S_{NN}(q, \omega)$. Needless to say, computer simulations will be essential to the interpretation of any scattering data. In particular, simulations on mixtures now routinely calculate not only the basic correlation functions between the conserved microscopic quantities, but also between many of their derivatives (Bryk and Mryglod, 2002). In this way, Bryk and Wax (2009) have determined for the cases of Li_4Pb and Li_4Tl mixtures exactly how the fast- and slow-sound modes emerge out of the hydrodynamic region, and what their character is once they become visible in the density–density correlation functions.

6.5   Fast sound versus viscoelastic behavior

The transition from sound modes propagating at the hydrodynamic speed of sound to modes propagating at a considerably higher speed (Fig. 6.7) is not something unique to binary mixtures. Instead, it is a very common phenomenon in liquid metals, as we will discuss in Chapter 7. In fact, sometimes it is very hard to distinguish between fast sound and viscoelastic behavior. We will discuss the findings on molten NaCl and water to illustrate this point.

Demmel *et al.* (2004) demonstrated the existence of sound modes propagating in a liquid NaCl mixture in a recent X-ray-scattering experiment. These sound modes show up as shoulders in the dynamic structure factor, and their propagation frequencies and damping rates are shown in Fig. 6.11. The propagation frequencies of these sound modes are ~ 70% higher than the hydrodynamic frequencies (Demmel *et al.*, 2004). On the one hand, this finding is in agreement with the viscoelastic model displaying the transition from viscous relaxation at low q to elastic behavior at the higher q values. During this transition the speed of propagation of the short-wavelength sound waves will increase from $[f_{un}(q)^2 + f_{uT}(q)^2]^{1/2}/q$ to $[f_{un}(q)^2 + f_{u\sigma}(q)^2]^{1/2}/q$, and a 70% increase is entirely within the realm of possibilities for NaCl. We draw the reader's attention to Fig. 2.19 as an illustration of such an increase, and to the various combinations of forces and damping rates shown in the figures of Appendix D.

On the other hand, close inspection of the propagation frequencies shows that they are very similar to those of pure Na (Demmel *et al.*, 2004) in the range $0.4 < q < 1.2$ Å^{-1} where the viscoelastic model describes the dispersion. Thus, in this binary ionic liquid the observed transition from hydrodynamic sound to faster than hydrodynamic sound could equally well be a transition very similar to the hydrodynamic to fast-sound

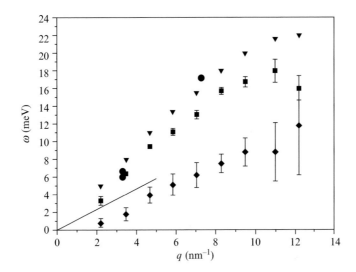

Fig. 6.11 Figure adapted with permission from Demmel *et al.* (2004). The excitation frequencies (squares) of liquid NaCl at 1170 K, as determined from X-ray-scattering experiments, and their associated damping rates (lozenges). The excitation frequencies correspond to short-wavelength sound modes that are propagating at a speed in excess (70%) of the hydrodynamic sound velocity (solid line). This excess can be caused by the liquid responding in a viscoelastic manner, or it can be caused by the lighter sodium ions undergoing a fast-sound oscillation in which the heavier Cl-ions do not partake. The triangles denote the peaks of the current–current correlation function, and the circles are the results for a molecular dynamics computer simulation.

transition we have described for our 65–35% He–Ne mixture. From the dispersion alone one cannot tell which description is the correct one, however, inclusion of the amplitudes of the modes should be able to shed light on this. After all, once the eigenvalues of the matrix $G(q)$ have been determined through model fitting, then the amplitudes of the modes also follow from this matrix (see appendices D and F). Following this procedure, one should be able to tell the difference between a sound mode in which all the particles partake, or one in which only half of them partake. This discussion also underscores the strength of the effective eigenmode formalism where the sum rules for the relative strengths of the various modes are automatically built into the model.

Short-wavelength sound modes propagating at a speed far in excess of the hydro-dynamic sound velocity have also been observed in water. Whereas normal sound travels at a speed of 1.5 km/s in water, the short-wavelength sound waves travel at speeds in excess of 3 km/s. These fast modes were given the name fast sound, according to the premise that these modes could represent a wave that would involve the light hydrogen atoms only. The origin of this mode has been the subject of many scattering experiments and computer simulations. Ruocco and Sette (2008) have given an historical account of this fast-sound mode that, most likely, turned out not to be a

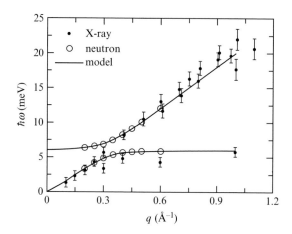

Fig. 6.12 Figure reproduced with permission from Sacchetti *et al.* (2004). Neutron- and X-ray-scattering results for the two branches of excitations in water at 293 K. The solid lines describing both the X-ray- and neutron-scattering results is the model by Sacchetti *et al.* (2004) based upon a flat dispersionless mode interacting with a dispersing mode. The slope of the dispersing mode for $q > 0.4$ Å^{-1} is 3 km/s.

fast-sound mode as observed in mixtures, rather it most likely reflects the viscoelastic response of water to short-wavelength perturbations.

The identification of this fast mode in water required the combination of neutron, X-ray, Brillouin and inelastic ultraviolet light-scattering techniques. The data interpretation was made more difficult by the existence of a second, almost non-dispersive mode present in the scattering spectra (see Fig. 6.12). This second mode (again most likely) originates from oxygen-bending modes associated with the tetrahedral structure of the liquid. In fact, Sacchetti *et al.* (2004) achieved a very good phenomenological description for the dispersion of these two modes based upon the branch repulsion that happens when a dispersing and a dispersionless mode interact with each other. This repulsion model is shown in Fig. 6.12. Such a branch repulsion is also observed in superfluid helium (see Chapter 9). Sacchetti *et al.* (2004) were able to describe the scattering data on water from 273 K to 450 K using this model, and in this temperature range it appeared that the transition from hydrodynamic sound to fast sound was strongly influenced by this branch repulsion. However, subsequent experiments showed (Santucci *et al.*, 2006) that this branch repulsion could not be the ultimate cause of the transition; it was discovered that the q value at which the transition to higher propagation speeds takes place is in fact strongly dependent on temperature, especially so in the supercooled regime. In contrast, the q value at which the branch repulsion occurs should only be weakly temperature dependent. Nonetheless, even though this branch repulsion may not prove to be the underlying cause of fast sound, it does appear to modify the shape of the dispersion.

Not only do the experiments by Santucci *et al.* (2006) show that branch repulsion is unlikely to be the full explanation of the fast-sound-like dispersion in water, out of

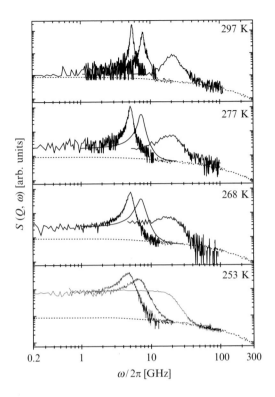

Fig. 6.13 Figure reproduced with permission from Santucci *et al.* (2006). Shown are Brillouin light-scattering spectra on water at the temperatures indicated in the figure. Three q values are shown at each temperature ($q = 0.00229$, 0.00326 and 0.00896 Å^{-1} for the top three panels, and $q = 0.00229$, 0.00326 and 0.00712 Å^{-1} for the bottom panel). Note the change from ordinary hydrodynamic sound modes at 297 K, to strongly damped, viscoelastic modes in the supercooled regime, demonstrating the strong temperature dependence of the q value at which the cross-over occurs in the supercooled regime.

the many scattering experiments that have been performed on water during the past 4 decades, their experiments provided the most conclusive evidence for the interpretation of the fast mode being related to viscoelastic relaxation. They were able to follow the behavior of the fast mode into the supercooled regime (see Fig. 6.13), and witness the transition from ordinary hydrodynamic sound to a viscoelastic excitation upon cooling. By lowering the temperature into the supercooled regime and by measuring at fixed momentum transfers $q < 0.01$ Å^{-1} they could see the onset of glass-like behavior characterized by excitations moving at increased speeds with decreasing temperature, even with the onset of heavy damping. The spectra in the supercooled regime were all very well described by a viscoelastic model that has excitations moving at a speed of 3 km/s ($= c_\infty = [f_{un}^2 + f_{u\sigma}^2]^{1/2}/q$). These measurements into the supercooled regime established that the fast modes visible in the neutron-scattering spectra were indeed due to the elastic-like response of the liquid, a response that is of course well known in the glass-forming regime (Balucani and Zoppi, 1994). Thus, water was shown to behave very similar to any other fluid when it comes to short-wavelength sound propagation, with the speed of propagation increasing because of the solid-like response of a liquid when probed at high frequencies.

On a final note in this discussion of the excitations in water, most results for the propagation frequency are presented in terms of the undamped frequency. This frequency gives the peak position in the current–current correlation function, as explained in Section 3.2.4, rather than the poles of the dynamic susceptibility. However, in this instance it does not affect the interpretation of the fast-sound mode as being a viscoelastic effect, as the comparison between the different temperatures was carried out across the literature (on water) using this undamped frequency. As long as one bears in mind that the fast-sound mode in water does not actually propagate at 3 km/s, but rather at a lower speed because of the damping affecting these modes, then there is little cause for confusion.

We will encounter this viscoelastic behavior also in liquid metals (see Chapter 7) where the increase in sound-propagation frequency can be of the same order of magnitude as is observed in water and NaCl. Thus, the change from ordinary hydrodynamic sound to a region where the sound modes propagate at a much higher velocity seems to be a very common feature in all classical liquids. Upon increasing q, all liquids seem to enter a region where the frequency of the short-wavelength sound wave is so high that the liquid responds much more like a solid, and the response is determined to a large extent by the instantaneous structure present in the liquid, resulting in a propagation frequency that is close to the propagation frequency of the solid state, albeit that these modes are much more strongly damped than in the solid sate because of the fact that the local structure present in a liquid is not as perfect as that present in the solid state.

7
Liquid metals

Liquid metals can be very successfully described by the effective eigenmode formalism. Surprisingly, the formalism for single-component fluids appears to be adequate in describing the dynamics of the ions, even though liquid metals are manifestly a two-component system consisting of the ions and the electron sea. The reason why a single-component description appears to work so well is that one can see the ionic motions as taking place through an interaction that is mediated by the conduction electrons; the resulting interaction can be captured in a pseudo-potential that depends upon the density of the system. However, a potential nonetheless, and consequently we find the same phenomena in liquid metals that we have already encountered in simple fluids, such as propagating short-wavelength sound modes and diffusive heat modes.

Nonetheless, there also exist quite a few differences between the behavior of simple liquids and that of liquid metals. First, the short-wavelength extended sound modes are much more prominent in liquid metals, and in contrast to the case for simple fluids, they can easily be seen with the naked eye (without the aid of a decomposition into extended modes). Inelastic X-ray-scattering experiments on a range of liquid metals have now shown that extended sound modes with wavelengths as short as twice the interparticle spacings are very prominent features of the scattering spectra. In fact, had the European Synchrotron Radiation Facility ESRF been in existence during the 1980s, then there would have been no debate as to whether short-wavelength sound modes exist in simple liquids, or not.

The prominence of the extended sound modes occurs for two reasons. First, the speed of propagation of these modes is very high, resulting in a distinct separation (in frequency) from the quasi-elastic features of the scattering spectra. Secondly, the amplitude of the (extended) heat mode is much weaker in comparison to simple liquids because the ratio of the specific heats γ is very close to 1, and the relative amplitude of the heat mode is given by $(\gamma - 1)/\gamma = 1 - c_V/c_p$ (see Table 2.3). Since $c_p - c_V \sim \alpha^2$, with α the coefficient of thermal expansion, we can expect small heat-mode amplitudes given the smallness of the thermal expansion coefficient in liquid metals. The reason for the smallness of the latter is that the number density of the metal is to a large extent determined by the electrons, and the electronic properties are only weakly temperature dependent since they are governed by the Pauli exclusion principle. Note that unlike very cold liquids, γ is not identical to 1, just fairly close to 1.

At some level in the decay tree of the effective eigenmode formalism we can expect to see the direct effects of the coupling between the electronic motions and the ionic ones. At this level, a disturbance that originated from a disturbance that ultimately originated from a density fluctuation will decay on the same time-scale as that of the electronic motions. The effective eigenmode formalism will not be able to distinguish whether the power put into this decay channel dissipates because of the electronic degrees of freedom, or because of the ionic ones; the only thing that we will find is that the decay tree can be cut off at this level, and that we can treat this part of the decay tree as if the disturbances instantaneously relax back to equilibrium. See also our discussion following Fig. 2.22. As we will discuss in this chapter (Section 7.4.1), computer simulations may steer the interpretation of the scattering data at this level of the decay tree.

In most of the recent literature reporting on scattering experiments on liquid metals (e.g., Scopigno and Ruocco 2005) one finds that the excitations are described using $S_{nn}^{\mathrm{sym}}(q,\omega)$ as opposed to $S_{nn}(q,\omega)$, albeit sometimes somewhat apologetically, even though it is the correct function to apply the effective eigenmode formalism and the memory formalism to. While liquid metals are measured at fairly high temperatures, the excitations extend up to very high energies, so that the asymmetries arising from the pre-factor in eqn 2.2 are still visible as lineshape distortion. And given the high accuracy of inelastic X-ray experiments, using $S_{nn}(q,\omega)$ as opposed to $S_{nn}^{\mathrm{sym}}(q,\omega)$ would lead to perceptible errors in the interpretation of the scattering results. As a final note, traditionally one uses the memory function formalism in the description of liquid metals as opposed to the effective eigenmode formalism; however, as the two are fully equivalent – in principle at least (see Appendix E) – we shall use both methods in this chapter in order to stay as close to the literature as we can.

7.1   Extended sound modes in liquid metals

Since the prominent extended sound modes are the most obvious features in the spectra of liquid metals, we will start our discussion with those.[1] We show the impressive level of detail and accuracy that inelastic X-ray-scattering experiments can achieve in Fig. 7.1. These spectra (Scopigno *et al.*, 2000*a*) represent the state-of-the-art in inelastic X-ray-scattering technology in 2000; not only is the energy resolution so good that it is very straightforward to see the extended sound modes as separate entities, the statistical accuracy of the data points demonstrates that even a system like lithium with only two bound electrons scatters strongly enough to do X-ray-scattering experiments. This figure also nicely illustrates why X-ray scattering has been so successful in the study of liquid metals over the past decade. The extended sound modes propagate at such high speeds that even at relatively low momentum transfers they are located at high-energy transfers. This is a region of (q,ω) space that is virtually impossible to access by means of inelastic neutron scattering while maintaining a decent energy resolution.

The solid line in Fig. 7.1 is the best fit based upon the extension of the viscoelastic model given by eqn 2.57. Depending on exactly how this model is implemented in the fitting procedure it is identical to the decay tree shown in Fig. 2.22(a), or to a good approximation of it. Also shown in this figure is the best fit as obtained from the viscoelastic model (Fig. 2.15). Clearly, this latter model fails to reproduce all

---

[1] This by no means implies that the central line does not provide us with information on the dynamics; see for instance Fig. 4.15.

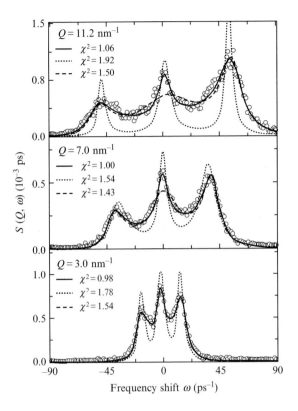

Fig. 7.1 Figure reproduced with permission from Scopigno *et al.* (2000*a*). Shown are the spectra for lithium at 475 K at three momentum transfers listed in the figure. The dashed and dotted lines that either underestimate or overestimate the central feature are the results using the decay tree depicted in Fig. 2.15, the solid line is the result for using, more or less, the model shown in Fig. 2.22(a). The improvement of using the extended model is most apparent for the quasi-elastic feature.

the features of the measured $S_{nn}(q, \omega)$ – especially for the quasi-elastic features – demonstrating the level of sophistication that models need to have in order to fully describe the measured spectra. Or alternatively, all this demonstrates the degree of information that one can distil from the spectra regarding the microscopic dynamics of liquid metals. There is one caveat that we have to make: in comparing the various models, the parameters relating to the thermal part of the decay tree ($f_{uT}(q)$ and $z_T(q)$) were kept fixed at their hydrodynamic values (Scopigno *et al.*, 2000*a*). This is standard practice in the analysis of scattering data on liquid metals in order to keep the number of free parameters down when fitting a model.

These lithium spectra (Fig. 7.1) are fairly typical for liquid metals close to their melting points. The alkali metals all support very well defined propagating short-wavelength sound modes (see for instance, Torcini *et al.* 1995; Scopigno *et al.* 2002; Bove *et al.* 2003), and the non-alkali metals also display reasonably well defined

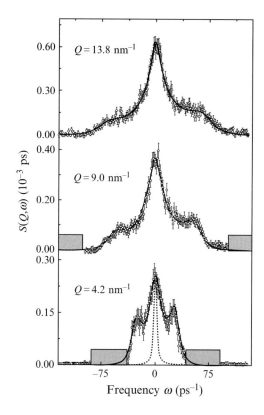

Fig. 7.2 Figure reproduced with permission from Scopigno *et al.* (2000*b*). The normalized dynamic structure factor (symbols) for liquid Al at $T = 1000\,\text{K}$ measured at the q values given in the figure. The solid lines are the fit according to the extended viscoelastic model (eqn 2.57). The extended sound modes are visible as side peaks or distinct shoulders. These modes are not as prominent as they are in liquid Li (Fig. 7.1), however, they are still much better defined than the extended sound modes in simple liquids at comparable momentum transfers. The energy resolution of the spectrometer is given by the sharp central feature in the lower dataset.

extended sound modes, although not as prominent as the alkali metals. We show the spectra for liquid Al measured by means of inelastic X-ray scattering in Fig. 7.2, and we reproduce (Scopigno *et al.*, 2000*b*; Scopigno and Ruocco, 2005) the comparison between the propagation speed of the extended sound modes of Li and Al in Fig. 7.3. The extended sound mode-propagation frequencies for cesium and silicon are shown in subsequent sections of this chapter.

Overall, the behavior of the extended sound modes in liquid metals is very similar to the behavior of these modes in simple liquids, at least at first sight. First, a direct comparison to the literature is made somewhat more difficult because propagation frequencies in liquid metals tend to be identified by the peaks of the current–current correlation function $C_L(q, \omega) \sim \omega^2 S(q, \omega)$, which as we have shown in Section 3.2.4 overestimate the actual propagation frequencies corresponding to the poles of the

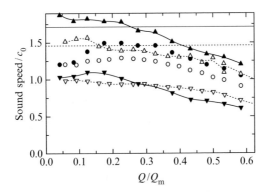

Fig. 7.3 Figure reproduced with permission from Scopigno *et al.* (2000*b*). Shown are the scaled sound velocities as estimated from the peaks of the current–current correlation function for Al (filled circles) and Li (open circles). As mentioned in Section 3.2.4, this slightly overestimates the actual speed of propagation of the extended sound modes; given the smallness of the damping, this effect is fairly minimal. The data have been scaled to the position Q_m of the main peak of $S(q)$, and to the isothermal sound velocity $c_0 = \lim_{q \to 0} \sqrt{f_{un}(q)}/q$. Also shown are the fit parameters of the viscoelastic model $c_\infty(q) = [f_{un}(q)^2 + f_{uT}(q)^2 + f_{u\sigma}(q)^2]^{1/2}/q$ (upward-pointing triangles), and the isothermal sound velocity $c_0(q)$ (downward-pointing triangles).

dynamic susceptibility. For extended sound modes that are well separated from the heat mode, the maxima of the current–current correlation function correspond roughly to $\Omega(q) = f_{un}(q)$, rather than to the propagation frequencies of the extended sound modes, which are under the above restrictions (roughly) given by $\sqrt{\Omega(q)^2 - \Gamma(q)^2}$. For example, for a damping rate $\Gamma(q) = 0.5\Omega(q)$ this leads to an overestimation in propagation frequency of 15%. Secondly, liquid metals are characterized by a value of the specific heat ratio γ that is relatively close to 1, and as such, we can expect changes in the mechanism that determines the propagation speed of the extended modes outside of the hydrodynamic region. We will return to this issue in the following sections.

The direct comparison between the extended sound modes of liquid Li and liquid Al (Fig. 7.3) shows that these two liquids are qualitatively very similar. The full spectra for both liquids are described very well by the extended viscoelastic model of eqn 2.57, and even the parameters of this model are rather similar for both systems. This is encouraging, given that Li is expected to more closely resemble a simple liquid than Al, given the fact that Li is essentially a fluid of charged He atoms embedded in an electron sea. This overall similarity between Li and Al appears to be norm, with the dynamics pertaining to the extended sound modes being very similar in most metals that have been studied. We refer the reader to the review by Scopigno and Ruocco (2005) for a detailed comparison between the various liquid metals. The most noticeable difference between simple liquids and liquid metals is at the lowest q values; the transition between ordinary hydrodynamic behavior and extended hydrodynamic behavior beyond which the transport coefficients become q dependent occurs at much smaller q values in liquid metals. We will discuss the hydrodynamic $q \to 0$ limit of the extended modes in Section 7.2.

7.1.1   The fine print

The scattering spectra of liquid metals are clearly distinct from their counterparts for simple liquids, given the prominence of the sound modes in the presence of an anemic heat mode, but how different are the extended sound modes of the two systems? First, the speed of propagation over the entire q range is not all that different. When the propagation frequencies are presented in both cases as the eigenvalues of the dynamic matrix $G(q)$ – as opposed to the peaks in the current–current correlation function as is done in the field of liquid metals – then the behavior of the two systems is almost identical. The propagation frequencies are given by the velocity of first sound at the lower q values, with a (modest) upwards curvature. What is different is the underlying propagation and damping mechanism. The extended sound modes in liquid metals are better defined (visible) because the ratio Γ_s/ω_s is much smaller than it is in simple liquids. In addition, the propagation frequency is determined almost solely by the coupling $f_{u\sigma}$ between 'u' and 'σ' in liquid metals, while the thermal decay channel f_{uT}–f_{Tq} plays a (much) larger role in simple liquids. This difference is because of the smallness of the coupling parameter f_{uT} in liquid metals. The strength of the coupling parameter $f_{u\sigma}$ compared to f_{un} is fairly similar between both systems. Hence, the main difference between the extended sound modes in the two systems is attributable to the difference in f_{uT}; all other parameters including the damping rates of the momentum flux are more similar than they are different.

Using the memory function formalism as expressed in eqn 2.58 yields an excellent description of the very accurate X-ray experiments on liquid metals (Scopigno and Ruocco, 2005; Bafile *et al.*, 2006). As explained in the discussion accompanying Fig. 2.22, there are some (minor) aspects in which this memory function differs from the decay tree of the effective eigenmode description. Here, we look at those minor differences and we show that one should not lose track of these (minor) issues.

We start by saying some mean things about the memory function of eqn 2.58, even though we do not really mean it. The decay tree of hydrodynamics as shown in Fig. 2.11 is cut off at the 'u' and 'T' branches. On the one hand, the results for the simple liquids discussed in Chapter 4 demonstrated that this decay tree needed to be expanded by including one more variable at the end of *each* branch. This yielded the decay tree shown in Fig. 2.12; this tree can be viewed as the generalization of the hydrodynamics tree by including one more level of detail. Both branches have been expanded by inclusion of the time derivative of the microscopic variables, allowing one to probe and describe the faster decay associated with fluctuations of shorter wavelengths.

On the other hand, the generalization as embodied in the memory function formalism of eqn 2.58 (Fig. 2.22) lacks this symmetry. Instead, the 'T' branch is left unaltered, while the 'u' branch is expanded by including two levels ('σ' and 'ν' in Fig. 2.22) of time derivatives. Thus, liquid metals appear to require a level of detail that is not needed in the description of simple liquids. Given that the liquid-metal ions move just as fast as the atoms in simple liquids this seems to be unjustified, unless it reflects the much faster motion of the electrons. The latter would be good news because at some level we would expect to observe the coupling between the ionic and electronic degrees of freedom. However, the standard interpretation of the two decay constants are in terms of the ionic motion; they do not include any coupling of the ionic–electronic motion. Thus, either we have an inconsistency in the behavior of simple liquids versus that of liquid metals, or we have an inconsistency in our description of either simple liquids or liquid metals.

The determination of the relevant parameters from the experimental results might also be slightly influenced by the minor differences between the decay tree shown in Fig. 2.22 and the memory function employed in eqn 2.58. We illustrate this in Fig. 7.4 where we have calculated the spectrum corresponding to Fig. 2.22(a) by employing a set of parameters that yields a spectrum similar to that of liquid lithium. We also show the spectrum corresponding to the memory function of eqn 2.58 using slightly different values (5–10% higher) for the coupling constants f_{un} and $f_{u\sigma}$. Both descriptions (probably) yield an equally satisfying description of experimental data when one takes into account the resolution function of X-ray spectrometers, and therefore, these experiments by themselves would not be sufficient to study the minor differences between the two descriptions. As discussed in Section 2.4, the main difference is in interpretation of the decay rates. In the memory function of eqn 2.58 the decay rates z_1 and z_2 are independent of each other and they characterize two different processes. In the effective eigenmode formalism both decay rates are determined by only one

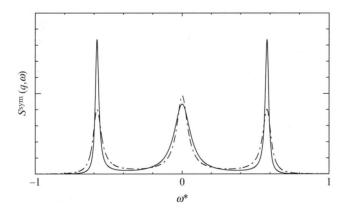

Fig. 7.4 The calculated (symmetric) scattering for the model shown in Fig. 2.22(a) as a function of scaled frequency ω^* using the following (scaled) parameters: $f_{un} = 0.4$, $f_{uT} = 0$, $f_{u\sigma} = 0.4$, $f_{\sigma\nu} = 0.24$, $z_\nu = 0.6$ (solid line). This figure can be roughly compared to the two top panels of the measured scattering in lithium, as displayed in Fig. 7.1. Also shown are the results for the memory function of eqn 2.58, with $z_1 = 0.28$, $z_2 = 0.075$, $\alpha = 0.3$, $f_{un} = 0.42$, $f_{uT} = 0$, $f_{u\sigma} = 0.45$ (dashed curve). Given the resolution obtained in real scattering experiments (Fig. 7.1), both curves would yield a very good description of the experimental data.

process, namely the decay characterized by the decay constant z_ν. Hence, there is a subtle difference in interpretation that comes with using either the effective eigenmode formalism, or a memory function that is known to involve one more approximation as well as a number of tacitly assumed restrictions (see Section 2.4).

What is really needed here are a set of molecular dynamics computer simulations in which the relevant correlation functions are determined directly from the simulation. As mentioned, there is a discrepancy between the behavior of simple liquids and liquid metals, or there is a discrepancy in our interpretation of one of them. Computer simulations can be used to determine the full set of independent correlation functions, similar to what was done in the study of simple liquids (Alley and Alder, 1983; de Schepper *et al.*, 1988), in order to assess whether the decay tree looks like the one shown in Fig. 2.22, or more like the one shown in Fig. 2.12. Such simulations can be made to be very accurate by explicitly calculating correlations using the microscopic variables 'u' and 'σ'. In addition, these computer simulations would tell us about the decay of temperature fluctuations, even when they might not be (indirectly) probed by experiment in the absence of (substantial) $u - T$ coupling, thereby also lifting the practical restriction of keeping f_{uT} and z_T fixed at their hydrodynamic values. The decay of temperature fluctuations would yield information on cage diffusion, as explained in Section 2.4.

Lastly, a technical point, although it might be a moot one. Even in the case where there is no direct coupling between the microscopic variables 'u' and 'T' ($f_{uT} = 0$), this part of the decay tree can be relevant when $z_{q\sigma}$ (see Fig. 2.12) is different from zero. The dependence of the effective eigenmodes on $z_{q\sigma}$ is shown in Appendix D for the case of a non-zero coupling parameter f_{uT}. In the case of $f_{uT} = 0$ and $z_{q\sigma} \neq 0$ we can anticipate the behavior of the eigenmodes by inspecting eqns 4.3 and 4.5. The generalized damping rate $z_\phi(q, z)$ does appear to have the structure consistent with the memory function of eqn 2.57 that is so successful in describing the dynamics of liquid metals:

$$z_\phi(q, z) = \frac{f_{u\sigma}(q)^2}{z + z_\sigma(q) + z_{q\sigma}(q)^2/[z + z_q(q)]}. \tag{7.1}$$

Depending on the ratio of $z_q(q)$ and $z_\sigma(q)$, the above expression could yield two distinct damping rates as desired to justify the use of eqn 2.57. However, this victory comes at a price. Even though f_{uT} is identical to zero, we now must also consider the generalized coefficient $\Delta(q, z)$ of eqns 4.3 and 4.5. This coefficient will not be identical to zero as there is absolutely no reason to expect that the coupling constant f_{Tq} will be small (see Appendix B.2). As a consequence, the generalized damping rate $z_T(q, z)$ (eqns 4.3 and 4.5) would also show up in the decay mechanisms with a similar strength as those related to $f_{u\sigma}$, given the symmetry of this part of the matrix given in eqn 4.3. However, this has not been observed in liquid metals. Therefore, despite the success in describing the scattering data on liquid metals using the memory function formalism as embodied in eqn 2.57, some degree of caution is required in the interpretation of the fitting parameters. As mentioned, computer simulations for all independent correlation functions are needed at this junction for an unambiguous interpretation.

7.2   The hydrodynamic limit

The hydrodynamic limit has not been reached in scattering experiments on liquid metals. Based upon the scattering results, it would appear that the hydrodynamic region does not extend much beyond $q = 0.01$ Å^{-1}. In simple liquids, the eigenvalues corresponding to the extended heat and sound modes when measured by means of scattering experiments or determined by computer simulations in the range $q > 0.3$ Å^{-1} allow for an extrapolation to $q \to 0$ using a linear q dependence for the coupling constants f and a q^2 dependence for the damping rates. The extrapolated values yield transport coefficients that tend to agree reasonably well with the measured, macroscopic values for the transport coefficients; perhaps the deviations are of the order of 10–20%. For liquid metals this is absolutely not the case, as we show in the following paragraphs. We illustrate this for liquid cesium, as measured by means of inelastic neutron scattering (Bodensteiner *et al.*, 1992). The findings for liquid cesium are typical for liquid metals when it comes to the hydrodynamic limit, or rather, the lack thereof in scattering experiments.

The dispersion of the extended sound modes in liquid Cs is very similar to that of simple liquids. We reproduce the findings of Bodensteiner *et al.* (1992) in Fig .7.5. Note that the dispersion displayed in this figure is given by the peaks of the current–current correlation function, which overestimates the actual propagation frequency of the extended sound modes. At small q values, the dispersion is linear in q, and the speed of propagation is very close to the hydrodynamic speed of propagation. For q values near $q\sigma = 2\pi$ the sound dispersion displays a minimum. This is all very similar to simple liquids, however, the damping rate of the extended sound modes is

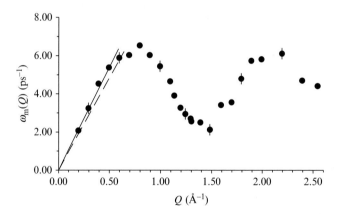

Fig. 7.5 Figure reproduced with permission from Bodensteiner *et al.* (1992). The maxima of the current–current correlation function for liquid Cs at $T = 308$ K are given by the dots, the hydrodynamic speed of sound is given by the dashed line (corresponding to $c_s = 965$ m/s), and the instantaneous, undamped viscoelastic speed of sound is given by the solid line ($c_\infty = 1061$ m/s). Note the linear $\sim q$ behavior at the smaller q values. Also bear in mind that the maxima of the current–current correlation function overestimate the actual propagation speed of the extended sound modes.

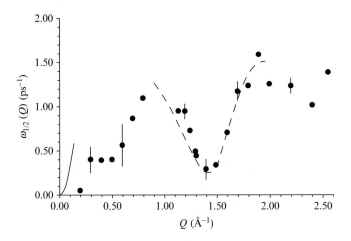

Fig. 7.6 Figure reproduced with permission from Bodensteiner *et al.* (1992). The observed width of the quasi-elastic feature of the spectra for liquid Cs at $T = 308$ K are given by the dots, the dashed line is the fit to eqn 4.9. The hydrodynamic prediction for the width of the heat mode aq^2 is given by the solid line; note that this prediction overestimates the width by an order of magnitude, demonstrating that even at the lowest q values probed in this neutron-scattering experiment the hydrodynamic limit is nowhere close to being reached.

very dissimilar: even at $q = 0.3$ Å^{-1} the damping rate in Cs is smaller by an order of magnitude (Bodensteiner *et al.*, 1992) than the hydrodynamic prediction $\Gamma_{\mathrm{s}} = \phi q^2$.

The same holds true for the extended heat mode. Qualitatively, the characteristic decay time of the heat mode[2] $\tau_{\mathrm{heat}} = [D_T(q)q^2]^{-1}$ has a similar q dependence (Fig. 7.6) as the characteristic width of the spectra for simple fluids. In particular, the minimum in characteristic width at $q\sigma \simeq 2\pi$ is well described by the hard-sphere expression given in eqn 4.9; this is not too surprising as Cs^{1+} has the outer electronic shell configuration of krypton. However, the width at the smallest q values (Fig. 7.6) is smaller by an order of magnitude than the predicted hydrodynamic width aq^2. It is of course the smallness of the damping rates of the extended sound and heat modes that make the sound modes so easily visible in the spectra, however, why this would occur in liquid metals as opposed to simple liquids, and how and when the hydrodynamic limit is reached in liquid metals is a different proposition altogether.

Before we discuss one of the options of how the hydrodynamic limit might be reached, we want to stress that the above observations are by no means restricted to the alkali metals. As an example of the ubiquity of these results, we reproduce the measurements by Hosokawa *et al.* (2003) on liquid Si in Fig. 7.7. Si is probably as different from alkali metals as one can imagine, with the effects of covalent bonding and electron localization detectable in the scattering spectra at higher q values (Hosokawa *et al.*, 2003). Nonetheless, the low-q behavior is qualitatively similar to that of Cs, not

---

[2] We identify the quasi-elastic feature reported on by Bodensteiner *et al.* (1992) as the extended heat mode.

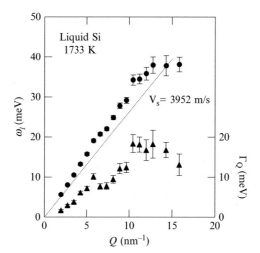

Fig. 7.7 Figure reproduced with permission from Hosokawa *et al.* (2003). Shown are the maxima of the current–current correlation function (dots) for liquid Si as determined from a fit using the damped harmonic oscillator model (Fig. 2.16) to $S_{nn}^{\mathrm{sym}}(q, \omega)$. These data have been determined by means of inelastic scattering experiments using the SPring-8 facilities in Hyogo, Japan. The damping rates $\Gamma(q) = z_u(q)/2$ are also shown in the figure (triangles). Within this model, the maxima of the current–current correlation function are given by the undamped frequency $\Omega(q) = f_{un}(q)$.

only for the sound-propagation frequency, but also for the damping rates of the sound and heat modes (Hosokawa *et al.*, 2003).

It is clear from the outset that the hydrodynamic limit occurs at smaller q values in liquid metals than in simple liquids because of the large thermal conductivity. Since both the width of the central line (the heat mode) and the widths of the extended sound modes depend on the thermal conductivity, we can expect the hydrodynamic criterion $\Omega(q) \gg aq^2$ to be violated at small q values, somewhere in the range of $0.01 < q < 0.1$ Å^{-1} (Balucani and Zoppi, 1994; Scopigno and Ruocco, 2005). As is clear from the experimental results, the linewidths actually become much smaller than the hydrodynamic predictions. This opens up the possibility of a new type of region between the hydrodynamic region characterized by the adiabatic sound propagation velocity and the viscoelastic region characterized by $c_\infty = \lim_{q \to 0}[(f_{un}^2 + f_{u\sigma}^2)^{1/2}/q]$. This new region would be characterized by an isothermal speed of propagation of the sound waves $c_T = c_s/\sqrt{\gamma} = \lim_{q \to 0}[f_{un}/q]$, where these waves would no longer be able to propagate in an adiabatic manner. In order to observe such a potential transition, one needs a liquid metal with a reasonably large value of γ. Note that it is sufficient to observe the viscoelastic to isothermal transition to demonstrate the existence of such a region, it is not necessary to observe the isothermal to hydrodynamic transition (which occurs at q values that are experimentally out of reach for the foreseeable future).

Liquid nickel is a prime candidate to look for an isothermal region, given its relatively large value of γ. Recently, this system has been investigated by means of

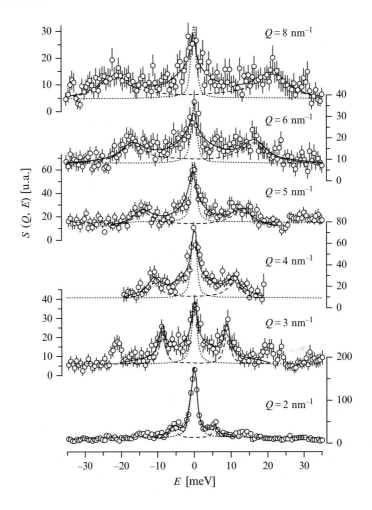

Fig. 7.8 Figure reproduced with permission from Cazzato *et al.* (2008). Shown are inelastic X-ray spectra for liquid Ni at $T = 1767$ K, measured at the q values indicated in the figure. The fit to the data (solid line) consists of a central line (heat mode) and two extended sound modes, which were represented by a damped harmonic oscillator contribution (dashed curves). The energy-resolution function of the spectrometer is given by the dotted line. The sapphire sample cell is responsible for the inelastic features beyond the extended sound-mode peaks. The propagation frequencies of the extended sound modes are easy to determine from the experiment and are shown in Fig. 7.9, and the damping rates of the extended heat and sound modes are shown in Fig. 7.10.

inelastic neutron- (Bermejo *et al.*, 2000) and X-ray (Cazzato *et al.*, 2008) scattering experiments, with the X-ray experiments extending the accessible q range down to $q = 0.2$ Å^{-1}. We show these X-ray extensions in Fig. 7.8, and the results for the extended sound and heat modes in Figs. 7.9 and 7.10. While the spectra for the X-ray scattering experiments are perhaps not of the same accuracy as those presented earlier

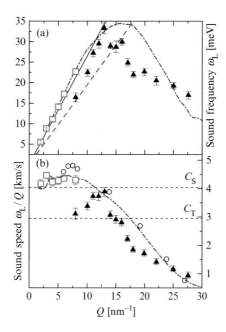

Fig. 7.9 Figure reproduced with permission from Cazzato *et al.* (2008). Shown are the undamped sound frequencies $\Omega(q)$ for liquid Ni as determined by inelastic X-ray-scattering (Cazzato *et al.*, 2008 – open squares) and by inelastic neutron-scattering (Bermejo *et al.*, 2000 – closed triangles) experiments. The dashed line in the top panel represents the isothermal frequency for sound propagation, the solid line the adiabatic one. The bottom panel is the same as the top panel, but now with the observed dispersion divided by q. The values shown in this figure are overestimates of the actual propagation frequency of the extended sound modes, since the latter is given by $\omega_s(q) = \sqrt{\Omega(q)^2 - \Gamma(q)^2}$, with the sound damping $\Gamma(q)$ shown in Fig. 7.10. The equivalent findings calculated using a molecular dynamics computer simulation are shown by the dashed-dotted lines (Ruiz-Martin *et al.*, 2007). Note the good agreement between the X-ray data and the results of the simulation. The difference between the neutron data and the X-ray data near $q = 0.9$ Å^{-1} is likely caused by the limited kinematic range in the neutron-scattering experiment at these q values.

in this chapter, they are accurate enough to determine the speed of propagation of the extended sound modes, and to determine whether an isothermal region does exist in liquid Ni in the q range probed.

The X-ray experiments (Cazzato *et al.*, 2008) indicate that an isothermal region does not exist in the behavior of the extended sound modes, while neutron-scattering experiments (Bermejo *et al.*, 2000) leave open the possibility that it does. Given the more extended q range probed in the X-ray scattering experiments combined with the larger kinematic region accessible to X-ray scattering experiments, it would appear that an isothermal to viscoelastic transition does not take place in liquid nickel. Instead, the X-ray scattering data strongly suggests that the speed of propagation reaches the (hydrodynamic) adiabatic speed of propagation at $q \sim 0.8$ Å^{-1}, while

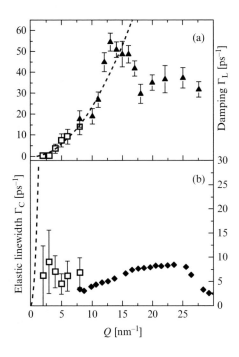

Fig. 7.10 Figure reproduced with permission from Cazzato *et al.* (2008) covering the same experiments shown in Fig. 7.9. The open symbols are the results of the fit to X-ray scattering experiments, the filled symbols are those of neutron-scattering data. The top panel displays the results for the damping rate of the extended sound modes, the bottom panel is the result for the extended heat mode. The dashed line in the top panel is a quadratic fit to the calculated damping rates (which differs from the hydrodynamic prediction by an order of magnitude), the dashed line in the bottom panel is the sad prediction of hydrodynamics.

the damping rate of the extended sound modes is nowhere near its hydrodynamic equivalent. The same can be said for the damping rate of the extended heat mode, which is smaller by an order of magnitude compared to its hydrodynamic equivalent. Thus, while the speed of propagation has reached its hydrodynamic value, the damping rate has not. Therefore, the existence of an isothermal region in Ni is very unlikely, albeit not impossible.

Even though an isothermal region in Ni might not exist, it is certainly possible within the framework of the effective eigenmodes for such an isothermal region to exist. In Fig. 7.11 we show the results for a particular set of parameters (f's and z's from the matrix $G(q)$ in eqn 2.36) that would yield an isothermal region beyond the hydrodynamics region. The parameters in this figure correspond to a ratio of the specific heats similar to Ni. The change from adiabatic to isothermal propagation is clearly visible in this figure, however, such a change is not an isolated feature within the effective eigenmode description. In order to achieve a transition to isothermal behavior, the ratio of the forces and the damping rates has to be chosen within a certain range. Note, however, that this range actually produces two sets of propagating

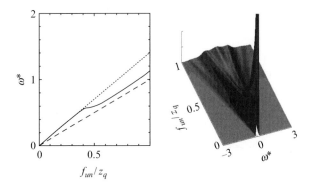

Fig. 7.11 The peak positions of the current–current correlation function $(\sim \omega^2 S_{nn}^{\mathrm{sym}}(q,\omega))$ mark the transition from hydrodynamic behavior to isothermal behavior (solid line, left panel) for a simple liquid with specific heat ratio $\gamma = 2$. This figure is similar to Figs. 2.18 and 2.19. $S_{nn}^{\mathrm{sym}}(q,\omega)$ is shown in the right panel. This figure has been calculated with the following combination of coupling constants f: $f_{un} = f_{uT} = f_{u\sigma} = f_{Tq}/2$. The two damping rates have been kept constant at $z_q = z_\sigma = 1$. The dotted line in the left panel corresponds to the adiabatic propagation of sound modes $([f_{un}^2 + f_{uT}^2]^{1/2})$, the dashed line corresponds to isothermal behavior (f_{un}). On close inspection it can be seen that the transition from hydrodynamic to isothermal behavior coincides with the appearance of an additional set of propagating modes in $S_{nn}^{\mathrm{sym}}(q,\omega)$, as indicated by the five peaks for the higher f_{un}/z_q values (right panel). These five peaks correspond to the quasi-elastic extended heat mode, two extended sound modes and two modes corresponding to propagating temperature fluctuations.

modes, rather than just one. The reason for this is that the force f_{Tq} is now so large compared to the associated damping rate z_q that oscillations in this channel are possible. After all, when the damping rate is small compared to the driving force $(z_q/2 < f_{Tq})$, then the oscillations will not be overdamped. Because of the coupling between all the variables $(f_{Tq} \to f_{uT} \to f_{un})$ these oscillations will show up in the density–density correlation function. Thus, a transition to isothermal behavior should be marked by the observation of more than one set of propagating modes. Besides the (standard) extended sound modes, we would now also expect to find a set of modes corresponding to propagating temperature fluctuations.

The behavior of the five effective eigenmodes is better visualized when we also inspect their damping rates and amplitudes in $S_{nn}^{\mathrm{sym}}(q,\omega)$, as we did in Section 2.3 (Figs. 2.20 and 2.21). We show the results corresponding to Fig. 7.11 in Fig. 7.12. One observes in this figure that the transition from a hydrodynamic speed of propagation to isothermal behavior at $f_{un}/z_q \approx 0.3$ (see Fig. 7.11) is accompanied by a mixing in of the kinetic mode corresponding to the heat flux 'q'. At this point, the character of all five modes changes very rapidly as a function of f/z. The heat mode, which started off at its hydrodynamic intensity of $(\gamma - 1)/\gamma = 0.5$ initially increases, but at $f_{un}/z_q \approx 0.3$ its large positive amplitude is negated by a large negative amplitude of an extended kinetic mode. The amplitude of the extended sound mode decreases from its hydrodynamic value of $1/2\gamma = 0.25$ to close to zero, while at the same time a new set of

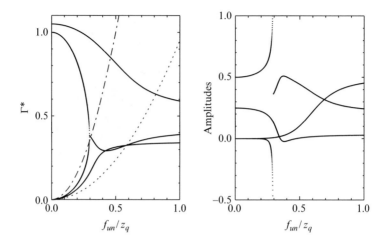

Fig. 7.12 The damping rates (left panel) and the relative amplitudes (right panel) for the five modes in $S_{nn}^{\mathrm{sym}}(q,\omega)$ whose spectra are shown in Fig. 7.11. The dotted line in the left panel corresponds to the hydrodynamic damping of the sound modes $f_{u\sigma}^2/z_\sigma$, the dashed line is given by f_{Tq}^2/z_q, which is closely related to the hydrodynamic damping rate for the heat mode: $f_{Tq}^2/\gamma z_q$ with $\gamma = 2$. The parameters in this figure are identical to the ones in Fig. 7.11 except that z_σ has been increased slightly to 1.05 to better visualize the distinction between the two kinetic modes at $q = 0$.

propagating modes appears (propagating at an isothermal speed of f_{un}/q). This new set of modes makes up virtually all of the intensity in $S_{nn}(q,\omega)$ at $f_{un}/z_q = 0.4$. With increasing q the relative intensities of the two sets of propagating modes gradually change with the isothermal mode accounting for almost half of the total intensity (2×0.25 at $f_{un}/z_q = 1$). The other propagating modes are fairly weak (2×0.025 at $f_{un}/z_q = 1$) but they still give rise to rather distinctive features in $S_{nn}(q,\omega)$ because of the fact that they are well separated in frequency from the other modes (right panel of Fig. 7.11). The central extended heat mode accounts for the remainder of the intensity (0.45 at $f_{un}/z_q = 1$). It is hard to determine what exactly the character is of the extended heat mode at this point, but it would appear that the kinetic mode 'σ' is mixed in to a high degree so that the central feature in this case would correspond to a viscous response of the liquid.

The behavior of the extended sound modes can be even more complex than a transition into an isothermal region upon probing the system at wavelengths shorter than hydrodynamic length scales. We show an example of such behavior in Fig. 7.13. With the set of parameters that was chosen for this figure we now see a transition from adiabatic behavior to isothermal behavior, followed by an increase back to adiabatic behavior. In addition, we also see the appearance of propagating modes associated with temperature fluctuations. Of course, whether such a set of parameters is even remotely physical is a different matter altogether, but it does illustrate that the standard observed behavior of the extended sound modes increasing in propagation

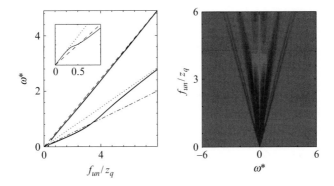

Fig. 7.13 Shown are the eigenvalues corresponding to the propagating modes of the 5×5 dynamical matrix $G(q)$ (solid lines, left panel). These eigenvalues mark the transition from ordinary hydrodynamic behavior consisting of three modes (one heat mode and two sound modes propagating at $[f_{un}^2 + f_{uT}^2]^{1/2}$ – dotted line) to the existence of five observable modes. A new set of propagating modes becomes observable around $f_{un}/z_q \approx 0.4$ where the propagation speed starts to deviate from its adiabatic value. The inset of the left panel shows the peak positions of the current–current correlation function ($\sim \omega^2 S_{nn}^{\mathrm{sym}}(q, \omega)$) at the smallest q values. This figure is somewhat similar to Fig. 7.11 and has been calculated with the following combination of coupling constants f: $f_{un} = f_{uT} = f_{u\sigma}/1.5 = f_{Tq}/2$. The two damping rates have been kept constant at non-identical values: $z_q = 0.25$ and $z_\sigma = 2$. The dashed-dotted line in the left panel corresponds to the isothermal speed of propagation f_{un}, the dashed line capturing the speed of propagation of the temperature fluctuations corresponds to $[f_{un}^2 + f_{uT}^2 + f_{Tq}^2]^{1/2}$. $S_{nn}^{\mathrm{sym}}(q, \omega)$ is shown from a bird's eye perspective in the right panel. Note the rather curious behavior of the extended sound mode in the left panel for the higher values of f_{un}/z_q where its propagation speed starts to approach the adiabatic speed once more.

speed upon leaving the hydrodynamic region is not an implicit feature of the decay tree; rather, it is associated with a particular range of forces f and damping rates z.

The damping rates and amplitudes corresponding to this rather esoteric (and so far hypothetical) liquid show the full merging of all five modes. This is depicted in Fig. 7.14. The transition from the hydrodynamic region to the isothermal region ($f_{un}/z_q \approx 0.2$–0.3) is characterized by a mixing in of the kinetic heat-flux mode 'q'. The subsequent transition to the region where the speed of propagation is once again given by $[f_{un}^2 + f_{uT}^2]^{1/2}$ (for $f_{un}/z_q > 3$) is marked by a gradual mixing in of the longitudinal momentum flux mode 'σ'. Note that $S_{nn}(q, \omega)$ consists of five clearly distinguishable modes at the highest f/z values. As before (Fig. 2.21), the transition where two modes change from diffusive to propagating ($f_{un}/z_q \approx 0.2$–0.3) is marked by a discontinuity in the amplitudes of the modes.

It is possible, given the absence of firm experimental evidence of behaviors such as the ones shown in Figs. 7.11 and 7.13, that there is a restriction on the values that the parameters f and z might take on. At present, such a restriction has not been shown to exist from a theoretical point of view. The main difference – from an experimental

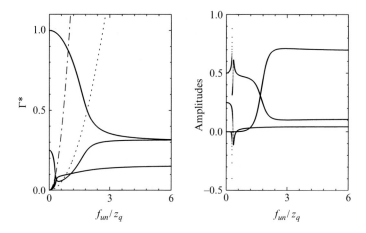

Fig. 7.14 This shows (roughly) the damping rates (left panel) and the relative amplitudes (right panel) for the five modes in $S_{nn}^{\mathrm{sym}}(q,\omega)$ whose dispersion is shown in Fig. 7.13. The dotted line in the left panel corresponds to the the hydrodynamic damping of the sound modes $f_{u\sigma}^2/z_\sigma$, the dashed line is given by f_{Tq}^2/z_q. The parameters in this figure are identical to the ones in Fig. 7.13 except that the rate z_σ has been decreased from 2 to 1 for the sake of visual clarity and hence we used the term roughly in the first line of this caption to indicate that the correspondence between this figure and the preceding one is only approximate.

point of view – between liquid metals and simple liquids appears to be the behavior of the damping rates of the extended heat and sound modes. This difference is almost without doubt caused by the presence of the conduction electrons. It might even turn out to be the case that the hydrodynamical damping rates of the heat and the sound modes involve the contribution of the electrons, whereas by the time the liquid is probed on neutron- and X-ray-scattering length scales, that the dynamics of the ions and the electrons have decoupled to such an extent that we are only observing the damping rates associated with the ions. Such damping rates would be very similar to those of simple liquids once the conductive properties of the electrons have been taken out of the picture.

In conclusion, we do not know exactly how the dynamical behavior of liquid metals changes from the hydrodynamic region to the observed behavior. On the bright side, the results obtained (Ruiz-Martín *et al.*, 2007) by molecular dynamics computer simulations (Fig. 7.9) are very close to the observed experimental values; conceivably, such simulations might be able to provide us with the answer. This hope is not solely based on the agreement obtained in the case of liquid Ni, but also on the agreement observed in other liquid metals, even in liquid metals as complex as Ga (Tsai *et al.*, 2007).

7.3   Excitations with $q\sigma \simeq 2\pi$

In the preceding section we reproduced the findings by Bodensteiner *et al.* (1992) that the characteristic width of the central feature in $S_{nn}(q,\omega)$ in liquid Cs can be well described by hard-sphere theory in the region $q\sigma \approx 2\pi$, as shown in Fig. 7.6.

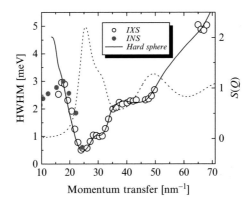

Fig. 7.15 Figure reproduced with permission from Scopigno *et al.* (2005*a*). Shown are the characteristic widths of liquid gallium spectra $S_{nn}^{\text{sym}}(q,\omega)$ as measured by X-ray and neutron scattering. The X-ray-scattering results have later been reproduced by neutron-scattering experiments (Bermejo *et al.*, 2005*b*). The solid line is the hard-sphere prediction for the heat mode given by eqn 4.9, using a hard-sphere diameter that is too large to describe a single Ga ion. The static structure factor $S(q)$ for liquid gallium is given by the dashed line on the scale shown on the right-hand side of the figure.

Given the fact that liquid alkali metals are believed to resemble simple fluids in many respects, this is perhaps not that surprising. However, from experiments it would appear that such hard-sphere behavior is also relevant to the dynamics of liquid metals such as Ga that are not generally considered to be hard-sphere fluids. However, the correspondence between the observed dynamics and that of a hard-sphere system is not as straightforward in the case of liquid Ga as it is for the case of liquid Cs.

In Fig. 7.15 we show the measured characteristic widths of the liquid gallium spectra (Scopigno *et al.*, 2005*a*) as determined by means of inelastic X-ray-scattering experiments. The measured width displays a high degree of structure with q. At the position of the main peak of the static structure factor $S(q)$ the width reaches a minimum. Around $q \sim 3$ Å^{-1}, in the region of the shoulder of the main peak of $S(q)$ the width reaches a plateau, followed by a marked increase until it reaches a second plateau around $q \sim 4$ Å^{-1} (Scopigno *et al.*, 2005*a*; Bermejo *et al.*, 2005*b*). Scopigno *et al.* discovered that this structure could be remarkably well reproduced by the predicted behavior of a hard-sphere fluid, as given by eqn 4.9. This was remarkable for two reasons. First, even though eqn 4.9 predicts the characteristic width of the heat mode, it apparently describes the characteristic width of the full spectra very well, even in a liquid metal where the extended sound modes are much more prominent than in hard-sphere fluids. Secondly, the essentially perfect agreement was obtained (Scopigno *et al.*, 2005*a*) by employing an equivalent hard-sphere diameter for liquid Ga that is unphysically large.

The ability to describe the characteristic spectral width using a hard-sphere theory outside its range of validity gives food for thought. The effective hard-sphere diameter used ($\sigma_{\text{hs}} = 2.79$ Å^{-1}) in the description shown in Fig. 7.15 is so large that it would not be possible to pack the hard spheres in the volume corresponding to the density of the liquid. Scopigno *et al.* (2005*a*) interpreted this as a precursor of ions bonding with their neighbors according to the arrangement in the solid phase. The effective hard-sphere diameter would then represent an averaged size of such 'clusters', and the resulting hard-sphere diameter should be viewed more as a dynamic entity than as a static number. This is a reasonable explanation, especially given the complex, non-close-packed structure that Ga crystalizes in (see also Fig. 4.8).

Of course, this explanation (Scopigno *et al.*, 2005*a*) could still be wrong. For instance, the fact that the function $d(q\sigma)$ (see eqn 4.9) predicts the correct oscillatory behavior could just imply that the correct theory would yield a function that behaves very similarly to $d(q\sigma)$. At this point it is good to recall the results for liquid Hg (Badyal *et al.*, 2003), shown in Fig. 4.15, where the characteristic widths are also rather well described by the hard-sphere expression (eqn 4.9) using $\sigma_{hs} = 2.81$ Å^{-1}. This latter value is an entirely reasonable equivalent hard-sphere diameter for the case of liquid Hg. Bearing in mind that both Hg and Ga crystallize in a similarly complex structure, the fact that both descriptions mandate a very similar behavior of $d(q\sigma)$ might indicate that something else besides cluster formation is happening in liquid Ga. In Section 10.2 we revisit this issue where we look into potential quantum-diffraction effects present during the binary collision between two particles that are approaching each other at very low relative velocities, and where we discuss what influence such diffraction effects might have on the characteristic width of the scattering spectra.

Whatever the ultimate explanation might be for the excellent agreement between the characteristic widths of the gallium spectra and the functional form given in eqn 4.9, it is very encouraging to see that the data are so accurate (Scopigno *et al.*, 2005*a*; Bermejo *et al.*, 2005*b*) that the discussion on the various interpretations (Bermejo *et al.*, 2005*a*; Scopigno *et al.*, 2005*b*) takes place at a level of great detail. As an example, the plateau in the characteristic width at $q \sim 3$ Å^{-1} (Fig. 7.15) is clearly associated with the shoulder of the main peak of $S(q)$, a shoulder not visible in simple liquids. Tsai *et al.* (2007) link this shoulder to excitations unique to the metallic phase, as we will discuss in the following section.

7.4 New excitations and old warnings

Given the fact that there are so many similarities between the microscopic dynamics of liquid metals and of simple liquids, one question one might ask is whether there are any excitations that are truly unique to liquid metals. This is a different issue from trying to explain the peculiarities of the dispersion of the extended sound mode in terms of a two-fluid model (Bove *et al.*, 2008). In this section we discuss two such novel excitations; whether they actually exist or not is still an open question, but it illustrates the type of behavior to be on the lookout for in liquid metals now that we have access to highly accurate experimental results.

7.4.1 Friedel oscillations

One very noticeable feature in Ga is the shoulder in the static structure factor (dashed line in Fig. 7.15), at slightly higher q values than the main peak. The origin of this shoulder is still under debate, but since there are no truly elastic features in a liquid, this shoulder is most likely associated with a type of excitation that is not seen in simple liquids. Tsai *et al.* (2007) identify this shoulder as being caused by the Friedel oscillations induced by the conduction electrons. While the short-range interaction between the ions is largely determined by the repulsive part of the interparticle potential, as it is in simple liquids, the conduction electrons mediate a long-range interaction. The resultant long-range interaction is oscillatory in nature, with the period between attraction and repulsion determined by the Fermi wavevector k_F. It is the same type of interaction that goes by the name of Ruderman–Kittel–Kasuya–Yosida (RKKY) interaction in solids.

In a series of very elegant computer simulations, Tsai *et al.* (2007) demonstrate that this long-range electron-mediated interaction is indeed responsible for the shoulder in $S(q)$ in Ga, as well as for the plateau in the characteristic width (Fig. 7.15). First, they show that the shoulder in $S(q)$ and the plateau in the characteristic width can be (qualitatively) reproduced when the long-range part of the interparticle potential is included (see Fig. 7.16). Next, they removed the long-range part from their interaction, and they found (Fig. 7.16) that both the shoulder as well as the plateau disappeared. Since the position of the shoulder occurred at the expected wave number $q = 2k_F$, Tsai *et al.* concluded that the interplay between ionic motions and the conduction electrons influences the overall dynamics. In particular, they argued that if the wavelength of a density fluctuation matches the wavelength $\lambda_F = \pi/k_F$ of the Friedel oscillations, then the resulting coherent nature of the ionic and electronic motions makes it harder for these density fluctuations to decay. As a result, the characteristic width is smaller than what can be expected without the presence of the conduction electrons.

If Friedel oscillations are indeed responsible for the change in dynamics around $q = 2k_F$, then it will not be a sinecure to properly describe this using the effective eigenmode formalism. Of course, there will not be a problem of obtaining a satisfactory description of the measured spectra by including more and more Lorentzian lines, but this would defeat the purpose of the formalism in trying to describe the system with the smallest possible number of adjustable parameters. Instead, one would have to set up – from the outset – the dynamic matrix $G(q)$ so as to accommodate the two-component nature of the liquid metal, very much like the matrix described in Chapter 6. While this is by no means a trivial task, the effort put into such an approach might pay off when one tries to understand the propagation speed of short-wavelength sound modes in terms of plasma oscillations (Bove *et al.*, 2008). On a final note, one way to verify or to disprove the presence of Friedel oscillations is to compare the spectra of normal and expanded metals. By heating (and expanding) the fluid it is possible to take the system through the metal–non-metal phase transition (Sumi *et al.*, 1999). X-ray- or neutron-scattering experiments on either side of the transition should then be able to reveal the role that Friedel oscillations play in the relaxation of density fluctuations.

7.4.2 Fluctuating magnetic moments

When it comes to excitations unique to the metallic phase, another very intriguing possibility is the existence of magnetic excitations. There are hints that magnetic

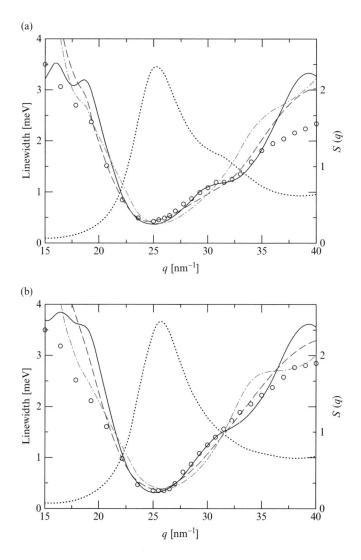

Fig. 7.16 Figure reproduced with permission from Tsai *et al.* (2007). Shown are the calculated $S(q)$ (dotted lines, scale on right-hand side of the figure) and the characteristic width (dots, scale on the left-hand side) of $S_{nn}(q,\omega)$ for liquid Ga at 323 K, as determined by means of molecular dynamics computer simulations. Inclusion of the long-range, oscillatory part (top panel) of the effective two-particle interaction potential reproduces the shoulder of $S(q)$ for $q \sim 3$ Å^{-1}, as well as the plateau in the characteristic width. Only keeping the repulsive, short-range part of the interaction removes both the shoulder and the plateau (bottom panel), making it highly plausible that these two features are a direct result of the coupling between the ionic and electronic degrees of freedom (Tsai *et al.* 2007).

excitations are indeed present in liquid metals, but hard evidence has not been forthcoming. In liquid metals, the electronic configuration of most ions is such that they have a closed outer shell. This fully filled shell does not have to correspond to the principal quantum number, it can also be a subshell. For instance, Ga ions have a valency of 3+ in the liquid, corresponding to an empty 4s shell, an empty 4p shell, and a closed 3d shell and hence, Ga ions are diamagnetic. Ni and Co ions on the other hand will always have an unfilled outer shell and hence, magnetic ions can be present in these systems.

In principle, even systems composed of diamagnetic ions can possess fluctuating magnetic moments, provided these systems are in the liquid phase. Badyal *et al.* (2003) suggested a mechanism through which magnetic moments can be induced through collisions. In their picture, during a binary collision between ions an electron of a closed shell would be ejected into the Fermi sea of electrons, leaving the ion with a magnetic moment. A short time after acquiring this magnetic moment, another electron would be captured and the moment would vanish. Alternatively, an ion could find itself relatively far away from its neighbors, allowing it to capture an electron into one of its outer shells, again only to give it up a short time afterwards. If these processes were indeed to happen, then on a timescale comparable to the time between collisions, ions could possess an unpaired electron, and hence a magnetic moment. In turn, such a magnetic moment would produce a magnetic cross-section for neutrons (Squires, 1994) that would be seen to fluctuate (that is, to appear and to disappear) on the same timescale as the rattling motion of an ion in the cage formed by its neighbors. The above process is sketched in Fig. 7.17. Note that the timescale of these fluctuating magnetic moments (ps) is very different from the timescales (fs) in the solid associated with spontaneously fluctuating moments within the Fermi sea. Once more, we stress that currently there exists no direct observations of such collision-induced magnetic excitations; only indirect evidence exists and therefore, this section should be read with a certain degree of levity.

Unlike the case for a liquid where there are permanent magnetic moments, fluctuating magnetic moments would give rise to a new signal. Permanent magnetic moments

(a)   (b)   (c)   (d)

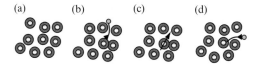

Fig. 7.17 Figure reproduced with permission from Patty *et al.* (2006). Sketch of how a fluctuating magnetic moment can come into existence as a result of a binary collision. (a) A non-magnetic ion is locked up in the cage formed by its neighbors, its outer shell (depicted as the dark gray region) is completely filled. (b) A collision with a neighbor during the rattling motion associated with cage diffusion brings two ions in such close contact that an electron is ejected from the outer shell into the Fermi sea. (c) The unpaired electron gives rise to a magnetic moment. (d) After the collision the ions separate and the outer shell is filled up again when an electron is captured from the Fermi sea, annihilating the magnetic moment.

would just augment the incoherent scattering contribution by adding a paramagnetic cross-section to the overall neutron scattering cross-section. This would merely provide another way of following the motion of a single particle, with the loss of correlation at very short times determined by the rattling motion, and over long times by Fickian diffusion (eqn 2.15). As with normal incoherent scattering, the loss of correlation resulting from the rattling motion will be restricted by the size of the cage, as shown in Fig. 2.23. Fluctuating magnetic moments would give rise to an additional, broad in energy signal whose width Γ is characteristic of the collision time between particles: $\Gamma = 1/\tau_{\text{collision}}$. A sketch of such a signal is shown in Fig. 3.4. The strength of this signal is determined by the size of the collision-induced moment. Thus, the shape of this signal looks very much like the shape of the signal that characterizes the fast rattling motion of a particle in its cage that represents the very small amount of correlation loss in the self-correlation function (Fig. 3.11), but its strength in scattering experiments can be many times larger owing to the new magnetic scattering mechanism.

Fluctuating magnetic moments would show up in neutron-scattering experiments through the electromagnetic interaction between the neutron spin and the ionic magnetic moment, but they would not be visible in X-ray-scattering experiments. By performing polarized neutron-scattering experiments one should be able to verify whether the origin of this signal is because of the magnetic scattering mechanism, or whether it is caused by the nuclear scattering mechanism (provided a broad signal were observed in neutron-scattering experiments). We show the observed scattering by liquid mercury in Fig. 7.18 (Badyal *et al.*, 2003). The scattering in this experiment was not carried out using polarization analysis, so the only conclusion that could be drawn was that a broad (in energy) signal was present in liquid Hg that showed all the tell-tale signs of scattering caused by fluctuating moments, with the width of the signal given by $\Gamma = 1/\tau_{\text{collision}}$.

Subsequently, Patty *et al.* (2006) carried out a literature review of scattering experiments investigating whether neutron-scattering experiments on liquid metals other than Hg exhibited an enhanced cross-section compared to X-ray-scattering experiments, and whether this enhanced cross-section displayed the hallmarks of fluctuating magnetic moments as described above. They found that many liquid metals appear to display an enhanced cross-section for neutrons. As an example, we show in Fig. 7.19 the difference in scattering between when liquid Pb is probed by X-rays (Dzugutov *et al.*, 1988) and by neutrons (Reijers *et al.*, 1989). These data strongly seem to suggest that fluctuating magnetic moments could indeed be present in liquid Pb. However, unlike the case for liquid Hg, no inelastic data were available in the case of liquid Pb, so that the requirement that the magnetic scattering should show up as a broad contribution in the liquid Pb spectra could not be verified.

The literature search by Patty *et al.* (2006) also suggested that liquid Ga might possess magnetic moments, even though this might not be an obvious candidate given the relative lightness of the ions.[3] A polarized neutron-scattering experiment carried out by Patty *et al.* (2009) on liquid Ga showed that while the broad mode is indeed present in liquid gallium just above the melting point, its origin could not

---

[3] The literature search revealed that the heavier elements appeared to possess a larger magnetic cross-section.

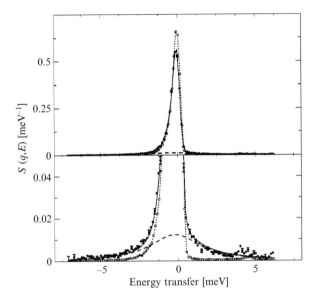

Fig. 7.18 Figure reproduced with permission from Badyal *et al.* (2003). Shown are the spectra (filled dots) for liquid Hg at 295 K and $q = 0.45$ Å^{-1}; the bottom panel is an enhancement. One observes the presence of a broad (in energy) signal (dashed line) in the Hg spectra, while such a broad component is absent in a vanadium reference standard (dotted line with open circles) that serves to indicate the asymmetric energy resolution of the MARI spectrometer at ISIS. This broad component could not be attributed to any coherent or incoherent signal since the sum rules for these signals were already fulfilled by the remainder of the scattering, and hence, this broad component could be indicative of a magnetic contribution to the scattering.

be magnetic. Since this represents the only magnetic experiment on liquid metals in this book and since this experiment illustrates the various contributions – including the unwanted contributions – that make up the total scattering, we elucidate this experiment in the following.

In half of magnetic scattering events the neutron flips the orientation of its spin, in the other half of the cases the neutron spin remains unchanged (Moon *et al.*, 1969; Squires, 1994). For this particular experiment, this ratio was the same ratio as that calculated for incoherent scattering given the particular combination of spin-incoherent scattering and isotope-incoherent scattering in Ga. In contrast to these types of scattering that can give rise to spin-flip events, scattering that takes place because of the strong nuclear force never results in a flipping of the neutron spin. Therefore, when one performs a polarized neutron-scattering experiment that keeps track of whether the spin of the neutron was flipped, or not, then one can determine whether magnetic scattering actually took place.

We show the scattering by liquid gallium just above the melting point in Fig. 7.20, separated into scattering events in which the spin of the neutron flipped and events in which the spin of the neutron remained unchanged. These experiments were performed

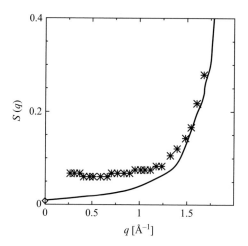

Fig. 7.19 Figure reproduced with permission from Patty *et al.* (2006). The static structure factor for liquid Pb at 623 K (solid line; X-ray-scattering results Dzugutov *et al.*, 1988) and at 613 K (symbols; neutron-scattering results Reijers *et al.*, 1989). The X-ray-scattering data approach the thermodynamic limit $S(q = 0) = 0.008$ (open diamond) while the neutron-scattering data show a signal well in excess of the X-ray scattering. Since Pb does not possess an incoherent cross-section for neutrons, and since the neutron-scattering data were corrected for multiple-scattering, Patty *et al.* (2006) concluded that liquid Pb must possess fluctuating magnetic moments, even though an energy analysis of the scattered signal was not available.

(Patty *et al.*, 2009) at the Canadian Neutron Beam Centre that houses one of the best triple-axis spectrometers in the world for doing polarized neutron scattering. Shown in this figure are the measured intensities of liquid gallium in a sample container, as well as the scattering by the empty container. The first thing one observes in this figure is how weak the actual scattering by liquid gallium is at low momentum transfers. This is of course not helped by using a polarized setup that typically results in a reduction of the neutron flux by an order of magnitude. Nonetheless, one can observe scattering exceeding that of the empty cell for the case where the spin of the neutron does not change (a broad signal that extends over the entire energy-transfer range), as well as for the case where the spin of the neutron does get flipped (a very weak signal restricted to energy transfers close to the elastic channel $\omega = 0$ and depicted by the shaded region). This latter quasi-elastic signal is the result of the motion of individual ions; the width in energy of this very weak quasi-elastic signal – whose strength is given by the small incoherent cross-section of gallium – is limited to the resolution of the spectrometer, as one would expect for a self-diffusion process at $q = 0.45$ Å^{-1}.

The outcome of the polarized scattering experiment is unequivocal: magnetic scattering by unpaired electrons does not take place in liquid gallium, ruling out the existence of fluctuating moments in this system. A broad component to the scattering is visible in the case where the spin of the neutron is not flipped. Had the origin of this signal been magnetic (or incoherent), then one should have observed the same signal – with identical amplitude – in the spectra corresponding to a flipping of the

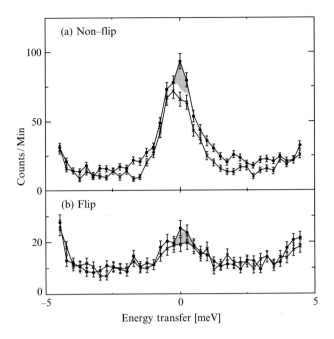

Fig. 7.20 Figure reproduced with permission from Patty *et al.* (2009). Shown are the spectra (dots) for liquid gallium at 303 K at $q = 0.5$ Å^{-1}. The bottom panel is the signal corresponding to neutrons that had their spin flipped during the scattering event, the top panel is for those that scattered without suffering a flipping of the spin. The only scattering that is visible above the empty-cell scattering (crosses) for neutrons that have their spin flipped (bottom panel) is a very weak and narrow signal around zero energy transfer. This signal (indicated by the light shading) corresponds to standard Fickian diffusion of individual particles. Neutrons that do not have their spin flipped (top panel) produce – in addition to this weak incoherent contribution once again suggestively indicated by the light shading – a broad signal at non-zero energy transfers. This is the broad signal whose shape is characteristic of that of fluctuating magnetic moments. However, since this broad component does not show up in the scattering where neutrons change their spin orientation, it cannot be associated with magnetic or with incoherent scattering. Hence, the origin of the broad signal in gallium must be coherent in origin.

neutron spin. Clearly, this is not the case, and therefore, the broad signal observed in liquid gallium cannot be caused by magnetic or by incoherent scattering. Hence, its origin must be linked to the coherent cross-section. Therefore, this broad mode could be a new collective excitation such as the one proposed by Bermejo *et al.* (2005*b*) who suggested that Ga could exhibit strongly damped solid-like phonon excitations. This would be in keeping with the suggestion by Scopigno *et al.* (2005*a*) that clusters might be present in the liquid phase of this unusual metal. The overdamped phonon excitations would then correspond to excitations of these clusters.

Having established that the excess scattering did not have a magnetic origin, Patty *et al.* (2009) performed an extensive series of neutron-scattering measurements using various sample geometries in order to investigate to what degree the excess scattering could be caused by second- and even third-order multiple scattering. We illustrate this in Fig. 7.21 by reproducing some of their data where the anticipated single-scattering signal (only representing the heat mode) is compared to the multiple-scattering contributions. As can be seen in this figure, the shape and amplitude of the multiple-scattering signal is very similar to the broad contribution that we discussed in the preceding paragraphs, demonstrating that teasing out the single-scattering contributions (at the lowest q values) associated with the heat mode and with potentially new modes is anything but a sinecure. In particular, the fact that third-order multiple scattering should be taken into account might be somewhat surprising, but it demonstrates the overall weakness of the scattering signal at the

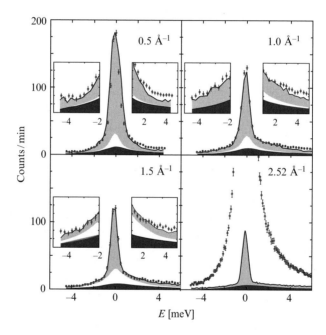

Fig. 7.21 The inelastic neutron-scattering spectra of liquid Ga (solid circles) for various q values shown in the figure. This experiment was carried out on the TRIAX spectrometer at the Missouri Research Reactor on a 3-mm thick slab of gallium. This experiment was set up in order to gauge under what experimental conditions a broad mode could be observed in the presence of multiple scattering and strong background scattering. The sum (solid line) of the calculated single scattering (coherent + incoherent, unshaded), multiple scattering (dark shading) and empty-cell scattering (light shading) is very close to the observed data points for all q and E. Thus, in order to distill a broad mode out of the experimental data, one needs a better experimental setup than this one. For reference, the data at $q = 2.52 \text{Å}^{-1}$ correspond to the peak in the static structure factor; here, the details of multiple-scattering corrections and empty-cell scattering are far less important. This figure has been adapted from Patty *et al.* (2009).

smaller wave numbers q.[4] For the experiment shown in Fig. 7.21 it would not be possible to say anything meaningful about the existence of a broad mode at low momentum transfers; the level of background scattering arising from the sample cell as well as from time-independent scattering is simply too large. The most one would be able to conclude based upon these experiments is that the overall intensity of such a broad mode would be weak. For experiments of this type one is much better off at a spallation source (Bermejo *et al.*, 2005*b*) where the overall level of background scattering is much less, and equally important, is much more uniform over the entire q range.

The data shown in Fig. 7.21 demonstrate the need for highly accurate multiple-scattering corrections during the data-analysis stage, as well as for the careful setup of the experiments in the first place in order to prevent background scattering from dominating the sought-after signal. Interestingly, inclusion of third-order multiple-scattering effects did not remove the excess scattering observed (Badyal *et al.*, 2003) in liquid Hg (Fig. 7.18), and it failed to bring the scattering down to the level expected from the values (measured by X-ray scattering) of $S(q)$ in this q range; hence, it might well still be possible that fluctuating moments do exist in the heavier metals like Hg (Bove *et al.*, 2001; Calderin *et al.*, 2009) and Pb.

Irrespective of whether fluctuating moments exist in liquid metals or not, the presence of excess scattering in neutron-scattering experiments (Fig. 7.19) should serve as a warning of how easy it is to infer the wrong behavior from the scattering functions. When excess scattering originates from a magnetic cross-section, or even from some leftover corrections, once we start modelling it as being part of $S_{nn}(q,\omega)$ then we impose the wrong intensities on, for instance, the heat mode. And since the relative intensities of the modes are governed by the same parameters that yield their eigenvalues, we would make the wrong inference. Any leftover intensity would quickly show up when employing the effective eigenmode formalism with its built-in sum rules since the results of the fit simply would not look very good. Conversely, fitting the spectra to a sum of Lorentzian without sum rules governing their relative amplitudes would produce a deceptively good fit, but a physically not very meaningful one. Lastly, should fluctuating moments actually turn out to exist, then this would provide us with a neat little check on our picture of what the rattling motion of particles inside a cage looks like in reality and perhaps even test whether the motion associated with the initial part of cage diffusion is similar between elemental liquids and charged colloidal suspensions (see Chapter 5).

---

[4] Roughly speaking, for a sample that scatters 10% of the incoming radiation, we can expect that 10% of the scattered particles scatters once more, and of those particles that have scattered twice, 10% will scatter a third time. Typical values for $S(q)$ at small momentum transfers are 0.01–0.05. On this intensity scale, second-order multiple scattering would show up at a level of 0.1, and third-order ones with an intensity of 0.01 (comparable to the single scattering in intensity, but spread out over a larger energy range).

8

Very cold liquids

When liquids are cooled, their excitations become slightly better defined since decay rates decrease slightly with decreasing temperature because of thermal population effects,[1] and because of fewer possible decay channels. Liquids that stay liquid down to the lowest temperatures do something much more special: the density fluctuations no longer couple to the temperature. As a result, the extended sound modes become the dominant features of the dynamic structure factor and their behavior as a function of momentum transfer can very easily be observed in experiment. As it turns out, the behavior of the extended sound modes in very cold liquids is very similar to the behavior of the extended sound modes in liquids at elevated temperatures, lending credence to the analysis of simple liquids that inferred the existence of short-wavelength sound modes through extensive data-fitting procedures. Another interesting development at low temperatures is that one can now also observe the difference between Bose statistics and Fermi statistics, even in the normal-fluid phase. In this chapter we restrict ourselves to cold fluids in the normal phase, the superfluid phase is the topic of the following chapter.

8.1 Prominent sound modes

When liquids are cooled to very low temperatures, the coupling between density fluctuations and temperature fluctuations weakens. To what extent density fluctuations give rise to temperature fluctuations depends on the value of $\gamma - 1 = c_p/c_V - 1$ (eqn 2.33). Since $c_p - c_V \sim \alpha^2$, with α the thermal expansion coefficient, we find that $\gamma - 1 \to 0$ when $\alpha \to 0$. For a liquid to stay liquid down to the lowest temperatures, we must have that the zero-point motion of the atoms is so large that it overcomes the crystallization tendencies. Thus, the volume a liquid occupies at these low temperatures will be determined by the amplitude of the zero-point motion; raising or lowering the temperature by a small amount will not make the liquid expand or contract. Hence, $\gamma - 1 \approx 0$, and density fluctuations cannot decay into temperature fluctuations. The damping of the sound waves is now entirely through the viscoelastic channel, while the heat mode will be largely absent since its amplitude is proportional to $\gamma - 1$ (Table 2.3).

We illustrate this effect in Fig. 8.1 for liquid ^{4}He at 2.3 K and at ambient pressure, corresponding to normal fluid ^{4}He just above the superfluid transition point $T_\lambda =$

---

[1] In quantum-mechanical terms we say that there are fewer quasi-particles excited for other quasi-particles to collide with. In general, this decrease is only a very small effect since most cold liquids do not exhibit an energy gap and therefore, the temperature will always be large compared to the minimum excitation energy. The glaring exceptions are of course the superfluids, to be discussed in Chapter 9.

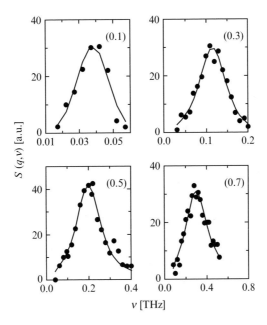

Fig. 8.1 Figure reproduced with permission from Montfrooij *et al.* (1997). Shown are the neutron-scattering data on liquid ^{4}He at saturated vapour pressure and $T = 2.3$ K as measured by Woods *et al.* (1976*b*). The solid line is the fitted result of a two-pole approximation for $S_{nn}(q, \omega)$ (eqn 3.9). Note that the quasi-elastic (extended) heat mode is entirely absent from the spectra. The q values are given in the figure in Å^{-1}.

2.172 K. Figure 8.2 shows the data for liquid ^{4}He at 4.2 K at elevated pressure ($p = 25$ bar), corresponding to a cold normal fluid. Both these thermodynamic states are characterized by a value of γ being very close to 1 (Sychev *et al.*, 1987), with the result that the extended heat mode is entirely absent from the spectra. The extended sound modes, that now make up the entire spectra at these low q values, can still be seen to broaden with increasing q. Thus, the sound modes still decay through the viscoelastic channel. This is in contrast to the case of superfluids, where the sound modes last forever and show up as very sharp (in energy) features in scattering experiments.

There is a subtle difference between the response of helium at low temperatures, and the response of liquid metals. In both cases, the macroscopic value of γ is close to 1, leading to the absence of the heat mode at low momentum transfers. However, when probing the liquid on shorter and shorter length scales, the extended heat mode remains absent in helium, but a central feature is present in liquid metals. This process is easily observed in scattering experiments on liquid metals where one needs to use a three-pole approximation of $S_{nn}(q, \omega)$ to adequately describe the spectra for q values near the peak of the static structure factor. The underlying difference in behavior traces back to the ratio of coupling parameter to damping rate $f_{u\sigma}/z_\sigma$. In liquid helium this ratio is found to be small, with the consequence that this part of the decay tree manifests itself as a damping rate $z_u = f_{u\sigma}^2/z_\sigma$. In metals, the above ratio

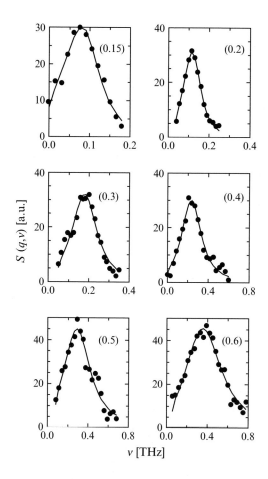

Fig. 8.2 Figure reproduced with permission from Montfrooij *et al.* (1997). Depicted is $S_{nn}(q,\omega)$ for liquid ^{4}He at 25 bar and $T = 4.2\,\text{K}$. The data were measured by Woods *et al.* (1976*a*). The solid line is the result of a fit using the damped harmonic oscillator expression of eqn 3.9. As is the case in Fig. 8.1, the extended heat mode is entirely absent from the spectra. The q values are given in the figure in Å^{-1}.

is not small and one has to specifically include this decay channel and use expressions such as the viscoelastic model given in eqn 2.48, or extensions thereof. This results in at least three modes in liquid metals, one of which will be the quasi-elastic feature that we have identified as the heat mode in this textbook. In Appendix E we depict the transition from one model to the other.

8.2 Lineshape distortion

The above disappearance of the heat mode in liquids at low temperature represents a real change in the liquid. In contrast, an artifact can also become visible in scattering experiments at low temperatures. This artifact manifests itself as a lineshape distortion that occurs when the width (in energy) of an excitation is larger than or comparable to the temperature: $\Gamma > k_B T$. What is meant by lineshape distortion is that the observed peak position in $S_{nn}(q, \omega)$ no longer corresponds in a straightforward manner to the underlying pole of the dynamic susceptibility $\chi_{nn}(q, \omega)$. As a result, it is relatively easy to arrive at the wrong conclusion regarding the dynamics of the system at low temperature. We note that this is not restricted to low-temperature liquids, but it is a potential source for mistakes in all systems where the linewidth is no longer negligible compared to the temperature. One encounters a good example of the latter in the study of soft modes in magnetic phase transitions at low temperatures.

We illustrate the effects of lineshape distortion and the fallacious conclusions that can be inferred for the case of the superfluid transition in ^{4}He under pressure (Svensson *et al.*, 1996). We will discuss the details of the normal fluid to superfluid transition in the next chapter, for our purposes in this chapter it suffices to know that ^{4}He at a constant density of 0.1715 g/cm^3 (p $\approx$ 20 bar) undergoes a second-order phase transition at $T = T_\lambda = 1.9202$ K. In Fig. 8.3 we plot $S_{nn}(q, \omega)$ as a function of temperature, both below and above T_λ. It is clear from this figure that nothing spectacular, or even particularly noteworthy happens when going through the phase transition; the lineshape looks the same on either side of the transition (right-hand column in Fig. 8.3), consisting of a broad feature centered around $\hbar\omega \approx 0.4$ meV.

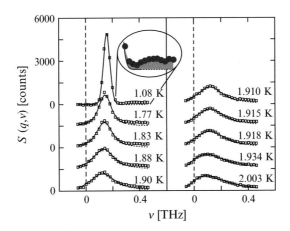

Fig. 8.3 Figure adapted with permission from Svensson *et al.* (1996). The dynamic structure factor $S_{nn}(q, \omega = 2\pi\nu)$ of liquid ^{4}He for $q = 2.0$ Å^{-1} at constant density $\rho = 0.1715$ g/cm^3 is shown at multiple temperatures (given in the figure). The solid lines are a fit to the two-pole approximation for $S_{nn}(q, \omega)$. These data do not immediately reveal the transition from a superfluid to a normal fluid that takes place at $T = T_\lambda = 1.9203$ K. In fact, the opposite appears to be the case; there is very little difference between superfluid ^{4}He and normal fluid ^{4}He when investigated by means of neutron scattering. The inset at $T = 1.08$ K shows the multiphonon component, a feature unique to superfluids (see Chapter 9).

It certainly does not look like we are witnessing the most spectacular phase transition known in liquids.

However, the superfluid to normal-fluid phase transition is clearly visible to the naked eye when plotting $S_{nn}^{\mathrm{sym}}(q,\omega)$:

$$S_{nn}^{\mathrm{sym}}(q,\omega) = \frac{(1 - e^{-\beta\hbar\omega})}{\beta\hbar\omega}S_{nn}(q,\omega). \tag{8.1}$$

It is clear from just looking at Fig. 8.4 that there is a marked change in lineshape between the spectra just below T_λ, and those just above T_λ. The peak position of the spectra are seen to migrate to $\omega = 0$, indicating the softening of the modes exactly at T_λ. Of course, since employing the above equation simply amounts to a replotting of the results, there is nothing in this procedure that would force modes to soften exactly at T_λ (Montfrooij and Svensson, 1994; Svensson *et al.*, 1996). This example illustrates that at low temperatures one should not draw too many conclusions based upon the peak position of the dynamic structure factor $S_{nn}(q,\omega)$, rather one should correct the data first for thermal population effects that distort the lineshapes.

The reason why a simple replotting of the data can directly reveal an otherwise obscured phase transition is fairly straightforward. The symmetrized relaxation function $S_{nn}^{\mathrm{sym}}(q,\omega)$ can be expressed in terms of the dynamic susceptibility $\chi_{nn}(q,\omega)$ and memory function $M(q,\omega)$ (eqn 2.54) as

$$S_{nn}^{\mathrm{sym}}(q,\omega) = \frac{8\pi^3 m}{\beta hq^2}|\chi_{nn}(q,\omega)|^2 M'(q,\omega) \sim |\chi_{nn}(q,\omega)|^2. \tag{8.2}$$

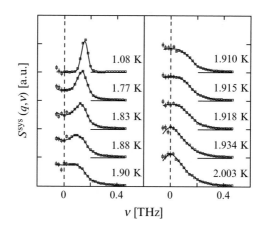

Fig. 8.4 Figure reproduced with permission from Svensson *et al.* (1996). $S_{nn}^{\mathrm{sym}}(q,\omega = 2\pi\nu)$ of liquid ^{4}He for $q = 2.0$ Å^{-1} at constant density $\rho = 0.1715$ g/cm^3 is shown as a function of temperature (given in the figure). The data in this figure are obtained through a simple replotting of the data in Fig. 8.3 (see text). The solid lines are guides to the eye. The data have not been corrected for the resolution of the spectrometer. It is now possible to follow by eye the complete softening of the extended sound mode at $T = T_\lambda = 1.9203$ K.

Thus, the poles of $\chi_{nn}(q, \omega)$ will be very prominent features in $S_{nn}^{\text{sym}}(q, \omega)$, especially in the case where $M(q, \omega)$ is only weakly dependent on energy in the energy range of interest, such as is the case in liquids at low temperatures, where $M(q, \omega)$ is almost independent of ω. For instance, in the damped harmonic oscillator model that does an excellent job at capturing the dynamics of liquids at low temperature, we have that $M(q, \omega) = z_u$.

8.3 Low-temperature versus high-temperature excitations

We can now make a comparison between the behavior of liquids and gases at high temperatures versus the response at very low temperatures. The obvious difference is of course the absence of the heat mode. Even in the case of liquid ^{4}He at 4.2 K under its own vapor pressure where γ has the large value of 2.08 (Woods *et al.*, 1976*b*), its amplitude rapidly diminishes with increasing wave number and it disappears altogether for $q > 0.8$ Å^{-1}. We show this in Fig. 8.5. As mentioned in earlier sections, the increasing equivalence with decreasing temperature of c_p and c_v is a direct consequence of the zero-point motion of the atoms; however, it does not explain why in the case of

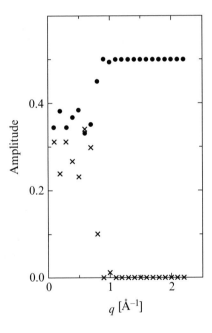

Fig. 8.5 Figure reproduced with permission from Montfrooij *et al.* (1997). Shown are the amplitudes of the extended heat mode (crosses) and one of the two extended sound modes (solid circles) for liquid ^{4}He at 4.2 K and 1 atm. Since $\gamma = 2.08$ the hydrodynamical value for these amplitudes would be 0.52 (heat mode) and 0.24 (sound mode). The experimental values already depart from the hydrodynamical predictions at the lowest q value probed in these experiments ($q = 0.1$ Å^{-1}). With increasing q the heat mode weakens in intensity until at $q \approx 0.9$ Å^{-1} it has disappeared from the spectra altogether.

^{4}He at 4.2 K the heat mode displays such a strong q dependence. We are not aware of any watertight explanation. It may be that the heat mode is fundamentally different at low temperature compared to its high-temperature equivalent. Or it may be that we are seeing a manifestation of the wave nature of the atoms at these temperatures in the sense that individual atoms are behaving less and less like individual entities because of the overlap of their wavefunctions. As such, a heat mode that resembles self-diffusion near $q\sigma \approx 2\pi$ (see Section 4.4.1) has to disappear from the spectra since the self-diffusion mechanism changes in character once the particles no longer move as individual ones. Since helium is the only system that we can investigate at these low temperatures in the liquid state we cannot really tell why the heat mode is absent. Perhaps future path-integral Monte Carlo (see Section 3.3.2) simulations modelling a hypothetical higher-mass system might be able to settle this issue.

What remains is a comparison of the extended sound modes between high and low temperatures. At first sight there is no difference between the high- and low-temperature datasets when it comes to the propagation of sound waves. We show the comparison between ^{4}He at 4.2 K (Montfrooij *et al.*, 1992*b*; Crevecoeur *et al.*, 1995) and Ar at 120 K (van Well *et al.*, 1985) in Fig. 8.6. Both systems are virtually identical when it comes to the details of the behavior of the extended sound modes. At low q values the speed of propagation is given by the adiabatic sound velocity. With increasing q the propagation speed increases, very much akin to the behavior of liquid metals when the propagation frequency starts to incorporate $f_{u\sigma}$, reflecting the elastic-like response of a liquid on short length scales. For $q > q_{\max}/2$ ($q_{\max}$ being the position of the maximum of $S(q)$) the propagation frequency rapidly decreases and

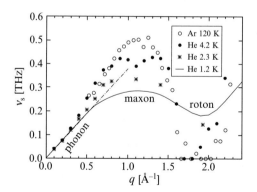

Fig. 8.6 Figure reproduced with permission from Montfrooij and Svensson (1997). Shown are the propagation frequencies $\nu_s = \omega_s/2\pi$ for ^{4}He at 4.2 K and saturated vapor pressure (SVP) (solid circles) (Woods *et al.*, 1978), for ^{4}He at 2.3 K and SVP (stars) (Woods *et al.*, 1976*b*), for superfluid ^{4}He at $T < 1.3$ K and SVP (Donnely *et al.*, 1981) (solid line), and for Ar at 120 K and 270 bars (van Well *et al.*, 1985) (open circles). The dashed line is the hydrodynamic dispersion relation $\nu_s = cq/2\pi$, with c the velocity of first sound ($c = 238$ m/s) in the superfluid phase. The frequencies for Ar and normal-fluid ^{4}He have been scaled to those for superfluid ^{4}He by using the known first-sound velocities.

a propagation gap develops near $q\sigma \approx 2\pi$. Thus, in this respect the low-temperature behavior is virtually identical to the high-temperature behavior.

Subtle differences between high and low temperatures develop when we lower the temperature below 4.2 K. The results for ^{4}He at 2.3 K (still in the normal fluid phase, see Chapter 9) and 1.2 K (in the superfluid phase) are also displayed in Fig. 8.6. At the very smallest q values the speed of propagation is still determined by the adiabatic sound velocity. With increasing q, there still exists an upward curvature to the propagation speed, but it is much less pronounced than is the case for ^{4}He at 4.2 K and for Ar at 120 K (van Well *et al.*, 1985). For instance, the maximum upward curvature in the superfluid phase is only 4% (Donnely *et al.*, 1981; Stirling, 1983) and this weak curvature happens under circumstances where the damping of the sound waves (at 2.3 K) is weaker than at high temperatures, or even completely absent (in the superfluid phase, see Chapter 9). Of course, in the case of weak damping we would expect the opposite, namely that the propagation frequencies would be higher than in the case of strong damping since the propagation frequencies ω_s are given by $\omega_s = \sqrt{f_{un}^2 - \Gamma^2}$ (in the damped harmonic oscillator model that describes the data well). Thus, this subtle difference between low and very low temperatures is more marked than a cursory glance at Fig. 8.6 might indicate.

As with the disappearance of the heat mode discussed above, there also is more than one explanation for the lack of vigorous upward curvature to the sound-mode dispersion. In fact, the following is hardly an explanation at all, but in our defense, textbooks that only list what is known beyond any doubt are hardly worth buying. One explanation could be that the upward curvature is caused by mode-coupling effects, as discussed in Section 2.2.2. The difference between the high- and low-temperature results would then be attributable to the mode-coupling mechanism becoming less effective.[2] Since mode-coupling theory does not come with an upper bound (in q) for its validity (Balucani and Zoppi, 1994), we cannot conclude for sure whether the upward curvature for Ar at 120 K is due to mode coupling, or not (see also Fig. 4.3. The 4% upward curvature for superfluid helium can certainly be explained in terms of mode-coupling effects provided frequency-dependent transport coefficients are used in the calculations (Kirkpatrick, 1984), so the demise of mode-coupling effects could well be a good explanation for the weakening upward curvature.

Another possible explanation could be that we are looking at the demise of the $f_{u\sigma} - z_\sigma$ decay channel (see Fig. 2.14). As explained in Appendix E, once z_σ becomes large then the coupling constant $f_{u\sigma}$ becomes irrelevant and no longer features in the determination of the speed of propagation of the extended sound modes. An increase in z_σ is certainly what could be expected for liquid ^{4}He on its way to becoming superfluid since a superfluid – by definition – does not support the build up of stress and hence, z_σ should be increasing upon lowering the temperature. Therefore, the lack of vigorous upward curvature at the lower temperatures could equally well be explained as the consequences of the approach to the superfluid state.

---

[2] The large-scale backflow pattern at the heart of mode coupling at intermediate densities involves coupling to the transverse, viscous modes of the fluid. When the viscosity diminishes, this backflow mechanism should be more difficult to establish.

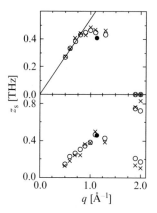

Fig. 8.7 Figure reproduced with permission from Montfrooij *et al.* (1997). Shown is a comparison of the extended sound modes for two thermodynamic states of normal fluid helium (open circles: $T = 4.21$ K, $\rho = 0.1640$ g/cm^3 and $p = 20.3$ bar; crosses: $T = 2.35$ K, $\rho = 0.1678$ g/cm^3 and $p = 16.7$ bar) that were chosen for having an identical speed of propagation of (hydrodynamic) sound. The top panel shows $\omega_s(q)$, the bottom panel $\Gamma_s(q)$. The two solid circles correspond to $T = 1.90$ K and $p = 20$ atm (Talbot *et al.*, 1988).

Despite the subtle differences in speed of propagation of the extended sound modes, there does not appear to be an underlying difference in mechanism between the lower and the higher temperatures. Rather, the differences we are seeing appear to be the consequence of the increasing influences of quantum effects. The same seems to hold true for the damping rate of the extended sound modes. At the normal to superfluid phase transition there is a rapid change in the damping rates of the extended sound modes; this rapid change is a natural side-effect of lowering the temperature in a quantum system when a gap opens up (as explained in Chapter 9). Away from the transition, the damping rate seems to be largely determined by the density and pressure of the system, and to a far lesser extent by the temperature. This effect was illustrated in normal-fluid ^{4}He (Montfrooij *et al.*, 1997) where the extended sound modes were compared between two thermodynamic states that had the same speed of sound propagation (and almost the same density), but that differed in temperature by a factor of two (roughly). We show the comparison between the two states in Fig. 8.7.

The identical behavior, from a quantitative point of view, demonstrates that also at very low temperatures the damping of sound waves in the normal fluid phase is largely determined by the density, with higher density corresponding[3] to higher damping rates (Montfrooij *et al.*, 1997). An example of this effect has already been shown in Fig. 4.4 where the damping rate for the extended sound modes was discussed for $q\sigma = 2\pi$. Not shown in that figure was that the damping rate for $q\sigma = \pi$ showed an identical behavior, namely $z_s \sim 1/l_{\text{free}}$.[4] Therefore – some quantum effects

---

[3] A good example of higher densities corresponding to higher damping rates in liquid ^{3}He can be found in the work of Scherm *et al.* (1987).

[4] The fact that τ_{free} does not appear to depend strongly, if at all, on temperature might seem somewhat counterintuitive, however, this is a consequence of the fact that most cold liquids do not display an excitation gap.

aside – we conclude that the extended sound modes behave virtually identically at high and at low temperatures. As an added bonus, at these low temperatures the results of the scattering experiments can be interpreted without ambiguity from knowledge of the density–density correlation function alone, in contrast to the so-called simple liquids and the liquid metals that we have discussed in the preceding chapters.

8.4 ^{4}He versus ^{3}He

^{3}He also stays liquid under its own pressure down to the lowest temperatures. And similar to the case for ^{4}He, the large zero-point motion that prevents the liquid from solidifying also ensures that the specific heat ratio γ is very close to 1. Thus the extended sound modes in ^{3}He cannot decay through coupling with the temperature, and the extended heat mode should be largely absent in the spectra. In addition to the standard heat and sound modes, ^{3}He shows a new type of excitation since it obeys Fermi–Dirac statistics. At low temperatures (but still above the superfluid transition temperature of ~ 3 mK) it is possible to excite atoms out of the Fermi sea, leaving a hole behind. These particle–hole excitations have indeed been observed (Scherm *et al.*, 1987; Fåk *et al.*, 1994; Glyde *et al.*, 2000*b*).

The microscopic response for ^{3}He has been measured by means of inelastic neutron and X-ray experiments (Albergamo *et al.*, 2007). Both types of experiment have proven to be a tour de force; the neutron-scattering experiments are greatly hindered by the extremely large neutron absorbtion cross-section of the ^{3}He atoms, whereas the X-ray experiments are hindered by the fact that the X-ray cross-section of such a light element is low. Nonetheless, with the increase in experimental capabilities at X-ray sources, accurate data now exist on the excitations that are present in the normal-fluid state of ^{3}He at low temperatures.

The particle–hole excitations are visible in inelastic neutron-scattering experiments through the incoherent cross-section, but they are absent in the inelastic X-ray experiments that are only sensitive to the coherent cross-section. Of course, both types of experiments measure the extended sound modes through the coherent cross-section. As such, the two types of experiments together offer insight not only into the behavior of the extended sound modes in ^{3}He, but they also allow one to determine whether the damping rate of the extended sound modes is influenced by the additional decay channel that the particle–hole excitations provide. In Fig. 8.8 we reproduce the inelastic scattering data taken by Albergamo *et al.* (2007) at $T = 1.1$ K. These data show the extended sound mode excitations (as mentioned, particle–hole excitations are absent in X-ray scattering experiments). Albergamo *et al.* have fitted their data using the two-pole approximation to the dynamic structure factor (eqn 3.9). This approximation yields an excellent description of $S_{nn}(q, \omega)$ over the entire range of momentum transfers probed in the experiment. Thus, ^{3}He is a very good example of a system where density fluctuations give rise to a velocity fluctuation. The velocity fluctuation does not give rise to a temperature fluctuation since $\gamma = 1$; therefore, the velocity fluctuation (mainly) decays through the coupling constant $f_{u\sigma}$. (see Fig. 2.14). In addition, since the damped harmonic oscillator model is adequate in describing the data, we also must have that the ratio $f_{u\sigma}/z_\sigma$ is small (see Appendix E).

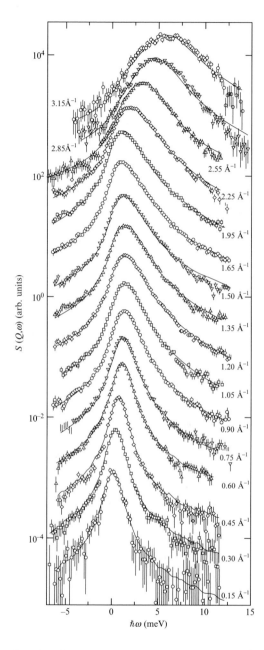

Fig. 8.8 Figure reproduced with permission from Albergamo *et al.* (2007). The figure shows inelastic X-ray-scattering spectra (symbols) of liquid ^{3}He at $T = 1.1$ K as a function of q (given in the figure). These data at saturated vapor pressure are well described by the damped harmonic oscillator model (solid lines – eqn 3.9). Note the logarithmic scale of the vertical axis. For clarity each spectrum has been multiplied by a factor of 3 with respect to the preceding one in the figure.

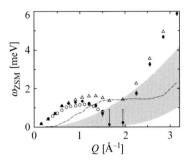

Fig. 8.9 Figure reproduced with permission from Schmets and Montfrooij (2008). The propagation frequency ω_{ZSM} of the extended sound mode in ^3He [X-ray: filled circles (Albergamo *et al.*, 2007); neutron: open circles (Fåk *et al.*, 1994)]. In the region of the particle–hole continuum (shaded area) the damping Γ (dashed line) becomes comparable to f_{un} (triangles) and the extended sound mode undergoes considerable softening.

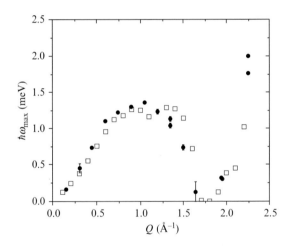

Fig. 8.10 Figure reproduced with permission from Albergamo *et al.* (2008). Shown are the extended sound-mode propagation frequencies for the ^3He experiments (solid symbols) depicted in Fig. 8.8 compared to the extended sound mode excitation energies for ^4He (open symbols) shown in Fig. 8.6. Both liquids have been measured at saturated vapor pressure. The similarity between the two datasets is such that one can easily mistake one system for the other.

The interesting question regarding the damping rate of the extended sound modes in ^3He is whether the particle–hole excitations provide enough coupling into an additional decay channel in such a way that the damping rate is increased compared to other liquids (Fåk *et al.*, 1994; Glyde *et al.*, 2000b). Inelastic neutron-scattering data seems to support this idea, whereas the inelastic X-ray-scattering data does not appear to show any obvious indication of an increased damping rate. We show the propagation frequency and the damping rate of the extended sound modes corresponding to the

X-ray data from Fig. 8.8 in Fig. 8.9. The extended sound modes can be seen to be very strongly damped in the region where decay into a particle–hole excitation is possible (Schmets and Montfrooij, 2008), and well defined outside this region. Note that while the extended sound modes are very strongly damped, or perhaps not propagating at all in the region $1.65 < q < 1.95$ Å^{-1}, because of the low-temperature lineshape distortion the peak of $S_{nn}(q, \omega)$ is not centered around $\omega = 0$ (See Fig. 8.8).

While this strong damping is certainly consistent with the extended sound modes interacting with the particle–hole decay channel, Albergamo *et al.* (2008) show that their values are in fact very similar to the extended sound modes in the normal fluid phase of ^{4}He. We reproduce their findings in Fig. 8.10. Given the similarity displayed in this figure between this low-temperature boson fluid and its fermionic equivalent, in order to assess whether the particle–hole channel is relevant to the decay of the extended sound modes we have to answer a much more subtle question: how unusual is it for a liquid with the number density of ^{3}He to show such strong damping of the extended sound modes? At present, this question cannot be answered. There are too few data points in the region of the propagation gap to be able to carry out such a detailed comparison. Perhaps even more importantly, we do not understand how the large zero-point motion of the helium atoms changes the range of densities over which a propagation gap can be observed. It is intuitively clear that a system in which the atoms have a large zero-point motion should mimic a system at higher density, but it is not possible to quantify this. Inelastic X-ray experiments on ^{3}He at elevated pressures might be able to answer all this, in particular if such pressures can be reached that the propagation gap is observed to disappear again. The latter scenario would yield a connection between the number density of the system, and the effective, increased density of a quantum fluid. Whatever the outcome might be of such experiments, the X ray data already show that even if the particle–hole continuum should provide an additional decay channel, the effects of such a channel upon the damping rate of the extended sound modes are likely to be fairly small.

The experiments on cold fluids show that the extended sound modes dominate the spectra. The extended heat modes are absent from the spectra because of the zero-point motion of the atoms. The details of the extended sound modes are by and large very similar to those of the extended sound modes at much higher temperatures in simple liquids. The sound-propagation frequency at low momentum transfers is similar, and so is the appearance of a propagation gap near the first peak of the static structure factor. All this demonstrates that the extended sound modes that are present in simple liquids over the entire q range covered in scattering experiments, even in regions where they can only be teased out of the dynamic structure factor through model fitting, are very much for real indeed. Liquids are most definitely capable of sustaining propagating density fluctuations of very short wavelength.

Thus far, we have illustrated the progression of the behavior of the extended sound modes from simple liquids, via the cases of binary mixtures and liquid metals where the extended sound modes are much more visible over an extended q range, to normal liquids at the very lowest temperatures where the extended sound modes dominate the spectra. In the next chapter we discuss what happens when the damping mechanism of the extended sound modes disappears, as is the case for superfluids.

9
Superfluids

Superfluidity is the property of a liquid to flow without friction through thin capillaries (Kapitza, 1938). This property is manifest in the Bose liquid ^{4}He below $T_\lambda = 2.17$ K. At this temperature, a small fraction of the ^{4}He atoms condenses into the lowest energy state and forms a Bose–Einstein condensate (London, 1938; Hohenberg and Martin, 1965). Other known superfluids are the Fermi liquid ^{3}He (Osheroff, 1997) and the recently discovered Bose–Einstein condensates that have been made in dilute systems at extremely low temperatures (Ketterle and Zwierlein, 2008). In this chapter we shall focus on the excitations in superfluid ^{4}He since it is for this system that a wealth of scattering data is available. We show the close similarity between the elementary excitations of this superfluid – the so-called phonon–roton curve – and the standard density fluctuations of normal fluids. In fact, it will be clear that the elementary excitations in superfluid ^{4}He are the natural extensions of the extended sound modes in normal fluids. In this chapter we focus entirely on density fluctuations, we refer the reader to the exhaustive texts by Griffin (1993), Glyde (1994) and Mayers (2006) on how the presence of the Bose-condensate in superfluid helium might change the character of the elementary excitations.

9.1   Superfluidity and Bose–Einstein condensation

As already mentioned, superfluidity is the property of a liquid to flow without friction through thin capillaries. In ^{4}He under its own vapor pressure the normal-fluid to superfluid transition occurs at $T_\lambda = 2.17$ K, the so-called lambda transition (named after the shape of the specific heat curve). Below 1 K, 100% of the liquid can flow without friction. Bose–Einstein condensation (BEC) on the other hand, is the property that a large fraction of the particles that make up a system condense into the same energy state. For instance, in an ideal Bose gas (a gas made up of bosons that do not interact with each other), 100% of the particles will condense into the state with the lowest available energy and form a Bose–Einstein condensate. However, an ideal Bose gas does not become superfluid. And conversely, in liquid ^{4}He only about 7% (Glyde *et al.*, 2000*a*) of the atoms actually do form a condensate, even when 100% of the atoms can flow without friction. In fact, there is no reason why a system could not become a superfluid even if only a very small fraction of the atoms were to form a condensate. All this nicely illustrates the fact that superfluidity and BEC are two different phenomena, even though the two are closely linked. The main difference between BEC and superfluidity is that BEC is a property of the ground state, while superfluidity is a property of the excited states. This is analogous to superconductivity, where the electrons condense into Cooper pairs (ground state), and where the interaction between the Cooper pairs (originating in the Pauli exclusion principle) introduces a finite energy gap between the ground state and excited states. In turn, this energy gap is responsible for the system becoming a superconductor. Thus, in both systems, it is the interaction between the particles that is responsible for the exotic behaviors, not necessarily how they arrange themselves in the ground state.

There exist both fermionic and bosonic superfluids. Whereas ^{4}He is the prototype bosonic superfluid, ^{3}He is the prototype fermionic superfluid. In ^{3}He the atoms condense into Cooper pairs at ~ 3 mK, and superfluidity takes place because of the finite energy gap between the ground state and excited states (Leggett, 1965, 1972). This is almost fully analogous to standard superconductivity; the main difference is that the Cooper pairs are bound by spin interactions rather than by exchanging phonons, as is the case in standard superconductivity. These two helium isotopes were the only two superfluid systems on Earth until 1995 when Bose–Einstein condensation was observed (Anderson *et al.*, 1995; Davis *et al.*, 1995) to take place in ultracold (~ 500 nK) systems of very dilute gases, such as Rb, K and Li. Both the bosonic and fermionic isotopes exhibit Bose condensation. Superfluidity in such Bose–Einstein condensates was confirmed (Zwierlein *et al.*, 2005) through the study of excitations and specific heat as well as through the observation of vortices signalling the quantization of angular momentum (see Fig. 9.1). Unfortunately, given the very cold temperatures required, and the limited number of particles (millions) that make up these condensates, there exist no scattering data on these systems, as of yet. As such, we shall only focus on ^{4}He in this chapter.

First, the fact that helium does not solidify at any temperature is a pure quantum effect. The weak van der Waals forces between the atoms are not strong enough to overcome the zero-point motion associated with confining a helium atom to a lattice site. The second aspect that makes helium stand out from other liquids is that it takes a finite amount of energy to create a disturbance in the liquid. This is shown in Fig. 9.2. The actual amount of energy required depends on the wavelength λ (or momentum

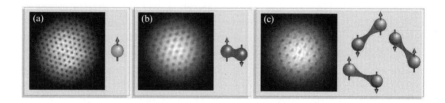

Fig. 9.1 Figure reproduced with permission from Grimm (2005). The observation (Zwierlein *et al.*, 2005) of vortices in three types of rotating Bose–Einstein condensates proves the property of superfluidity in all three of them. Panel (a) shows the vortex structure in a condensate formed from bosonic sodium atoms, panel (b) for the fermionic system where the condensate consists of tightly bound pairs of ^{6}Li atoms, and panel (c) for a fermionic gas of pairs of loosely bound ^{6}Li atoms.

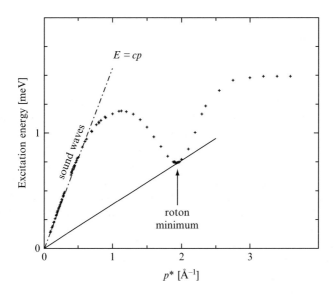

Fig. 9.2 The measured excitation energies $E_{ex}(p)$ in superfluid ^{4}He as a function of momentum transfer q ($= p^* = p/\hbar = 2\pi/\lambda$ in the notation of the figure). The slope of the curve at small q [dashed line] is given by the velocity of sound $c = 237.4$ m/s; the overall minimum slope [corresponding to a speed of 58 m/s] is given by the solid line that is tangent to the excitation curve near the so-called roton minimum ($q = 1.94$ Å^{-1}, $\hbar\omega_{ex}(q) = E_{ex}(q) = 0.743$ meV). The dispersion curve is given by a compilation of neutron-scattering results by Donnely *et al.* (1981) and Glyde *et al.* (1998).

$\hbar q = p = h/\lambda$) of this disturbance; the resulting dispersion curve was estimated from thermodynamic properties by Landau (1941; 1947) and this estimate turned out to be in remarkably good agreement with the early measurements of the excitation energies (Palevsky *et al.*, 1957, 1958; Yarnell *et al.*, 1958; Henshaw, 1958; Yarnell *et al.*, 1959). The measured values for the energy costs are shown in Fig. 9.2.

At low momentum transfers (long wavelengths) the energy disturbance is just a run-of-the-mill sound wave (a phonon). The energy of this excitation is given by $E_{ex}(p) = \hbar\omega_{ex}(q) = cp$, the standard hydrodynamics result. When we go to shorter wavelengths, such as the density disturbance pictured in Fig. 1.1, the energy cost starts to deviate from $E_{ex}(p) = cp$. For wavelengths comparable to the interatomic spacing d, the energy cost goes through a minimum, after which it goes up again. This minimum of the energy gap between the ground state and the excited state is commonly referred to in the literature on superfluid helium as the roton minimum or simply 'the roton', and it turns out to be the determining feature of superfluids.[1] As pointed out above, this situation is entirely analogous to superconductivity where

---

[1] Precisely what the roton excitation entails is not known. Many visualizations have been put forward over the years (Landau, 1947; Chester, 1967; Feynman, 1972; Glyde and Griffin, 1990; Williams, 1992); we shall only be discussing the close connection between the roton excitation and the excitations in normal fluids at similar wavelengths.

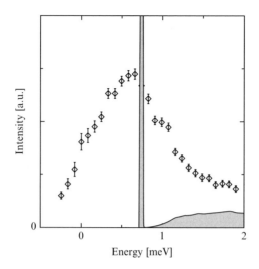

Fig. 9.3 Detailed view of the dynamic structure factor $S_{nn}(q, \omega)$ of ^{4}He for the q value corresponding to the roton minimum in the superfluid phase (shaded curve) and in the normal-fluid phase (points plus error bars). The data were collected during a neutron-scattering experiment on the MARI spectrometer at the ISIS facility. The neutron transfers energy $\hbar\omega = E$ to the liquid and an amount of momentum corresponding to the roton minimum (see vertical arrow in Fig. 9.2). When the amount of energy transferred exactly matches the energy difference between the ground state and the excited state, then a sharp resonance peak (at 0.743 meV) is seen in the superfluid. Note that there is no scattered intensity at energy transfers below this peak. In the normal fluid, however, even a small amount of energy transfer is sufficient to excite the liquid, and there is a clear signal even at $\omega = 0$. The signal at $\omega < 0$ implies that the normal liquid can transfer some of its energy to the neutron. The areas under both curves are the same, illustrating the colossal intensity of the roton excitation in scattering experiments.

the energy gap is the determining feature of superconductors. The presence of a gap also firmly sets superfluid ^{4}He apart from normal fluids, where nothing resembling an energy gap exists. This is shown in Fig. 9.3 where we compare helium in the superfluid phase and in the normal-fluid phase.

One can easily verify from Fig. 9.2 that the presence of a non-zero energy gap is synonymous with superfluidity (Landau, 1941, 1947; Feynman, 1972). The slope of a line that goes through the origin and a point on the excitation curve gives the (phase) velocity of the excitation. For instance, this slope at small momenta (where the group and phase velocities coincide) is given by the speed of sound (see Fig. 9.2). The smallest slope is encountered near the roton minimum, at the point where the group and phase velocities of the excitations are identical. The value of the slope at this point corresponds to the velocity below which the liquid can flow without friction, at least through small capillaries. After all, if the liquid is flowing at a lower speed, then the liquid cannot slow down because of the following restrictions due to the energy and momentum conservation laws. Take a liquid mass M that is flowing at speed v. For it to slow down to speed v' by creating an excitation of energy E_{ex} and momentum $\vec{p}_{ex}$ we have

$$Mv^2/2 = Mv'^2/2 + E_{\text{ex}}$$
$$M\vec{v} = M\vec{v}' + \vec{p}_{\text{ex}}. \qquad (9.1)$$

Eliminating v' we get

$$\vec{v}.\vec{p}_{ex} - p_{\text{ex}}^2/2M = E_{\text{ex}}. \qquad (9.2)$$

For the best case scenario – in which $\vec{v}$ and $\vec{p}_{\text{ex}}$ are parallel and in which M is very large – we find the lower bound on the flow velocity v for the liquid to be able to slow down:

$$v \geq E_{\text{ex}}/p_{\text{ex}}. \qquad (9.3)$$

In a normal liquid without an energy gap, liquid flow will always be damped because the minimum slope would be zero. This (eqn 9.3) is of course also the reason why an ideal Bose gas does not become superfluid. Here, the excitation energies are given by $E_{\text{ex}} = p_{\text{ex}}^2/2m$, and a parabolic curve does not have a minimum slope: no matter how slow an ideal Bose-liquid is flowing, it is always possible to transfer energy by creating an excitation, and the liquid will slow down. Also, note that even though it requires less energy to create a sound wave than it does to create a roton excitation, the roton minimum actually determines the critical flow velocity. Of course, it is well known that the actual critical flow velocity in bulk helium is much lower than the value inferred from the slope in Fig. 9.2 because of the formation of vortices in the fluid (the same type of vortices that were used to confirm the existence of superfluidity in the dilute Bose–Einstein condensed systems as shown in Fig. 9.1). However, when the fluid is made to flow through very small orifices (Varoquaux *et al.*, 1991, 1993) which inhibit the formation of vortices, then the critical flow velocity does indeed go up to values approaching those shown in Fig. 9.2.

The above also explains why the actual fraction of atoms in the condensate does not determine the critical properties. Bogoliubov (1947) showed for a fluid of weakly interacting bosons that the quadratic dispersion of the ideal Bose gas, $E_{\text{ex}} = p_{\text{ex}}^2/2m$, changes to a linear dispersion for small momentum values because of the presence of a condensate. This is sketched in Fig. 9.4. Because of the fact that the atoms interact with each other, the dispersion curve will exhibit some other changes as well, reflecting the fact that some disturbances cost more energy than others because the atoms have a preferred spatial arrangement owing to their excluded volumes. This is completely analogous to the case of simple liquids where we used the excluded volume argument to estimate the position of the minimum in the characteristic width of the spectra (Fig. 4.8). When the strength of the interaction increases even more, then the minimum slope will no longer be determined by the linear part of the dispersion, but instead it is given by the tangent to the overall dispersion curve. This is shown in Fig 9.4(d). Thus, while the formation of the condensate is essential in changing the quadratic part of the dispersion at low momentum values to a linear part so that superfluidity can take place in the first place, the actual critical parameters that have

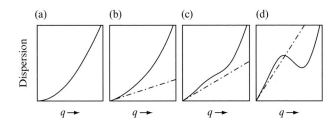

Fig. 9.4 The dispersion relation for an ideal Bose gas consisting on non-interacting particles is given by $E \sim q^2$ (part a). Bogoliubov (1947) showed that the low-momentum part of the dispersion renormalizes to a linear part $\sim q$ in the presence of a Bose-condensate in a dilute system (part b). This renormalization results in a minimum flow velocity, given by the slope of the dashed line in the panels. With increased density of the system, the dispersion curve will also reflect the structure of the liquid. It will cost more energy to create a hole in the liquid, and less energy to create a density fluctuation with a wavelength matching the interatomic separation d_{avg} (part c). For even higher densities, these structural effects will be such that the minimum slope to the dispersion curve is no longer given by the slope at low q, but rather it will be given by the minimum in the dispersion curve near $q = 2\pi/d_{\mathrm{avg}}$ (part d). It is for this reason that the properties of superfluid helium are determined by the roton minimum rather than by the condensate fraction.

to do with critical flow and by extension, with the damping of the excitations, are determined by the roton part of the dispersion. This part of the dispersion curve does not depend on the Bose-condensate directly.

In the following sections we apply the effective eigenmode formalism to the excitations in $^4\mathrm{He}$. The analysis will show how the extended sound modes of the normal fluid continuously change their character upon crossing the superfluid transition from above until they become undamped at the lowest temperatures. Note that since the damping itself does not depend on the condensate fraction *per se*, the effective eigenmode description does not yield any information about the condensate. Before we present an inventory of the excitations present in superfluid $^4\mathrm{He}$ at very low temperatures as seen in scattering experiments, we briefly revisit the arguments that Feynman presented to explain the property of superfludity in a Bose system at low temperatures (Feynman, 1953*a,b*, 1954, 1955, 1972).

9.2 Why Bose-liquids must become superfluids

Feynman explained in a beautiful argument why the energy gap between the ground state and the excited state is the unavoidable consequence of the fact that $^4\mathrm{He}$ atoms obey Bose statistics. We refer the reader to Feynman's 1955 account and 1972 textbook for details, but in a nutshell the argument is the following.

Assume that a certain configuration of the helium atoms represents the state with the lowest energy, the ground state. The quantum-mechanical wave function ϕ of this state depends on the positions of all atoms: $\phi(\vec{R}_1, \vec{R}_2, ..\vec{R}_N)$. The energy of this ground state consists of a kinetic-energy term that depends on the gradient of the wavefunction

$\sim |\nabla\phi|^2$ as well as on a potential term $|V\phi|^2$. The same holds for the wavefunction $\psi(\vec{R}_1, \vec{R}_2, ..\vec{R}_N)$ describing the excited state that is lowest in energy of all possible excited states. The potential operator has terms $\sim 1/|\vec{R}_i - \vec{R}_j|^n$ that tell us that the force between the atoms is strongly repulsive when they are too close together.

The exact details of the ground state are not important, but both the contributions of the kinetic and potential terms to the ground-state energy should be small. From this requirement we can expect that the atoms in a configuration that could represent the ground state are fairly well spread out (see Fig. 9.5(c)). After all, if they were to sit on top of each other (Fig. 9.5(a)), we would pay a high price in potential energy, and hence, we can assume the amplitude of the ground-state wavefunction ϕ to be zero for those cases where $\vec{R}_i \approx \vec{R}_j$. Also, the atoms will not be too close to each other (Fig. 9.5(b)), because this would correspond to a high gradient $\nabla\phi$ (and thus to a high kinetic energy), making it an unlikely choice of ground state. We can see that atoms almost touching each other would correspond to a high gradient as follows: if the amplitude of ϕ were not zero for a configuration where two atoms are very close, then we would have the situation that by slightly changing the coordinate of one atom in order to make it sit on top its neighbor (going from Fig. 9.5(b) to 9.5(a)), we would go from a non-zero to a zero amplitude for ϕ. This implies a steep gradient $\nabla\phi$ (large kinetic energy), and therefore, the amplitude of ϕ must also be zero for configurations where atoms are too close. Another way of saying the above is that if an atom actually were to move from Fig. 9.5(b) to 9.5(a), it must have had a large kinetic energy in the first place to be able to approach the other atom as closely as shown in Fig. 9.5(a). However, note that we do not actually 'move' atoms, we just compare the wavefunction for two different configurations $\vec{R}^N$. Thus, when we say 'move', we do not imply any dynamics.

We also know that ϕ will not have any nodes other than at the edge of the box confining the liquid, so without loss of generality we can assume that the amplitude of ϕ positive for all configurations. Also, the excited-state wavefunction ψ should have at least one more node. This is essentially the guitar string equivalent when comparing the first and second harmonic. This implies that half of the configurations

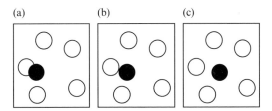

Fig. 9.5 Figure adapted from Feynman (1972). A depiction of various configurations representing different energies. Part (a) shows an unlikely ground-state configuration since two atoms being in the same spot implies a high potential energy. Similarly, part (b) shows an unlikely ground state because this configuration would represent a high kinetic energy (see text). On average, in the ground state the atoms will be spaced out as shown in part (c).

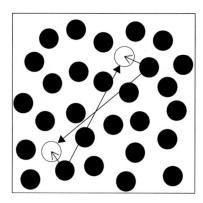

Fig. 9.6 Figure adapted from Feynman (1972). A reasonable guess of what a low-energy state in helium could look like. In order to flip the sign of ψ, atoms are 'moved' over large distances (long arrows) while smoothing out any holes left in the liquid to minimize the energy cost. However, since the Bose particles cannot be distinguished from each other and since permutation of particles does not affect the wavefunction, the outcome of the 'movements' indicated by the solid arrows are identical to the 'movements' indicated by the short arrows.

representing the excited state $\psi(\vec{R}_1, \vec{R}_2, ..\vec{R}_N)$ correspond to a positive amplitude, and half of the configurations correspond to a negative amplitude. We will now try to create an excited state that barely differs in energy from the ground state. In this way, the roton minimum would be very small (yielding a very small minimum slope in Fig. 9.2) and we would be able to stifle superfluidity. However, Feynman (Feynman, 1972) showed that this cannot be done; one always ends up with a sizeable energy difference between the ground state and the excited states.[2]

In Fig. 9.6 we have sketched a configuration for the excited state that we arbitrarily will take to correspond to a maximum positive amplitude for ψ. We do not really know what the configuration should look like, but we tried to make it look like the ground state, with atoms not sitting on top of each other. Next, we will rearrange the atoms to end up with a configuration that would correspond to a maximum negative amplitude. To achieve this, we should rearrange the helium atoms over large distances. We are not interested in short distances, because this would imply that ψ goes from a maximum to a minimum over short distances, which would correspond to a large gradient $\nabla\psi$ and therefore, to a high kinetic energy. We will also smooth out any holes or bumps that may materialize, otherwise we would end up with an excitation that looks like a phonon (see Fig. 1.1) and we already know that a phonon does not represent the minimum slope of the excitation curve (Fig. 9.2). The required changes to the configuration are shown in Fig. 9.6 by the long arrows, and we appear to have achieved our aim. We moved one atom over a large distance, followed by a second to fill in the hole we created.

---

[2] Phonons can be created at very little energy cost; however, phonons do not determine the minimum flow velocity in the superfluid phase (see Fig. 9.2) and hence we will only scrutinize excited states that do not represent phonon excitations.

However, the above approach does not work in a Bose-liquid since all the atoms are indistinguishable and interchanging two atoms does not lead to a change in the amplitude of the wavefunction. Thus, one could have obtained the same final configuration by simply 'moving' the atoms affected by the rearrangement over distances less than half the atomic separation (short arrows in Fig. 9.6). In fact, half the atomic separation is the best that one could achieve. However, such a rapid variation (from maximum to minimum over half the atomic separation) would represent a large gradient and signify a significant step up in energy. In other words, because of the Bose nature of the atoms, it is not possible to construct an excited state (which is not a phonon) that differs by a vanishingly small amount in energy from the ground state. Therefore, an energy gap must be present in a Bose-liquid and provided the liquid does not freeze, it must become a superfluid.

Whether a Bose–Einstein condensate forms or not is not relevant to the above argument since the argument links the property of superfluidity to the scarcity of excited states. Even a Bose-liquid where only a tiny fraction of the atoms condenses will become a superfluid when the temperature is low compared to the energy gap. Penrose and Onsager (1956) showed that Feynman's assumptions are a sufficient requirement for a Bose-condensate to form. Thus, we can view Feynman's argument as a convenient way of not only explaining the existence of the property of superfluidity, but also for the fact that a Bose-condensate will form upon cooling, even in a strongly interacting system like helium. While the actual value of the condensate fraction does not matter to the preceding argument, it does matter to what is observed in scattering experiments. It is possible for a neutron or photon to scatter a particle out of the condensate; if the condensate fraction is large, then this will produce a significant scattering cross-section. In the case of ^{4}He where the condensate fraction is below 10%, this scattering process only presents a minor contribution, and it is only visible for scattering near the recoil curve (which is shown in Fig. 9.7).

9.3   Density fluctuations in very cold superfluid ^{4}He

We can distinguish three types of excitations in scattering experiments on superfluid ^{4}He. They are the phonon–roton excitations or elementary excitations, the multiphonon component, and the recoil curve. We shall discuss these excitations (shown in Fig. 9.7) in some detail below. These three types seen in scattering experiments are by no means the only excitations present in the fluid. As mentioned, the fluid is capable of producing vortices in response to rotational flow. The fluid can be excited by knocking a particle out of the condensate. The fluid can also sustain propagating temperature waves, known as second sound. These temperature waves cannot be seen by means of scattering experiments since they are not accompanied by a density variation, rather they are an out-of-phase oscillation of the normal fluid and the superfluid component. The fact that these temperature waves do not produce a density variation is a direct consequence of the fact that $\gamma = 1$; there exists no coupling between the variables 'u' and 'T' ($f_{uT} = 0$). This lack of coupling makes it easy to follow the fate of the phonon–roton type of excitations when warming the system up through the superfluid transition. In addition, surface waves can also be

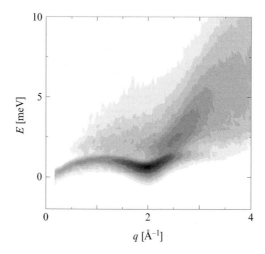

Fig. 9.7 The scattering in superfluid ^{4}He under its own pressure as a function of q and $E = \hbar\omega$, measured by means of inelastic neutron scattering on the MARI spectrometer at ISIS. The strongest scattering corresponds to the darkest bands. The scattering shows three components: the phonon–roton excitations, the (barely visible) multiphonon component and the scattering near the recoil curve $\omega_{\mathrm{recoil}}(q) = \hbar q^2/2m$ (eqn 2.27). The branch of phonon–roton excitations is smeared out in energy because of the non-perfect instrumental resolution of the spectrometer employed for this particular experiment. A vertical cut through the data at constant $q = q_{\mathrm{roton}} = 1.94$ Å^{-1} yields the data shown in Fig. 9.3 once they are corrected for the instrumental resolution. The intensity scale employed in this figure is non-linear and was chosen so as to bring out the three components. Subsequent figures yield better information on the relative intensities of the various features. See also cover illustration.

present in helium, and helium films also show excitation modes unique to the lowered dimensionality. Here, we will only deal with the density fluctuations in bulk helium.

The energies corresponding to the elementary excitations in superfluid ^{4}He are shown in Fig. 9.2, and their intensities in scattering experiments are shown in Fig. 9.7, and in more detail in subsequent figures such as Fig. 9.9 and Fig. 9.14. In contrast to normal fluids, these phonon–roton excitations are virtually undamped, and therefore show up as extremely sharp features in inelastic scattering experiments, such as the one shown in Fig. 9.3. The dispersion curve of the phonon–roton excitations flattens off beyond the roton excitation (Glyde *et al.*, 1998), and reaches a constant value of $\hbar\omega = 2\Delta$ (Fig. 9.2). The intensity of the excitations gets weaker and weaker with increasing q until they disappear from the spectra for $q > 2q_{\mathrm{roton}}$.

Close inspection of the scattered intensity shows that there is weak, yet highly structured scattering at energies higher than the phonon–roton excitations (Svensson, 1989; Crevecoeur *et al.*, 1996; Gibbs *et al.*, 1999). This component of the scattering, referred to as the multiphonon component, is shown in more detail in Fig. 9.8 for ^{4}He at 10 bar. This multiphonon component is unique to the superfluid phase (Juge and

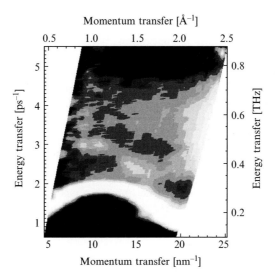

Fig. 9.8 Shown is the dynamic structure factor $S_{nn}(q,\omega)$ of superfluid helium ($T = 0.4$ K) at elevated pressure ($p = 10$ bar) measured on the IRIS neutron scattering spectrometer at the ISIS facility. In order to emphasize the multiphonon component of the scattering we have plotted $\omega S_{nn}(q,\omega)/q^2$ and we have used a grayscale that ignores the low (black) and high (white) scattering intensity. Note the rich structure present in this component, especially near twice the roton energy, twice the maxon energy and near the roton plus the maxon energy.

Griffin, 1994), no such excitations are present in the normal fluids discussed in this book up till this point.

The third component to the scattering consists of a broad (in energy) feature located (Woods and Cowley, 1973) near the recoil energy $E = \hbar q^2/2m$ (eqn 2.27). This recoil curve does not link up with the phonon–roton curve, a clear hybridization gap can be observed (see Fig. 9.9). As discussed in Section 2.2.3, this recoil scattering represents the scattering that can be linked to the motion of individual atoms. The width (in energy) of this feature is a measure of the kinetic energy of the atoms. In the superfluid phase, this scattering differs slightly from the scattering seen in the normal fluid phase because of the presence of the Bose-condensate. A neutron (or photon) can now scatter a particle directly out of the condensate, yielding a sharp peak superimposed on top of the broad distribution. It is from the intensity of this sharp, spectrometer-resolution-limited feature that the fraction of atoms that constitute the condensate has been determined (Sears *et al.*, 1982; Root and Svensson, 1991; Sokol, 1995; Glyde *et al.*, 2000*a*). The effective eigenmode formalism is not suited to deduce meaningful information from the scattering in this region. The lineshape of the response (in energy) is almost Gaussian, implying that we would need to include many Lorentzian lines to reproduce this shape. Thus, there would no longer be a clear distinction between the slow and fast decay channels, and hence the effective eigenmode description would lose its usefulness.

9.4   The normal-fluid to superfluid transition

The normal-fluid to superfluid transition in ^{4}He under pressure was displayed in Figs. 8.3 and 8.4 for one value of momentum transfer, corresponding to the roton excitation. These figures showed the continuous, yet very rapid transition upon changing the temperature through the superfluid transition. We now scrutinize these changes for all momentum transfers and for multiple pressures with the aid of the effective eigenmode formalism.

9.4.1   Global changes during the superfluid to normal-fluid transition

A blanket view of the transition is shown in Fig. 9.9, and the most significant changes have been described in great detail in the literature (see for instance, Svensson 1989, 1991; Stirling 1991; Ohbayashi 1991). At the lowest temperatures, the excitations in this figure appear broader (in energy) than they are in reality because of the instrumental resolution of the spectrometer. A cursory inspection of this figure shows that the excitations that make up the photon–roton dispersion curve broaden upon raising the temperature, and because of this spreading out, they appear to become less intense even though their energy integrated intensity barely changes. The temperature

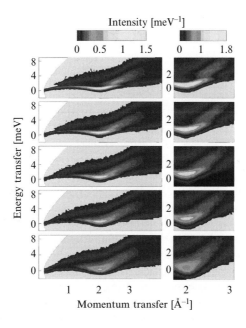

Fig. 9.9 The (resolution broadened) dynamic response of liquid ^{4}He at saturated vapor pressure for five temperatures ($T = 1.35$, 1.85, 2.0, 2.1 and 2.3 K, from top to bottom). These data were taken on the MARI neutron-scattering spectrometer at the ISIS spallation source. The superfluid transition temperature is at $T = 2.17$ K. The panels on the right show expanded views of the roton region where the recoil curve and the phonon–roton dispersion curve hybridize. The absolute, non-linear intensity scale is shown at the top of the figure.

dependence of the excitations at the high-q end of the dispersion curve is more difficult to follow because of their very weak intensities, even at the very lowest temperature. These latter excitations are only visible in the superfluid phase at the lower temperatures, they are not visible in the normal-fluid phase.

The hybridization gap between the recoil curve and the phonon–roton curve (Bedell *et al.*, 1984) can clearly be seen to disappear upon going from the superfluid phase to the normal-fluid phase. Thus, this gap must be associated with a property of the superfluid phase. The multiphonon component, located above (in energy) the phonon–roton dispersion curve can just be made out at the lowest temperature, however, its temperature dependence cannot be ascertained because of the broadening of the phonon–roton excitations. The effective eigenmode analysis will show that this component is only present in the superfluid phase. The recoil curve itself, away from the hybridization region, does not exhibit any temperature dependence in this figure. Thus, the disappearance of the Bose-condensate upon raising the temperature has such a small influence on the shape of this curve that it can only be observed through a very detailed inspection of this region. However, these more detailed experiments have been performed by various groups (Sears *et al.*, 1982; Root and Svensson, 1991; Sokol, 1995; Glyde *et al.*, 2000*a*), and they show that the Bose-condensate can indeed be seen to weaken upon approaching the superfluid transition from below, and to disappear upon going through the transition.

9.4.2 The roton region

The roton excitations are the most prominent excitations in the dynamic structure factor, and their temperature dependence upon going from the superfluid to normal-fluid phase is easiest to follow. The fate of the roton excitation at elevated pressure is displayed in Fig. 8.3 (Dietrich *et al.*, 1972; Svensson *et al.*, 1996). The solid lines through the data points in this figure are the result of the fit using the effective eigenmode formalism with two Lorentzian lines (see eqn 3.9). This corresponds to the situation where there is a very clear distinction between the decay of the velocity fluctuation that arose from the density fluctuation, and the very rapid decay of all other fluctuations. The absence of a third Lorentzian line is a direct result of γ being equal to one – which in turn was the direct result of the zero-point motion of the helium atoms – in combination with the extremely small ratio of $f_{u\sigma}/z_\sigma$ (Appendix E).

The two-pole approximation corresponding to the damped harmonic oscillator model of Fig. 2.16 gives a complete description of the scattering data for $T > T_\lambda$ (see Fig. 8.3), and a good description of the roton excitation below T_λ. It is also clear that, as expected, the two-pole approximation is inadequate in describing the multiphonon structure that can be seen to emerge in the spectra at higher energy transfers upon lowering the temperature into the superfluid phase (see inset Fig. 8.3). We show the fit parameters of the two-pole approximation for two pressures in Fig. 9.10. The damping rate $z_u/2 = \Gamma$ of the roton excitation shows a very marked temperature dependence (Fig. 9.10). Starting at the lowest temperature where the damping rate is essentially zero as befits an elementary excitation, z_u is seen to increase gradually upon warming up, followed by an increasingly more rapid change upon approaching the superfluid transition, after which it only increases slowly in the normal fluid phase. This behavior occurs both at the lowest pressure (Montfrooij *et al.*, 1997), as well as at the elevated pressure.

Most importantly, the change in the damping rate z_u is continuous. This proves that we are looking at a single excitation that rapidly broadens upon approaching the superfluid transition from below; what we are not seeing is an admixture of two types of excitations (Glyde and Griffin, 1990) where one excitation is only visible through (and because of) the presence of a Bose-condensate. This continuity in the changes as a function of temperature is also seen in the behavior of the coupling constant f_{un}. This constant, which is the ratio of two frequency moments, is almost independent of temperature for the lowest temperatures, and displays a significant and rapid increase upon going through the superfluid transition (Fig. 9.10). This change is attributable to the disappearing multiphonon component, as we will elucidate in the next paragraph. The most important observation at this point is that the fit parameters that describe the roton excitation part of the spectra vary rapidly, yet continuously as a function of temperature.

The reason why an increase in the fitted value of f_{un} – describing the roton excitation – is expected to occur when the multiphonon component disappears from the spectra is straightforward. In applying the effective eigenmode formalism to the roton excitation we are solely modelling the single excitations; the multiphonon component is left unmodelled. This implies that the model underestimates the first frequency-sum

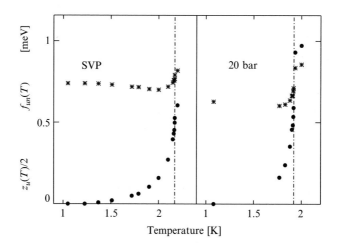

Fig. 9.10 Figure adapted with permission from Montfrooij and Svensson (2000). The temperature dependence of the damping rate z_u (solid circles) and the driving forces f_{un} (stars) of the roton excitation at saturated vapor pressure (SVP, left panel) and at elevated pressure (20 bar, right panel). The (pressure-dependent) superfluid transition temperature is indicated by the dashed-dotted vertical line. The data taken at SVP approach the superfluid transition to 0.7 mK, the data at 20 bar to 2.6 mK.

rule (f-sum rule), as well as the true static susceptibility [(−1)st frequency moment]. Therefore, f_{un} as given by the ratio of these two frequency moments is also underestimated by the model. A numerical (toy model) example of this is given by Montfrooij *et al.* (1997), which we repeat here. Assume for the sake of argument that all excitations are very sharp in energy in the superfluid phase at very low temperatures so that they can be approximated by delta functions in energy. We look at the example where we have the roton excitation located at ω_r with weight Z in $S_{nn}(q, \omega)$, a two-roton excitation at ω_{2r} with weight Z_{2r}, a roton–maxon excitation at ω_{rm} with weight Z_{rm} and a two-maxon excitation at ω_{2m} with weight Z_{2m}. The weights combine to give the static structure factor $S(q) = Z + Z_{2r} + Z_{rm} + Z_{2m}$. In this example, one finds for the ratio of the 1st and (−1)st frequency moments in the effective eigenmode description that only describes the roton excitation:

$$(f_{un}^{\text{model}})^2 = (\omega_r Z)/(\omega_r^{-1} Z) = \omega_r^2, \tag{9.4}$$

and for the true ratio as based on the exact frequency sum-rules:

$$(f_{un}^{\text{true}})^2 = \frac{\omega_r Z + \omega_{2r} Z_{2r} + \omega_{rm} Z_{rm} + \omega_{2m} Z_{2m}}{\omega_r^{-1} Z + \omega_{2r}^{-1} Z_{2r} + \omega_{rm}^{-1} Z_{rm} + \omega_{2m}^{-1} Z_{2m}} = \frac{\hbar q^2/2m}{\hbar \pi \chi(q)}. \tag{9.5}$$

The latter equation can be rearranged to read

$$(f_{un}^{\text{true}})^2 = \omega_r^2 \frac{1 + (\omega_{2r}/\omega_r)(Z_{2r}/Z) + (\omega_{rm}/\omega_r)(Z_{rm}/Z) + (\omega_{2m}/\omega_r)(Z_{2m}/Z)}{1 + (\omega_r/\omega_{2r})(Z_{2r}/Z) + (\omega_r/\omega_{rm})(Z_{rm}/Z) + (\omega_r/\omega_{2m})(Z_{2m}/Z)} > \omega_r^2. \tag{9.6}$$

Thus, in the presence of a multiphonon component the parameter f_{un} as determined from a fit to the roton excitation is an underestimate. When the system goes from being a superfluid to being a normal fluid, the static susceptibility barely changes (if at all) since the spatial arrangement of the atoms does not appreciably change. Thus, f_{un}^{true} does not significantly change upon going from the superfluid to the normal-fluid phase. However, when the weights Z_{2r}, Z_{rm} and Z_{2m} tend to zero upon approaching the transition from below, then f_{un}^{model} must necessarily approach f_{un}^{true} since the damped harmonic oscillator model gives a satisfactory description of the entire spectra in the normal-fluid phase. Hence, the rapid, yet modest (compared to z_u) increase of f_{un}^{model} upon approaching the superfluid transition from below is expected in a system where the multiphonon component disappears upon going through this transition.

9.4.3 Phonon and maxon excitations

A virtually identical behavior to that of the temperature dependence of the roton is observed for the phonon and maxon excitations (Cowley and Woods, 1971). This is displayed in Fig. 9.11. The damping rate z_u of these excitations shows an almost identical temperature dependence (Montfrooij and Svensson, 1997) to that of the roton upon approaching the superfluid transition. The same holds true for the driving force f_{un}, including the marked increase upon approaching T_λ from below. The effectiveness

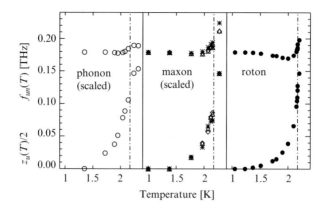

Fig. 9.11 Figure reproduced with permission from Montfrooij and Svensson (1997). The temperature dependence of the phonon and maxon parameters (damping rate z_u and driving force f_{un}) at saturated vapor pressure, compared to those of the roton. The forces f_{un} for the phonon and maxon excitation have been scaled to those of the roton to emphasize the very similar temperature dependence of all of them. The damping rates have not been scaled, their similarity reflects the Landau–Kalatnikov damping mechanism (Landau and Khalatnikov, 1949).

of the two-pole approximation in describing the non-multiphonon excitations in superfluid ^{4}He near the normal-fluid transition is illustrated in Fig. 9.12 for the maxon excitation. Because of the fact that the maxon excitation is much weaker (in intensity) than the roton excitation, the multiphonon component of the scattering at the lowest temperatures is more pronounced (Talbot *et al.*, 1988; Gibbs and Stirling, 1996). Again, it can be seen from this figure that the two-pole approximation gives a complete description of the spectra in the normal-fluid phase. This is another manifestation of the fact that the multiphonon component is restricted to the superfluid phase.

The fate of the elementary excitations with wave numbers $q > q_{\text{roton}}$ is less clear because of the weakness of the scattering. The best data thus far have been measured (Glyde *et al.*, 1998) by means of neutron scattering utilizing the IRIS spectrometer at ISIS. These data are reproduced in Fig. 9.13. As can be seen in this figure, the excitations show up as very sharp features on top of the tail of the recoil contribution. Their sharpness indicates that at the lowest temperature these excitations are essentially undamped, just like the other elementary excitations that make up the phonon–roton curve. Upon increasing the temperature, these excitations can be seen to broaden rapidly, and to disappear from the spectra. However, the excitations are so weak that it is not clear whether the broadening is similar to that of the roton excitation, or whether it is more rapid. What is clear, however, is that they are not present in the normal-fluid phase.

In summary, the analysis using the two-pole approximation of the effective eigen-mode description shows that the transition from the normal-fluid to the superfluid phase is marked by a rapid, yet continuous change in the damping rate $z_u(q)$ of the extended sound mode excitations of the normal-fluid phase. This change is

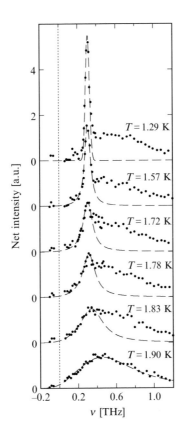

Fig. 9.12 Figure reproduced with permission from Montfrooij *et al.* (1997). The dynamic structure factor $S_{nn}(q, \omega)$ for ^{4}He at 20 atm for the maxon excitation ($q = 1.13$ Å^{-1}) is shown for six different temperatures T (given on the figure). The neutron-scattering data are given by the symbols (Talbot *et al.*, 1988) and the dashed curves show the result of the two-pole approximation (eqn 3.9) to the scattering. The bottommost panel corresponds to scattering in the normal-fluid phase where the two-pole approximation describing the extended sound modes of the normal-fluid phase gives an excellent description of the data. The differences between the model and the data in the other panels are attributable to the multiphonon component, which can be seen to weaken with increasing temperature, and to disappear once the superfluid transition has been crossed ($T_\lambda = 1.88$ K at this pressure).

accompanied by a much more modest change in the driving force f_{un}. There is no indication that the damped harmonic oscillator model needs to be expanded to include the $f_{u\sigma} - z_\sigma$ part of the decay tree (Fig. 2.14), implying that the ratio $f_{u\sigma}/z_\sigma$ is very small. At the lowest temperature, the extended sound mode excitations have evolved into the phonon–roton excitations characteristic of the superfluid phase. At the same time, a new component at higher energies than the phonon–roton excitations can be seen to emerge. This is the multiphonon component, which is unique to the superfluid phase. Other changes are that the hybridization gap between the phonon–roton

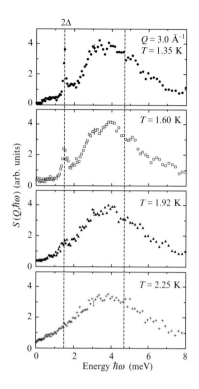

Fig. 9.13 Figure reproduced with permission from Glyde *et al.* (1998). The temperature dependence of the elementary excitations corresponding to the flat part of the phonon–roton dispersion curve. The sharp excitation located at twice the roton energy 2Δ (leftmost dashed vertical line) can be seen to broaden rapidly upon raising the temperature, and it is no longer present in the scattering in the normal-fluid phase ($T > T_\lambda = 2.17$ K). The temperatures are given in the figure, and the rightmost dashed line is the position of the recoil energy.

dispersion curve and the recoil curve opens up once the superfluid phase is reached upon cooling (Fig. 9.9). In the following section we explain how all of these changes are a natural consequence of what happens to the extended sound mode excitations in the absence of damping.

9.5   Perturbation theory for the extended sound modes in ^{4}He

The continuous change from ordinary density fluctuations to elementary excitations on lowering the temperature into the superfluid phase opens up the possibility of calculating the dispersion of the phonon–roton excitations using perturbation theory. What is required for this is a guess of what the excited-state wavefunction looks like as a function of q; we do not need to know the ground-state wavefunction. In the following, we denote the (unknown) ground-state wavefunction by $|0>$, the excited-state wavefunction by $|\psi_q>$, and the initial guess for this excited-state wavefunction

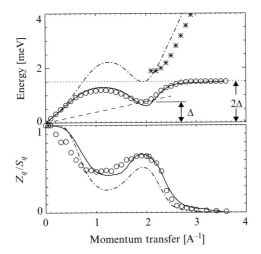

Fig. 9.14 Top: the phonon–roton dispersion curve of superfluid ^{4}He at $T < 1.3$ K (circles). The roton excitation is given by the minimum of the dispersion curve at $q_r = 1.94$ Å^{-1}, with energy gap Δ. For q values beyond the roton, the dispersion curve asymptotically approaches 2Δ (dotted line) and terminates at about $2q_r$. The stars indicate the peak positions of the recoil component of the scattering. The uncorrected excited state energies $\hbar\epsilon_F(q)$ are given by the dashed-dotted curve, and the corrected energies (self-consistent calculation) by the solid line. Bottom: corresponding weights Z_q of the phonon–roton excitations (circles: experiment; solid line: self-consistent calculation; dashed-dotted line: simple model)

by $|q>$. When the states are not properly orthonormalized, we shall denote them by $|...)$ as opposed to $|... >$.

Following Feynman (1972), our initial guess for the excited-state wavefunction is a density fluctuation created out of the ground state. We will also take $T = 0$ K, so as to keep matters tractable. Later on, we shall make qualitative statements about what is expected to happen upon raising the temperature. It can be shown that our initial guess is the proper choice based on the variational principle (Feynman, 1972), however, we have already seen that the extended sound modes (which are propagating density fluctuations) become the elementary excitations of the superfluid. Hence, a very natural guess for the excited state is

$$|q) = \hat{n}_q^+|0> \equiv \frac{1}{\sqrt{N}}\sum_{j=1}^{N} e^{-i\vec{q}.\hat{\vec{r}}_j}|0>. \tag{9.7}$$

Here, the operator $\hat{n}_q^+$ creates a plane-wave density fluctuation out of the ground state with momentum $\hbar q$ and energy $\hbar\epsilon_F(q)$. Normalization of eqn 9.7 yields the expression for $|q>$:

$$|q> = \frac{|q)}{\sqrt{(q|q)}} = \frac{|q)}{\{\sum_{j,l=1}^{N} <0|e^{i\vec{q}.\hat{\vec{r}}_j}e^{-i\vec{q}.\hat{\vec{r}}_l}|0> /N\}^{1/2}} = \frac{|q)}{\sqrt{S(q)}}. \tag{9.8}$$

This initial guess is not an eigenfunction of the Hamiltonian H, implying that this state would be allowed to decay, which cannot happen for an elementary excitation. The energy of the state $|q>$ is given by[3]

$$\hbar\epsilon_F(q) = <q|H|q> = <q|\sum_{k=1}^{N} \frac{-\hbar^2\nabla_k^2}{2m}|q> = \frac{\hbar^2 q^2}{2mS(q)}. \tag{9.9}$$

We show the energy corresponding to $|q>$ in Fig. 9.14, along with the measured phonon–roton excitation energies. The agreement at the lowest q values is rather good, reflecting the linear behavior in this q range of the static structure factor $S(q) \sim q$. In other words, phonons are well described by the initial guess for the excited state. This is not at all surprising given that long-wavelength phonons in liquids are very similar to their counterparts in solids, which in turn are well described as plane waves. With increasing q, the agreement between $\epsilon_F(q)$ and the measured dispersion deteriorates; $\epsilon_F(q)$ overestimates the observed excitation energies, which is of course to be expected for a variational guess of the excited-state wavefunction. Note that the actual dispersion and $\epsilon_F(q)$ both show a minimum at the same q value, reflecting that it costs less energy to create a density fluctuation with a wavelength that matches the average separation between the atoms. At the very highest q values, $S(q)$ approaches 1, and $\epsilon_F(q)$ reduces to the recoil energy $\hbar\omega_{\text{recoil}} = \hbar^2 q^2/2m$ (see eqn 2.27). As mentioned in Section 2.2.3, the peak positions (in energy) of $S_{nn}(q,\omega)$ of liquid helium do indeed (almost) approach this limit, lagging behind by a small amount $\hbar\omega_0/2$.

Overall, $\epsilon_F(q)$ represents a good description at the lowest and the highest q values, but in between it does not do a very good job. Moreover, it cannot account for the shape of the dispersion curve beyond the roton, namely the flat part of the dispersion shown in Fig. 9.14, nor can it account for the hybridization between the phonon–roton curve and the recoil line. The strength of this excitation in scattering experiments is given by the overlap between the excited state $|q>$ and the density fluctuation caused by the probing particle (such as a neutron). Since the density fluctuation at $T = 0\,\text{K}$ due to the probing particle is given by $\hat{n}_q^+|0> = |q)$ (Squires, 1994; Lovesey, 1971) we find for the initial guess for the strength of the excitation $Z^0(q)$ (Manousakis and Pandharipande, 1984):

$$Z^0(q) = |(q|q>|^2 = S(q). \tag{9.10}$$

We show the ratio of the measured strength of the phonon–roton excitations $Z(q)$, scaled by the static structure factor $S(q)$ in Fig. 9.14. Equation 9.10 predicts a ratio of 1, whereas the experiment shows that the ratio is only close to 1 at the lowest q values. It is also clear from this figure that the predictions at the highest q values only pertain to the recoil curve, there is no predictive value in Eqs. 9.9 and 9.10 when it comes to the flat part of the measured dispersion.

Using standard perturbation theory, we can improve our calculations for the excited-state energy and in doing so we end up with a remarkably good agreement between

---

[3] Partial integration of the integral contained in the ensemble average yields that only the kinetic term is important in determining the energy (Feynman, 1972).

calculations and experiment. This agreement not only extends to the excitation energies, but also to the strength of the signal in scattering experiments. In order to improve our guess of the excited-state wavefunction, we include states that have two density fluctuations in them: $|lm) = \hat{n}_l^+ \hat{n}_m^+ |0>$. This unnormalized state has a total momentum of $\hbar(\vec{l} + \vec{m})$. We orthonormalize this state to $|q>$ in order to be able to use it in our perturbation expansion:

$$|lm> = \frac{|lm) - <q|lm)|q>}{\{(lm|lm) - |(lm|q>|^2\}^{1/2}}. \tag{9.11}$$

We can now replace the single (quasi-)particle propagator $G_1^0(q,\omega)$ defined as

$$G_1^0(q,\omega) = \frac{1}{\omega - i0^+ - \epsilon_F(q)} \tag{9.12}$$

by the corrected expression

$$G_1(q,\omega) = \frac{1}{\omega - \epsilon_F(q) - \Sigma(q,\omega)}. \tag{9.13}$$

We sketch in Fig. 9.15 what the above correction amounts to. Since $|q>$ is not an eigenfunction of the Hamiltonian, we have that the matrix elements $<lm|H|q>$ are not zero. Hence, a one-quasi-particle state $|q>$ can decay into a two-quasi-particle state $|lm>$. An admixture of the states $|q>$ and $|lm>$ should produce a better estimate for the excited states, with a lower excitation energy than $\epsilon_F(q)$. This amounts to replacing the single particle propagator $G_1^0(q,\omega)$ associated with the state $|q>$ with $G_1(q,\omega)$ as shown in Fig. 9.15. The Dyson series shown in this figure is easily summed using energy and momentum conservation at all the vertices to give eqn 9.13, with the self-energy $\Sigma(q,\omega)$ given by

$$\Sigma(q,\omega) = \frac{1}{2} \sum_{l,m} \frac{\Delta(\vec{q} - \vec{l} - \vec{m})| <lm|H|q>|^2}{\omega - \epsilon_F(l) - \epsilon_F(m) - \Sigma(l,\omega - \epsilon_c(m)) - \Sigma(m,\omega - \epsilon_c(l))}. \tag{9.14}$$

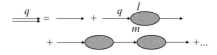

Fig. 9.15 The basic perturbation theory equation. The initial guesses for the excited states $|q>$ are modified through the inclusion of multiparticle states $|lm>$. This (graphical) equation can be solved once the function $<lm|H|q>$, which allows for the decay of the single quasi-particle state $|q>$ into the multi-quasi-particle state $|lm>$, is known. The single arrows denote the uncorrected single-excitation propagators of momentum $\hbar q$, the double arrow represents the corrected propagator of eqn 9.13. The shaded balloons are the self-energies given in eqn 9.14.

Here, the corrected excitation energies $\epsilon_c(q)$ that enter the equation through a recursive scheme are given by the poles of eqn 9.13. In this approximation, the new estimate $|\psi_q\rangle$ for the (unnormalized) exited-state wavefunction is given by

$$|\psi_q\rangle = |q\rangle + \frac{1}{2}\sum_{l,m} |lm\rangle\langle lm|H|q\rangle \,\Delta(\vec{q} - \vec{l} - \vec{m})G_2(lm, \omega = \epsilon_c(q)). \qquad (9.15)$$

Here, the two-particle propagator $G_2(lm, \omega)$ is implicit in eqn 9.14 (Manousakis and Pandharipande, 1986):

$$G_2(lm, \omega) = \frac{1}{\omega - \epsilon_F(l) - \epsilon_F(m) - \Sigma(l, \omega - \epsilon_c(m)) - \Sigma(m, \omega - \epsilon_c(l))}. \qquad (9.16)$$

The only thing that is left to do is to solve the above set of equations, for which we need expressions for the matrix elements $\langle lm|H|q\rangle$. Reasonable approximations for these (Chang and Campbell, 1976) can be obtained using the convolution approximation (Lee and Lee 1975) or the Kirkwood superposition approximation (Jackson and Feenberg, 1962). Using these approximations (and generalizations), remarkably accurate results have been obtained for the dispersion curve from the lowest q values up to the point where the dispersion flattens off at $q = 2.6$ Å^{-1} (Lee and Lee, 1975; Manousakis and Pandharipande, 1984). We show an example of such a calculation in Fig. 9.16. For higher q values, the perturbation results approach the recoil curve, while the flat part of the phonon–roton curve is not reproduced. In the following, we show how in fact the entire phonon–roton dispersion curve, including the flat part, can be reproduced from perturbation theory.

In order to understand the phonon–roton dispersion curve up to its termination point at twice the roton wave number and at twice the roton energy (Fig. 9.14) we first make some grossly simplifying approximations for $G_2(lm, \omega)$ that essentially remove all the difficult bits, but retain the overall structure of the solution. This simplification explains why one finds multiple solutions at the higher q values (the recoil curve and the flat part of the dispersion). Later on, we will return to the full solution. The poles of eqn 9.13 yield the excitation frequencies $\epsilon_c(q)$ and damping rates Γ_q:

$$\epsilon_c(q) + i\Gamma_q = \epsilon_F(q) + \Sigma(q, \epsilon_c(q) + i\Gamma_q). \qquad (9.17)$$

The self-energy, as given by eqn 9.14 will exhibit strong resonances for those frequencies where there are many combinations of l and m such that $\epsilon_c(m) + \epsilon_c(l)$ add up to the same energy. If this is the case, then there are many terms in the summation in eqn 9.14 that display a similar functional dependence near $\omega = \epsilon_c(m) + \epsilon_c(l)$. Of course, we expect many combinations of l and m adding up to the same total energy and producing resonances in $\Sigma(q, \omega)$ for l and m values originating from the minimum and the maximum of the dispersion. Thus, we would expect to find resonances in $\Sigma(q, \omega)$ at energies given by twice the roton energy (2Δ), twice the maxon energy ($2M$) and at the roton plus the maxon energy ($M + \Delta$).

In addition, at low temperatures the damping rates of the excitations are very small, so that to very good approximation we can approximate the imaginary part of the

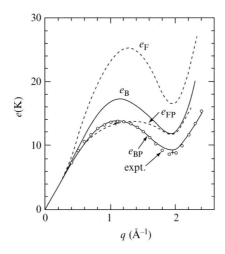

q (Å^{-1})

Fig. 9.16 Figure reproduced with permission from Manousakis and Pandharipande (1984). The measured phonon–roton dispersion curve (symbols) and the results from various perturbation theory attempts. The solid line that almost goes through the data points is the calculation by Manousakis and Pandharipande using the experimental $S(q)$ values as input. The agreement between calculation and experiment regarding the strength of the excitation $Z(q)$ is slightly less good (Manousakis and Pandharipande, 1984) than the excellent agreement for the excitation energies for $q < 2.6$ Å^{-1}. The dashed curve (labelled e_{FP}) is the result when $S(q)$ is also determined (from scratch) by a variational method. e_F is the uncorrected Feynman estimate of eqn 9.9, while e_B is the result of the calculations of Feynman and Cohen (1956) using an improved trial wavefunction that includes backflow.

self-energy of an excitation by Γ (independent of q^4), and as a result we have that $\Sigma(l, \omega - \epsilon_c(m)) \simeq \Sigma(l, \epsilon_c(l)) + i\Gamma$ since $\omega \simeq \epsilon_c(m) + \epsilon_c(l)$ for those regions where $\epsilon_c(m) + \epsilon_c(l)$ add up to the same energy. With this, we rewrite the two-particle propagator as

$$G_2(lm, \omega) = \frac{1}{\omega - \epsilon_F(l) - \epsilon_F(m) - \Sigma(l, \epsilon_c(l)) - \Sigma(m, \epsilon_c(m))}$$

$$= \frac{1}{\omega - \epsilon_c(l) - \epsilon_c(m) - 2i\Gamma}. \tag{9.18}$$

We are not done with our simplifications yet. Since the roton corresponds to the strongest resonance, we restrict the summation over all possible l and m to $l \approx q_{\text{roton}}$ and $m \approx q_{\text{roton}}$. In this very limited range, we take the matrix elements $< lm|H|q >$ to only depend on q: $< lm|H|q > \approx M_{\text{eff}}(q)$. And as a *coup de grace*, we anticipate the outcome of our calculations, and we replace the calculated excitation frequencies $\epsilon_c(q)$

[4] The independence of the damping rate on q for the phonon, maxon and roton excitations is a reflection of the Landau–Khalatnikov damping mechanism at low temperatures $T < 2$ K, and an illustration of this effect can be seen in Fig. 9.11.

by their measured counterparts $\epsilon_{\text{obs}}(q)$ in the evaluation of the self-energy. Thus, our resulting expression for the self-energy is

$$\Sigma(q,\omega) = \frac{1}{2} \sum_{l,m\approx q_{\text{roton}}} \frac{|M_{\text{eff}}(q)|^2 \Delta(\vec{q} - \vec{l} - \vec{m})}{\omega - \epsilon_{\text{obs}}(l) - \epsilon_{\text{obs}}(m) - 2i\Gamma}. \tag{9.19}$$

With all these simplifications, which turn out to actually be not too restrictive because of the overwhelming influence of the roton excitation, we can now determine the strengths of the poles of G_1 by solving a much simplified equation:

$$\omega + i\eta = \epsilon_{\text{F}}(q) + \Sigma(q, \omega + i\eta). \tag{9.20}$$

We show the graphical solution to this equation in Fig. 9.17 by plotting the real parts of both sides of eqn 9.20 for two q values beyond the roton excitation. From this figure we see that there are in fact two solutions to eqn 9.20. For large q, one of the solutions will be close to the initial guess of $\epsilon_{\text{F}}(q)$, while the other solution will be determined by the sharp feature in $\Sigma(q,\omega)$ that occurs at $\omega = 2\Delta$. This sharp feature

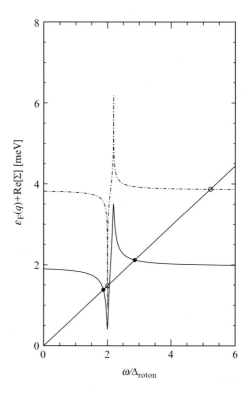

Fig. 9.17 Graphical solution of the roots of eqn 9.20 for q = 2.3 (solid line) and 2.7 Å^{-1} (dashed line). Only the real part of the equation is depicted here. In each case there are two solutions, shown by circles. The apparent third solution does not satisfy the imaginary part of the equation. Δ_{roton} stands for the roton excitation energy.

is the resonance in the self-energy attributable to the roton minimum of the dispersion curve. This latter solution yields the flat part of the dispersion curve, and its energy is virtually independent of the exact amplitude of $\Sigma(q, \omega)$ given the sharpness of the resonance at low temperatures. We mention the latter since in order to solve eqn 9.20 we need to know $M_{\text{eff}}(q)$. In our case, we fixed the value of $M_{\text{eff}}(q)$ in such a way that we would get the correct excitation energy for the lower (in energy) solution. While this last simplification may sound like it renders our efforts totally useless, bear in mind that we only use eqn 9.20 to explain the emergence of two solutions, and to qualitatively understand the flat part of the dispersion beyond the roton. As noted, the appearance of a solution at $\omega = 2\Delta$ is independent of the actual value of $M_{\text{eff}}(q)$.

The simple model demonstrates that the flat part of the phonon–roton dispersion curve originates from a two-roton resonance and it shows how two solutions will be obtained with the second (higher in energy) solution approaching the uncorrected excitation energy $\hbar \epsilon_F(q) = \hbar^2 q^2 / (2mS(q))$. It also makes it immediately clear why the phonon–roton curve not only flattens out at an excitation energy level of 2Δ, but also why it terminates at twice the roton wave number $2q_{\text{roton}}$. For larger q values, the requirement $\Delta(\vec{q} - \vec{l} - \vec{m})$ in eqn 9.14 can no longer be satisfied, and hence, the resonance will disappear from $\Sigma(q, \omega)$ and the dispersion will terminate at $2q_{\text{roton}}$.

There is one final thing that we can calculate using this simple model, and that is the strength $Z(q)$ of the pole in $S(q, \omega)$ (which gives the intensity in a scattering experiment). Within our approximations, including the one that fixes $M_{\text{eff}}(q)$ by imposing that the lower (in energy) solution matches the observed excitation energy, we can evaluate the excited-state wavefunction $|\psi_q >$ and the strength of the pole as (Manousakis and Pandharipande, 1984)

$$Z(q) = |(q|\psi_q >|^2. \tag{9.21}$$

We show the results in the lower panel of Fig. 9.14 by the dashed-dotted curve. The agreement between the simple model and the measured intensities is remarkably good given the utter simplicity of the model. It correctly predicts the overall q dependence of the intensity of the phonon–roton dispersion curve over the entire range, and it demonstrates that the weakening of $Z(q)$ with increasing q is a direct result of the difficulty of satisfying $\Delta(\vec{q} - \vec{l} - \vec{m})$ given that this is the only q dependence that is left in eqn 9.20.

We now repeat the calculations of the simple model, but we drop the restriction on the l and the m values that enter eqn 9.14, and we will use reasonable approximations for the matrix elements $< lm|H|q >$. In this way, we can solve eqn 9.14, and we can also calculate $\epsilon_c(q)$ since this solution is no longer fixed by an requirement on $M_{\text{eff}}(q)$. We do keep the simplification that we use the observed excitation energies in the determination of the two-particle propagator.[5] For $q < 2$ Å^{-1} we use the convolution approximation to determine $< lm|H|q >$, at higher q values we use the Kirkwood superposition approximation. The $q = 2$ Å^{-1} boundary was chosen because both approximations yield the same result at this q value. This procedure is very similar

---

[5] In other words, we solve the equations in a self-consistent manner, anticipating the outcome; we use the observed energies as a shortcut to a long, iterative solution process.

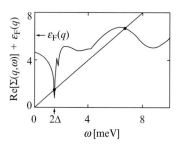

Fig. 9.18 Figure reproduced with permission from Montfrooij and Svensson (2000). Graphical solution of the real part of eqn 9.14 for $q = 3.3$ Å^{-1} at $T \sim 1$ K using the Kirkwood approximation for the matrix elements $< lm|H|q >$. The two solutions (circles) are given by the intersection of the line $\omega = \omega$ and the line $\epsilon_F(q) + \mathrm{Re}[\Sigma(q,\omega)]$. Note the appearance of a solution at the two roton resonance 2Δ, well away from the initial estimate $\epsilon_F(q)$ (horizontal arrow).

to what has been done in the literature (Jackson and Feenberg, 1962; Lee and Lee, 1975; Manousakis and Pandharipande, 1984). We show the results for this procedure in Fig. 9.18 for $q = 3.3$ Å^{-1}. From this figure we see that this much more realistic model that does not contain any adjustable or fudged parameters, is actually not that different from our simple model. We still find a solution to eqn 9.13 at $\omega = 2\Delta$ and one at $\omega \approx \epsilon_F(q)$. In Fig. 9.14 we show the corresponding results for the excitation energies, as well as for the strength of the pole (eqn 9.21).

The agreement between the model and the observed excitation energies and intensities of the phonon–roton curve is truly impressive. After all, helium is a superfluid that harbors a Bose-condensate. All we have done is to use perturbation theory, starting with density fluctuations that are present in the normal-fluid phase as our initial guess for the excited-state wavefunction, and apply straightforward perturbation theory to reach a better estimate of the wavefunctions. We short-circuited the iterative nature of the solution by simply evaluating the self-energy using the knowledge of the ultimate solutions, thereby solving the equations in a self-consistent manner. So, in effect, the only input necessary to the results[6] shown in Fig. 9.14 is the static structure factor $S(q)$.

Another thing we can understand based on the calculations using perturbation theory is the appearance of a multiphonon component in the scattering spectra. As shown in Fig. 9.8, the superfluid phase has excitations at energies higher than the phonon–roton dispersion curve, and this so-called multiphonon component displays quite a bit of structure in (q, ω)-space. The normal-fluid phase does not display a multiphonon component, at least not as far as we know. In Fig. 9.19 we sketch why this component appears in the superfluid phase. Detailed calculations of the shape of the multiphonon component have been carried out by Manousakis and Pandharipande (1986), here we merely wish to elucidate why it is present in the first place.

---

[6] The self-energy can also be evaluated using the uncorrected excitation energies as input. This would require more work, but from the results of Lee and Lee (1975) we already know that it can be done.

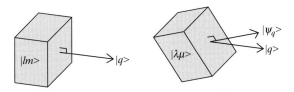

Fig. 9.19 Pictorial illustration of how an excited (single quasiparticle) state $|\psi_q>$, which does not coincide with a simple density fluctuation $|q>$, necessarily leads to the observation in scattering experiments of multi-quasi-particle states $|\lambda\mu>$.

The reader will probably find the fact that the multiphonon component displays energy bands related to 2Δ, $2M$ and $M+\Delta$ resonances not too far fetched. In the normal-fluid phase we observe the fate of density fluctuations $|q>$; these fluctuations make up the entire cross-section in a scattering experiment, the two-quasi-particle states $|lm>$ are perpendicular to what can be observed by means of scattering experiments (left panel of Fig. 9.19). The situation is different in the superfluid phase. As we have seen, the two-quasi-particle states $|lm>$ mix in with the one-quasi-particle states $|q>$ to give the excited-state wavefunction $|\psi_q>$. By the same token, the two-quasi-particle states $|lm>$ now also have an admixture of the $|q>$ states to produce two-quasi-particle states $|\lambda\mu>$ that are perpendicular to $|\psi_q>$. This rotation of the $|lm>$ states is sketched in the right-hand panel of Fig. 9.19. Because the $|\lambda\mu>$ states now have the $|q>$ states mixed in, they are directly visible in scattering experiments with a cross-section given by $|<\lambda\mu|q>|^2$. So ultimately, this multiphonon component is visible because of the rotation of the states in the space spanned by $\{|lm>, |q>\}$, and this rotation is caused by the resonances present in the self-energy. These resonances disappear once the excitations stop being elementary excitations since the sharpness of the resonances is directly linked to the sharpness of the underlying excitations. Therefore, once the underlying excitations acquire a substantial width (which occurs upon raising the temperature as shown in Figs. 9.10 and 9.11), then not only does this affect the flat part of the dispersion, it also rotates the two-quasiparticle states $|\lambda\mu>$ back to the states $|lm>$, rendering the multiphonon component invisible. It is this disappearance of the multiphonon component that shows up as an increase in the value for $f_{un}(q)$ upon raising the temperature through the superfluid transition, as shown in Fig. 9.11.

In fact, the character of the excitations close to the termination point of the phonon–roton dispersion curve at $\omega = 2\Delta$ and $q = 2q_{roton}$ is that of a multiphonon excitation. The strength of the poles is such that it produces a very weak signal in scattering experiments. This implies that the eigenstate for this solution has only a very small component on $|q>$ and therefore, it must consist almost exclusively of $|lm>$ states. Thus, we should view these states near the termination point as two-roton states, very similar to the two-roton states (Ruvalds and Zawadowski, 1970) that have been observed at low momentum transfers in neutron-scattering (Svensson *et al.*, 1976) and in light-scattering experiments (Murray *et al.*, 1975).

Finally, the presence of the gap between the phonon–roton excitations and the recoil curve at the lowest temperature also directly follows from perturbation theory. The resonance in the self-energy at $\omega = 2\Delta$ results in two solutions that repel each other, with one ending up at an energy below 2Δ, and one above it. One can see this branch repulsion even when using the simple model at $q = 2.3\,\text{Å}^{-1}$ as shown in Fig. 9.17. As before, once the resonance in the self-energy disappears upon raising the temperature, then this branch repulsion will cease, and the recoil curve will smoothly link up the extended sound mode dispersion, as borne out by the data shown in Fig. 9.9.

We summarize the findings of this chapter for the superfluid state of ^{4}He at $T = 0\,\text{K}$ in Fig. 9.20. Many ingredients come into play to give the superfluid phase its unique characteristics. First, because of Bose statistics an energy gap will be present in the fluid. This energy gap causes the fluid to become a superfluid at low enough temperatures since the Bose-condensate will renormalize the dispersion of the low-q excitations from a quadratic one into a linear one. Meanwhile, the gap at $q = q_{\text{roton}}$ ensures that a minimum flow velocity is required in order for liquid flow to slow down. Hence, the presence of a gap is synonymous with superfluidity. The gap also ensures that the excitations will become elementary excitations; at low temperatures there will be very few thermally excited fluctuations, and hence once excitations form they cannot decay though collision with other excitations. Had the gap not been there, then the fluid would have accommodated enough thermally activated excitations to limit the lifetime of the newly formed excitations. In other words, the excitations would have been damped, they would not have been elementary excitations.

Because of the fact that the excitations are virtually undamped at $T = 0\,K$ as a direct consequence of the presence of an energy gap that in turn is a direct consequence of Bose statistics, we can expect to see resonances appear in the self-energy. And because of the resonances in the self-energy, we can expect to see the existence of a flat part to the phonon–roton dispersion, a hybridization gap between this dispersion

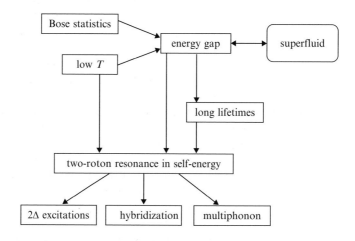

Fig. 9.20 Schematic diagram to illustrate how the extended sound modes become the elementary excitations of the superfluid phase, and how they give rise to the unique density fluctuations observed in the superfluid state. The defining feature is the appearance of the two-roton resonance, which in turn arises because of the presence of an energy gap.

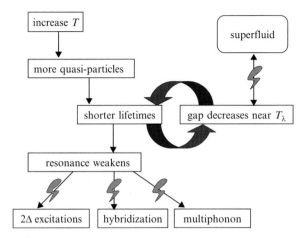

Fig. 9.21 Upon raising the temperature, more quasi-particles become excited. This leads to an increased damping, which will lead to a lowering of the energy gap. This starts the feedback mechanism where the smaller energy gap allows for yet more quasi-particles to be excited, and the damping rate increases exponentially until the normal-fluid phase is reached.

and the recoil curve, as well as the appearance of a multiphonon component to the scattering. In other words, superfluids are not that special, they merely house dressed-up density fluctuations. These fluctuations are the very same density fluctuations that we have been referring to as extended sound modes in non-superfluids except that in the case of superfluid ^{4}He these extended sound modes now exist in the absence of a damping mechanism. In short, we can feel vindicated for having focused so much on the extended sound modes in normal liquids, and for having paid special attention to their behavior in the region $q\sigma \approx 2\pi$ since these sound waves transform themselves in the all-determining roton excitation of the superfluid phase.

Based upon our findings at $T = 0\,K$ for the superfluid phase, and the observed behavior of the excitations in going from the superfluid to the normal-fluid phase, we can conjure up a qualitative picture of what drives the superfluid to normal fluid transition. We sketch this in Fig. 9.21. When the temperature is raised from the lowest values, then one observes an increase in damping rate of all the elementary excitations (Mezei and Stirling, 1983). In fact, the damping rate of the excitations does not show a significant q dependence, as shown in Fig. 9.11, and as known from the work by Woods and Svensson (1978). Highly accurate measurements of the linewidth of the roton excitation show that this linewidth (at low T) can be well understood as being caused by Landau–Khalatnikov damping (Landau and Khalatnikov, 1949; Bedell *et al.*, 1984), as shown in Fig. 9.22. This is a process where the damping rate of the quasi-particles is entirely determined by the number of thermally activated quasi-particles present in the liquid at a given temperature.[7] The quantitative predictions for the

---

[7] The damping rate is given by a four-quasi-particle process, in which the quasi-particle of interest collides with a thermally excited quasi-particle, resulting in two new quasi-particles. The

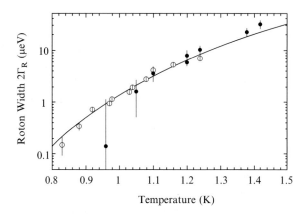

Fig. 9.22 Figure reproduced with permission from Andersen *et al.* (1996). Shown are the measured linewidths for the roton excitation [open circles – Andersen *et al.*(1996); closed circles – Mezei (1980)], compared to the Landau–Khalatnikov damping prediction [solid line – Bedell *et al.* (1984)]. The essentially perfect agreement demonstrates that the damping rate of the excitations in the superfluid phase is given by the number of quasi-particles that are thermally excited. Note the logarithmic scale on the vertical axis.

Landau–Khalatnikov damping mechanism are only valid when the number of excited quasi-particles is limited, however, we can qualitatively extend this damping process to temperatures increasingly closer to the superfluid-transition temperature.

When the temperature is raised, more quasi-particles will pop into existence adding to the damping rate z_u since this rate is given by the number of collisions between the quasi-particles. Once this damping rate becomes significant compared to the force f_{un}, then this will result in a lowering of the excitation energy $\omega_s = \sqrt{f_{un}^2 - (z_u/2)^2}$. This effect will be most pronounced for the roton excitation since this excitation corresponds to the smallest value of f_{un}, at least in the q region where the excitations are not ordinary long-wavelength sound modes that we know can propagate even in the normal-fluid phase. The lowering of the roton excitation energy because of the increased damping it experiences will have an amplifying effect on the damping rate. When the roton excitation energy is lowered, even more excitations (pre-dominantly rotons) will become thermally activated, producing in an even larger damping rate, producing a further lowering of the excitation energy. In other words, once the damping becomes significant, it will very rapidly result in a complete demise of the roton excitation. Thus, the excitation gap will close[8] (Fig. 9.10), and the roton excitation will become an ordinary extended sound mode.

accompanying prediction (Bedell *et al.*, 1984) of a very modest softening of the roton excitation caused by the increased damping rates is not borne out by experiment (Andersen *et al.*, 1996). It is possible that this could be a consequence of the anticipated increase in propagation frequency (Figs. 9.10 and 9.11) caused by the weakening of the multiphonon component.

[8] Whether the excitation energy actually softens all the way, or whether the excitation merely becomes very strongly damped is not all that relevant; the main effect is that the flowing fluid will have a substantial amount of (smeared out) states available even at the lowest flow velocities.

The multiphonon component and the flat part of the dispersion curve will be affected even more. The damping rate of the excitations in the flat part of the dispersion is actually twice that of the roton, as follows from the simple model (solution to eqn 9.20). Hence, these excitations will broaden very rapidly upon raising the temperature, and they will disappear from the spectra before the roton excitation becomes less well defined. This can be seen to happen in the data shown in Fig. 9.13. The disappearance of the multiphonon component at all q values is reflected in the increase of $f_{un}(q)$ upon approaching the normal-fluid phase (Fig. 9.11). Again, this is as anticipated in the presence of substantial damping since the visibility of the multiphonon component is directly related to the presence of the two-roton resonance in the self-energy, and this resonance rapidly disappears with increased damping. Finally, the hybridization gap, which is also a direct consequence of the presence of the two-roton resonance, is seen to disappear from the scattering spectra upon raising the temperature in Fig. 9.9.

In summary, the extended sound modes that we have followed in simple liquids and liquid metals, are also present in the superfluid phase. Not only do they dominate the spectra as is the case in very cold, but still normal, fluids; because of the emergence of the energy gap in the excited-state spectra they now acquire the character of elementary excitations. But deep down, they are still extended sound modes. The remainder of the features that make the superfluid phase unique (multiphonon component, hybridization gap and the flat part of the dispersion) are straightforward consequences of what happens to extended sound modes in the absence of damping. All observations are consistent with our assertion that the elementary excitations of the superfluid phase are extended sound modes. It would appear from the effective eigenmode analysis that the Bose–condensate is not directly visible in the excited states that make up the phonon–roton dispersion curve. This is unlike what one might expect based on the marked differences (at first sight) between the spectra of the normal–fluid and of the superfluid phases, and based upon an intuitively appealing qualitative scenario that was put forward by Glyde and Griffin (Glyde and Griffin, 1990; Glyde, 1992). However, note that the presence of the condensate is still required to ensure that the energy gap is relevant. Without the condensate, we would have excitations at low q with a quadratic dispersion that would allow for a damping mechanism, which in turn would prevent the two-roton resonance from forming.

10
Summary and outlook

10.1 Summary

We have mapped out the behavior of the extended eigenmodes from classical liquids to quantum fluids, from simple liquids to liquids metals. We have shown how to determine the characteristics of these modes out of the experimental data using the effective eigenmode formalism that is valid for classical liquids and quantum fluids alike. We have seen that the extended sound modes are always present outside of the hydrodynamic region. In simple liquids like Ar these extended sound modes are hidden (for most q values outside of the hydrodynamic region) in the broad quasi-elastic shape that is $S_{nn}(q, \omega)$. Depending on the density, the extended sound modes become strongly damped, or even diffusive and overdamped in the region where the static structure factor reaches a maximum. It is possible to make the sound modes visible to the naked eye by reducing the width of the central feature related to the heat mode. This can be achieved by mixing in heavier atoms, such as is done in He–Ne and Ar–Kr binary mixtures.

The extended sound modes are prominent features in liquid metals and very cold liquids because of the weakness of the coupling between the microscopic longitudinal velocity ('u') and the microscopic temperature ('T'). This coupling strength determines the strength of the quasi-elastic feature related to the transport of heat. This heat transport is of course not the only way in which a fluid can produce a central line; after all, even the simple viscoelastic model that has zero coupling between 'u' and 'T' predicts a quasi-elastic line. However, the central feature in this viscoelastic model is not as dominant. The overall behavior of the extended sound modes is very similar in all cases, from simple liquids to liquid metals. The sound modes propagate at a higher speed than the adiabatic sound velocity in the region $q_H < q < q_{max}/2$, with q_H the upper q limit of the hydrodynamic range and q_{max} the peak position of the main maximum of the static structure factor $S(q)$.

The causes behind this faster than hydrodynamics propagation can differ between the various liquids. First, mode-coupling effects result in a large-scale backflow, increasing the speed at which these modes can propagate. Such mode-coupling effects have been observed in simple liquids (van Well *et al.*, 1985), liquid metals (Pilgrim and Morkel, 2002) and superfluid ^{4}He (Woods and Cowley, 1973; Kirkpatrick, 1984). Secondly, the speed of sound propagation is expected to increase upon entering into the collisionless regime. This happens both in the viscoelastic model (Fig. 2.15), as well as in the extended hydrodynamics model (Fig. 2.12). We would expect to observe a transition from ordinary hydrodynamic sound with propagation frequency

$[f_{un}^2 + f_{uT}^2]^{1/2}$ to (undamped propagation frequencies) $[f_{un}^2 + f_{u\sigma}^2]^{1/2}$ or to $[f_{un}^2 + f_{uT}^2 + f_{u\sigma}^2]^{1/2}$, respectively. Note that these are the undamped frequencies; because of fairly substantial damping, the actual propagation frequencies can be considerably lower than, for instance, $[f_{un}^2 + f_{u\sigma}^2]^{1/2}$. The transition to a viscoelastic, collisionless region has been observed in most liquid metals, as well as in water. Also note that historically in these systems the undamped frequencies – as determined from the peaks of the current–current correlation function – are presented as opposed to the frequencies defining the poles of the dynamic susceptibility, rendering the interpretation of the scattering results a little more difficult.

In superfluid ^{4}He, the extended sound modes become the elementary excitations of the liquid. In the absence of a heat mode (absent because of the zero-point motion) and in the absence of damping (absent because of Bose statistics) the extended sound modes now show up as very sharp (in energy) features in $S_{nn}(q, \omega)$. This sharpness leads to the emergence of resonances, which in turn make the multiphonon component visible in scattering experiments. In all, the extended sound modes are very similar between all liquids.

The extended heat mode is less well understood, partly because it changes in character with decreasing wavelength of the fluctuations. In the hydrodynamic region, the heat mode represents heat transport. With decreasing wavelength, the character of the heat mode changes in such a way that the momentum flux gets mixed in through the coupling parameter $f_{u\sigma}$. From scattering experiments alone we can no longer tell exactly what this quasi-elastic feature represents, other than that it is a continuous extension of the mode associated with heat transport in hydrodynamics. By using computer simulations to determine the other basic correlation functions between density and temperature, we can make an unambiguous identification of the heat mode. At present, there is a need for more such simulations in order to interpret the scattering results in liquid metals.

When the wavelength of the fluctuations is comparable to the interparticle spacing, then the extended heat mode dominates $S_{nn}(q, \omega)$ in simple liquids. In this wavelength range, the approximate expression for the decay time of the heat mode as calculated for a fluid of hard spheres yields a very good description of the characteristic width of $S_{nn}(q, \omega)$ both in simple fluids as well as in liquid metals. This not only confirms that the heat mode is the most important decay channel for density fluctuations, it demonstrates that the packing of the particles at liquid densities is the most important feature determining the decay of disturbances. When the temperature of the liquid is lowered, and provided the liquid remains liquid during this process, we observe that the heat mode becomes less important because of the zero-point motion of the particles, however, it is still present. Only in superfluids close to 0 K is the heat mode entirely absent because of the symmetry requirements of the wavefunction.

The extended heat and sound modes give an excellent description of the collective excitations that make up $S_{nn}(q, \omega)$ over the wavevector range covering hydrodynamics $q \approx 0$, to wavevectors beyond the peak position of the main maximum in $S(q)$. In other words, only three modes are required to describe – within the experimental accuracy – the decay of density fluctuations with wavelengths ranging from very large, to smaller than the interparticle separation. To explain the character of these three modes so that we know through which decay channels the density fluctuations decay, we need more than three microscopic variables. In simple liquids five such variables are sufficient. The same holds for liquid metals, but note that these might not be the same five microscopic variables that are required in simple liquids.

For wavelengths much shorter than the interparticle separation the effective eigenmode formalism loses its usefulness, and we should instead focus on single-particle motion as opposed to collective motion. Under these circumstances, we can describe $S_{nn}(q,\omega)$ by the motions of individual particles corrected for the fact that they are surrounded by other particles. Classically, we do this by incorporating the effects of binary collisions, quantum mechanically we do this by taking into account that the possible energy states that a particle can have are quantized because of its confinement.

From computer simulations we know that a three-mode description is not entirely accurate. Even for density fluctuations of long wavelength, we can expect to see a very fast, but very small decay of the correlation function because of the rattling motion of the particles inside the cage formed by their neighbors. The amplitude of this decay is very small in scattering experiments (less than 1%), but given the increasing accuracy of such scattering experiments, we should be able to observe this part of the decay directly before too long.

In our opinion, the main question in the dynamics of simple liquids remains unanswered: why do three modes suffice (at least up to the 99% level) in describing the decay of density fluctuations? Given the relevance of at least five microscopic variables in describing the dynamics of simple liquids, and given the many ways in which the couplings between those microscopic variables can combine to yield more than one pair of propagating modes (see Appendix D), why do we not observe more than one pair of such modes? What is it in a simple liquid that appears to prevent the existence of more than one pair of propagating modes? We do not have the answer to this question yet, nor are we even sure for that matter whether three modes are actually sufficient for all liquids given the potential observation of a second pair of propagating modes in very dense helium at 13 K (Montfrooij *et al.*, 1992*a*). Clearly, we need more scattering data and simulation work on potential systems where multiple pairs could exist (such as on the aforementioned state of helium, or on liquid nickel). It is likely that the explanation behind this mystery will involve the damping rates of the microscopic heat and momentum flux (z_q and z_σ) on length scales comparable to the interparticle separation. As discussed in various sections of this book, the frequency and effectiveness of binary collisions determine the decay rate for both the heat and momentum flux. Both decay rates appear to be of a very similar magnitude (albeit that this still has to be investigated in computer simulations on liquids different from Lennard-Jones systems), and both decay rates depend only weakly on the wavelength of the density fluctuation. These observations may prove to be the guiding lights in solving the three-mode mystery, or they might just prove to be manifestations of a deeper truth that so far still eludes us.

It would be nice to think that the decay rates z_q and z_σ of the heat and momentum flux hold the key to a microscopic theory for liquids. After all, based on the weak variation with q of these decay rates we can picture those rates in terms of binary collisions, and then view the overall wavelength dependence of the decay of density fluctuations as being pre-ordained by the q dependence of the coupling parameters f_{un}, f_{uT}, f_{Tq}, and $f_{u\sigma}$. This would present the sought-after connection between the macroscopic transport coefficients and the microscopic collisions between the atoms. Such a picture might have some merit in the description of simple liquids and could certainly serve as a scaffold for interpreting the scattering data in terms of microscopic motions, however, it would appear that it would not suffice in the description of liquid metals. Notwithstanding this, a study that would compare these decay rates between simple liquids and liquid metals would be most welcome.

10.2   Outlook

Trying to predict where or when the most exciting new insights into the behavior of liquids will take place is a fairly useless endeavor. It is safe to say that many new discoveries will be made at X-ray sources, and that most questions that need to be answered are in the field of liquid metals. Of course, that is not to say that there are no remaining questions left to be answered when it comes to other liquids such as binary mixtures, it is just more likely that liquid metals will receive more attention when it comes to doing scattering experiments. Molecular dynamics will continue to play a very important part in furthering our overall understanding of liquids, especially when analysis and interpretation of scattering experiments is combined with simulations that also determine the other basic correlation functions, such as the density–temperature correlation function.

Instead of gazing into a crystal ball trying to see the outlines of the future in general, we discuss two topics in particular. One deals with quantum effects that might be lurking in classical liquids, rendering the interpretation of increasingly more accurate data more difficult; the other one deals with quantum liquids. In both cases, it should be possible to make rapid progress.

10.2.1   Quantum effects in classical fluids?

One quantum effect that is visible in the characteristic width and peak position of excitations at high momentum transfers in liquid helium are the so-called glory oscillations. These oscillations are easily identified in Fig. 2.6. The origin of these oscillations is understood to be diffraction effects associated with the wave-like nature of the particles, and their period can be calculated by solving the Schrödinger equation for two colliding particles given an interaction potential. In Fig. 10.1 we show the characteristic width of helium scattering spectra as measured by Andersen *et al.* (1997), as well as their comparison to the ^{4}He–^{4}He scattering cross section. As can be seen in this figure, there is a one-to-one correspondence between the characteristic width of the scattering spectra of ^{4}He and the anticipated quantum diffraction effects. This naturally brings up the question of whether such quantum effects would also be observable in more mundane liquids, such as liquid sodium, or even liquid gallium. In particular, could it be possible that the oscillations observed in the characteristic width of liquid Ga are in fact influenced by quantum diffraction effects? And if so, would this offer an alternative explanation for the peculiarities associated with this

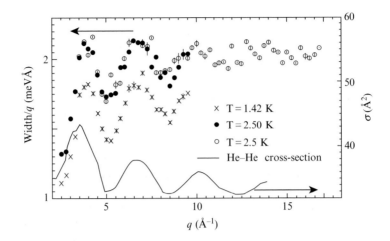

Fig. 10.1 Figure reproduced with permission from Andersen *et al.* (1997). Shown are the characteristic (reduced) widths of $S_{nn}(q, \omega)$ for liquid ^{4}He as measured by neutron scattering (symbols, scale on the left-hand side). Glory oscillations are clearly visible in these widths, and these oscillations are in phase with the calculated ^{4}He–^{4}He cross-section (solid line, scale on the right-hand side) as experienced by two helium atoms in a binary collision. Note that the relative amplitude of the oscillations in the characteristic width is larger than the relative oscillations in the ^{4}He–^{4}He cross-section, implying that even small oscillations in this cross-section have a marked effect on the characteristic width.

width (as detailed in Chapter 7 and shown in Fig. 7.15)? We discuss some of the particulars of glory oscillations in the following, and we give details in Appendix G.

Most binary collisions are not head-on collisions and therefore, angular momentum will play a role. Classically, the collision is expressed in terms of the relative velocity along the line of collision, and the impact parameter s, which is the distance between the line of collision and the center of the stationary particle, as shown in Fig. G.3. The collision results in a scattering angle θ, that depends both on the relative velocity and on the size of the impact parameter. The resulting angle-dependent differential cross-section $|f(\theta)|^2$ can be expressed as a function of impact parameter s (Appendix G) as

$$|f(\theta)|^2 = \frac{s}{\sin(\theta)} \frac{ds}{d\theta}. \tag{10.1}$$

This cross-section will diverge when $\sin(\theta) = 0$, as happens in forward and backward scattering, or when $d\theta/ds = 0$. The former goes under the name of forward and backward glory, the latter is referred to as rainbow glory. This nomenclature comes from wave optics, and the reader is most likely familiar with the backward glory. This backward glory, consisting of light waves backscattered to the observer and interfering with the incoming waves, can be observed as a bright ring of colored light around the shadow that a plane casts on the clouds below, or as a ring of light around the shadow of one's head (hence the name glory). Rainbow glories occur when a range of impact parameters results in the same deflection angle. We explain in Appendix G under what conditions these rainbow glories appear.

When we solve a binary collision using the Schrödinger wave equation, we also find that the cross-section is marked by glories, implying that some scattering angles are more likely to result from the collision than other scattering angles. Depending on this angle, the particle will finally escape the cage it has been trying to get out of for over nine chapters now, or it will stay within the cage. Given an interparticle potential, this cross-section (eqn 10.1) can be calculated in the quantum-mechanical case by solving the Schrödinger equation numerically. This is achieved by separation of variables (radial and angular dependence) and, bearing in mind that angular momentum is quantized, the problem can be solved using the partial waves expansion (Appendix G). The more classical the collision (heavier particles), the more terms are required in the partial wave expansion and the calculations become more tedious in the sense that more computer power is required. One also has to bear in mind that depending on whether the colliding particles are bosons or fermions, the solutions to the scattering problem will have to be symmetric or anti-symmetric upon interchanging the two colliding particles. Once the numerical solution has been calculated, then one can calculate, for instance, the total cross-section $\sigma = 2\pi \int \sin(\theta)|f(\theta)|^2 d\theta$. This calculated total cross-section, which still depends on the relative velocity v, is shown in Fig. 10.1 for binary collisions of ^{4}He atoms. In order to make the transition to scattering experiments one uses $mv = \hbar q$.

The oscillations in the cross-section are directly visible in scattering experiments, at least in the case of ^{4}He, as shown in Fig. 10.1. In fact, even relatively small oscillations in the total cross-section of ^{4}He–^{4}He collisions have a marked effect on the characteristic width of $S_{nn}(q,\omega)$. In the case of ^{4}He, the oscillations in the characteristic width and the cross-section are in phase. This implies that a large width – corresponding to a short residency time in the cage $\tau_{cage} = 1/\omega_H$ – goes hand in hand with a large ^{4}He–^{4}He cross-section. This makes sense provided the total cross-section is dominated by the contributions for $\theta < \pi/2$, allowing the particle to escape its cage. The latter is indeed borne out by calculations. A typical example of such a differential cross-section is shown in Fig. G.2, where one observes that most of the cross-section relates to scattering events with $\theta < \pi/2$.

While we expect to see quantum effects – related to the wave nature of the particles play a role in light atoms at low temperatures – we would not necessarily expect such effects to show up in heavier particles at room temperature. This is not correct, however. Whereas it is true that $mv_{thermal} \sim \sqrt{Tm}$ increases both with temperature and mass, this is not the quantity that is being probed in scattering experiments. Instead, mv is fixed by q, implying that at a fixed momentum transfer one probes a fixed relative velocity v that depends on the mass of the particles. For example, at $q = 3.3$ Å^{-1}, the relative velocity in the case of ^{4}He is ~ 500 m/s, in the case of Ga it is 30 m/s. Since the relative velocity is the only relevant velocity in a binary collision, in both cases it would correspond to a particle with a de Broglie wavelength of ~ 2 Å. Thus, quantum-diffraction effects should be expected in both cases.

In Fig. 10.2 we show the calculated Ga–Ga cross-section for $T = 300$ K. The details of such a calculation, including the positions of the maxima and minima, depend

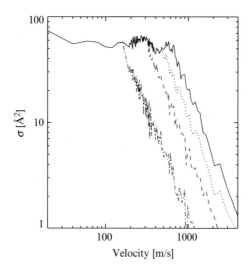

Fig. 10.2 Figure reproduced from Patty (2009). Shown is the Ga–Ga cross-section as a function of relative velocity v, calculated for the Bose isotope of Ga. Note the double log-scale. This cross-section was determined using the Ga interparticle potential published by Baskes *et al.* (2002). The solid curve is calculated taking 141 terms (l-values) into account in the partial wave expansion, the other curves that deviate at lower and lower relative velocity were calculated with 105, 70 and 35 terms, respectively. The range of interest to the scattering experiments shown in Fig. 7.15 is $v < 100$ m/s. The spacing of the v points in this range is 20 m/s, resulting in the jaggedness of σ in this range. Thus, we only take these calculations as an indication that quantum effects are important, we did not attempt a direct comparison to the experimental results for Ga.

sensitively on the interparticle potential used, and the calculation becomes increasingly more computer intensive with decreasing v. Nonetheless, even though only a few points were calculated (Patty, 2009) for $v < 100$ m/s (the range of interest to scattering experiments in liquid Ga just above the melting point), it is clear from this figure that oscillations in the cross-section do indeed take place. The amplitude of these oscillations is of the same order of magnitude as the amplitude in the ^{4}He–^{4}He cross-section. Thus, it is quite possible that the Ga–Ga oscillations might also be visible in the characteristic width of the Ga spectra. Should this indeed be the case, then the comparison between the observed width and the theoretical prediction becomes a very challenging problem indeed. One can easily imagine a scenario where the oscillation due to the Ga–Ga cross-section and those predicted by eqn 4.9 would interfere with each other, resulting in a hard-sphere diameter that is trying to also account for quantum interference effects.

10.2.2   What is next for superfluids?

It would appear that virtually all imaginable scattering experiments on superfluid ^{4}He have already been carried out. Probably the one experiment that still awaits completion with a high enough accuracy is to determine the temperature evolution of the linewidths of the phonon–roton excitations close to the termination point of the dispersion curve. Based on the results of perturbation theory presented in Chapter 9, we would expect these linewidths to be (at least) double that of the roton linewidth. The best data so far (Fig. 9.13) are still not accurate enough to determine this. Perhaps experiments at the next generation of neutron spallation sources will solve this; it is unlikely at this point that the energy resolution of the X-ray spectrometers will be good enough to determine the linewidth with a high enough accuracy. But then again, few people could have imagined the whirlwind development of X-ray sources during the past decade, so we will just have to wait and see.

Better data on the multiphonon component and its temperature dependence are also needed, if only to inspire more theoretical work along the lines of Manousakis and Pandharipande (1984; 1986). As welcome as these new data would be, what would really be needed at this point are scattering data on the other superfluid systems, namely ^{3}He and Bose–Einstein condensates. It is hard to guess when we will be seeing these data, but it might not be too long. For ^{3}He it is a matter of being able to do scattering experiments at very low temperatures, but there does not appear to be an insurmountable technical difficulty in doing so. Even with the somewhat limited energy resolution of X-ray sources, it would definitely be possible to follow the fate of the excitations upon transitioning between the normal-fluid and superfluid phase. Hopefully, these experiments will be forthcoming. For Bose–Einstein condensates, the main problem is the limited number of particles that make up such systems (about 10 million), rendering the scattering signal too weak to measure. Again, it is foreseeable that this might change in the not too distant future, although probably not within a decade, as we show next.

We do a back of the envelop calculation to see how many orders of magnitude we still need to find to be able to measure the excited states of a Bose–Einstein condensate (BEC) by means of neutron scattering. We chose neutron scattering because neutrons have a relatively easy time making it through the trap that holds the BEC atoms. On the IRIS spectrometer at ISIS it is possible to get a clear scattering signal corresponding to the roton excitation in superfluid helium in a sample consisting of 10^{23} atoms in about 5 min. The flux at IRIS on the sample is $\sim 10^7$ neutron/cm^2/s of neutrons of all wavelengths (white-beam flux). Only 10^3 neutron/cm^2/s will have the right energy to create rotons and be detected. On the cold triple-axis spectrometer that is being developed for the CG1 beamline at the HFIR reactor at the Oak Ridge National Laboratory, the flux at the sample position would be 2×10^8 neutron/cm^2/s, counting neutrons of the desired energy. This beam can be (horizontally) focused, resulting in a factor of 100 intensity increase. Next, the detector area can be increased (compared to IRIS) to yield another factor of 10–100. In all, this instrument will have an increase in neutron flux of $\sim 10^9$. Since we are seeking to scatter from a sample that has 10^{16} times fewer atoms in it, we are still 6–7 orders of magnitude short. Rubidium scatters 4 times more strongly than He, so we take the optimistic approach that we are 6 orders of magnitudes shy of being able to do scattering experiments on Bose–Einstein condensates in 5 min. If we manage to focus the CG1 beam in the vertical direction as well (from 2 cm to 2 mm), we could gain another factor of 10. And we do not have to measure for 5 min, we can measure for a full day, in all giving us 3 orders of magnitude. So in this rather optimistic scenario, we are off by a factor of 1000. Either we measure for a full year, or we wait until more atoms can be brought together to form a BEC.

Progress on the theoretical front is more within our grasp. The results of perturbation theory presented in Chapter 9 give us a very solid handle on the entire phonon–roton dispersion curve, both in terms of excitation energies, strength of the excitations, and its termination at twice the roton energy and momentum. It also provides a natural explanation for the presence of the multiphonon component in the superfluid phase, as well as for the hybridization of the recoil curve with the phonon–roton dispersion curve. What it does not do is say anything about the Bose-condensate, or the ground state in general. This is almost by definition as the perturbation theory presented describes the excited states. Also, it does not describe the recoil curve very well. The predicted position is close, but it does a poor job at predicting the linewidth of the recoil curve; this linewidth is largely determined by the motion of a single atom. Trying to capture the motion of a single atom using collective plane waves as a starting point is simply not going to work.

The latter problem is actually very similar to the problem of calculating the electronic states in a solid. Here, the initial plane wave guesses for the electronic states (Bloch states) need to be augmented when describing the states of the band electrons in the vicinity of the localized electrons of the ions. In solids, this augmentation can be done in several ways by folding in the localized electronic states into the initial guesses for the band electron states. These methods go by the names of augmented plane-wave method and orthogonalized plane-wave method, and they work because of the periodic structure present in a solid. In order to do the same for liquid helium, we need to determine the single-particle states, and we need to be able to convolute these states with the plane waves that we have taken as our starting point in perturbation theory. The former is not too hard, but the latter is rather problematic.

As we briefly discussed in Section 2.2.3 and as we showed in Figs. 2.5 and 2.6, the approach to ideal gas in liquid helium is well described by quantum ideal-gas behavior corrected for the effects of helium atoms being subject to a harmonic potential (Schimmel *et al.*, 2002). The only parameter that enters this harmonic potential is the zero-point energy $\hbar\omega_0$ in such a way that the kinetic energy of a helium atom at zero Kelvin is given by $3\hbar\omega_0/4$. Thus, in the limit where we are measuring the behavior of individual atoms, we can – to a very good approximation – model the wavefunctions of a single atom by those of the harmonic oscillator. The next challenge is to fold these harmonic oscillator functions with the plane waves $n_q^+|0>$. We sketch this procedure in Fig. 10.3. One can already see from this sketch that the excited-state trial function should incorporate the first excited states of the harmonic oscillator potential. Trying to simply fold the harmonic potential ground-state wavefunction with a plane wave does not result in a final product that is perpendicular to the ground-state wavefunction. The new trial wavefunction, corresponding (roughly) to the maxon excitation and sketched in the bottom panel of Fig. 10.3, does not differ enormously from the original plane-wave function guess for an excited state. However, it is easy to imagine that with decreasing wavelength the convoluted trial wavefunctions will start to deviate more and more from the simple plane-wave trial function.

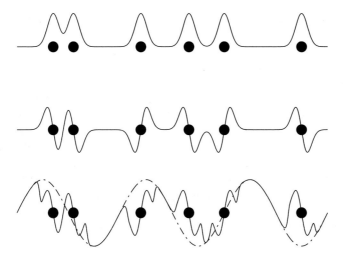

Fig. 10.3 Sketch of how the harmonic oscillator functions that represent the wavefunctions of individual atoms, which are confined to the cage formed by their neighbors, can be folded in with the plane waves that are being used in the perturbation theory discussed in Section 9.5. The top panel shows six atomic wavefunctions corresponding to six irregularly spaced atoms (big dots). The middle panel shows the corresponding (first) excited wavefunction for the harmonic oscillator. The bottom panel is an attempt at reconciling a plane wave excited state with the atomic wavefunctions in such a way that the resultant wavefunction (solid line) is perpendicular to the ground state, while already capturing the rapid variation of the wavefunctions near the atomic positions. The non-modified plane wave corresponding to a wavelength of twice the average atomic separation is shown by the dashed curve in the bottom panel.

This sketch makes the procedure look straightforward, but we have the problem that the atoms in liquids are not located at average lattice sites. In addition, trying to orthonormalize the single-particle states to the plane waves turns out to be a *tour de force*. We think the problem can be solved, but it will require some serious effort. Should this folding procedure be successful, then we would have obtained the wavefunctions for the excited states, describing the phonon–roton curve, the recoil curve and the multiphonon excitations. What would still be absent from this description is the Bose-condensate itself, whose presence is manifest in the lineshape of the high-q recoil-curve excitations. Since this component does not appear to have any direct influence on the excited states, we do not know at this point whether this component should be mixed in with the excited-state wavefunctions in order to achieve even better agreement with experiment, or whether we can simply consider it as a separate contribution that we can ignore in the determination of the excited-state energies.

Appendix A
Conversions

For historical and geographical reasons, the amount of energy transferred by the radiation to the sample is measured in various units such as meV, THz and ps^{-1}, as well as some less common ones such as cm^{-1}. The graphs in this book that have been reproduced from the literature employ all such units. The table below lists the conversions between the various common units, and the not so common ones. For example, if one wants to evaluate the exponential e^{-E/k_BT} with E expressed in meV and T in Kelvin, then this exponential would become $e^{-E*11.605/T}$.

The energy of a neutron is measured in meV, but it can also be characterized by its speed v, its wavelength λ or its wave number k. If we use the units [meV] for energy, [km/s] for speed, [Å] for wavelength and [Å^{-1}] for wave number, then the numerical conversions read:

$$E = 5.2267v^2 = \frac{81.799}{\lambda^2} = 2.072178k^2. \tag{A.1}$$

For example, a 4 Å neutron has an energy of about 5 meV and travels roughly at 1000 m/s.

Table A.1 The conversions between the energy units employed in scattering experiments. The conversions are carried out using : $E = h\nu = \hbar\omega = \hbar c k = k_B T = \mu_B H$.

		E	ν	ω	k	T	H
		[meV]	[THz]	[ps^{-1}]	[cm^{-1}]	[K]	[Tesla]
1 meV	=	1	0.24180	1.5193	8.0655	11.605	17.326
1 THz	=	4.1357	1	6.2832	33.356	47.994	71.655
1 ps^{-1}	=	0.65821	0.15912	1	5.3088	7.6384	11.404
1 cm^{-1}	=	0.12398	0.029979	0.18837	1	1.43883	2.1481
1 K	=	0.086170	0.020836	0.13092	0.69500	1	1.4930
1 Tesla	=	0.057717	0.013956	0.087689	0.46551	0.66980	1

Appendix B
Derivation of the effective eigenmode formalism

In this appendix we give a brief derivation of the basic equations of the projection operator formalism, which we use as the framework for the effective eigenmode description. We follow the exhaustive treatise by Forster (1975), and the articles by de Schepper *et al.* (1988) and by Montfrooij *et al.* (1997). We refer the reader to these publications for additional details as well as to the original papers outlining the projection formalism by Mori (1965) and Zwanzig (1961).

B.1 Projection formalism

This section follows Montfrooij *et al.* (1997) almost verbatim. The time evolution of an operator $A(t)$, in Heisenberg representation, is given by

$$\frac{\partial}{\partial t}A(t) = \frac{1}{i\hbar}[A(t), H] \equiv iLA(t), \tag{B.1}$$

where the brackets denote the commutator, H the Hamiltonian of the system and L the quantum analoge of the Liouville operator. We define the (imaginary part of the) response function $\chi"_{AB}(\vec{r}, t)$ pertaining to the two microscopic quantities A and B for N particles in a volume V by

$$\chi"_{AB}(\vec{r} - \vec{r}', t - t') \equiv \frac{1}{2\hbar} < [A(\vec{r}, t), B(\vec{r}', t')] >_{eq}, \tag{B.2}$$

with

$$< ... >_{eq} \equiv \frac{\mathrm{Tr}[e^{-\beta H}...]}{\mathrm{Tr}[e^{-\beta H}]} \tag{B.3}$$

the ensemble average and Tr the trace operator. Similarly, the generalized van Hove scattering functions $G_{AB}(\vec{r} - \vec{r}', t - t')$ are defined by

$$G_{AB}(\vec{r} - \vec{r}', t - t') \equiv < [A(\vec{r}, t)B(\vec{r}', t')] >_{eq}, \tag{B.4}$$

where the variables A (and B) are chosen in such a way that $< A(t) >_{eq} = 0$, which can be done without loss of generality. As a short-hand notation, we use $\chi(\vec{r}, t)$ to represent the matrix of all $\chi_{AB}(\vec{r}, t)$. The Fourier and Laplace transforms of $\chi''(\vec{r}, t)$ are given, respectively, by

$$\chi''(\vec{q}, \omega) = \int_{-\infty}^{\infty} dt'' e^{i\omega t''} \int_V d\vec{r}'' e^{-i\vec{q}\cdot\vec{r}''} \chi''(\vec{r}'', t''), \tag{B.5}$$

and

$$\chi(\vec{q}, z) = 2i \int_0^{\infty} dt'' e^{izt''} \int_V d\vec{r}'' e^{-i\vec{q}\cdot\vec{r}''} \chi''(\vec{r}'', t''), \tag{B.6}$$

for $\text{Im}[z] > 0$ (corresponding to 'physical', positive times $t'' > 0$), with $t'' = t - t'$ and $\vec{r}'' = \vec{r} - \vec{r}'$. A similar definition holds for $\text{Im}[z] < 0$ (corresponding to 'unphysical', negative times $t'' < 0$). Substituting eqn B.5 into eqn B.6 one finds, for $\text{Im}[z] \neq 0$,

$$\chi(\vec{q}, z) = \frac{1}{\pi} \int_{-\infty}^{\infty} d\omega \chi''(\vec{q}, \omega) \frac{1}{\omega - z}. \tag{B.7}$$

$\chi''(\vec{q}, \omega)$ is real, odd in ω, and only depends on $|\vec{q}| = q$ for an isotropic system. $\chi''(\vec{q}, \omega)$ and $S(\vec{q}, \omega)$, the latter being the Fourier transforms of the $G(\vec{r}, t)$, are related through the fluctuation dissipation theorem (cf. eqns B.1–B.4)

$$\chi''(\vec{q}, \omega) = \frac{1 - e^{-\beta\hbar\omega}}{2\hbar} S(\vec{q}, \omega). \tag{B.8}$$

We define a scalar product $< ... >$ that is related to the canonical ensemble average $< ... >_{\text{eq}}$ through the Kubo function

$$< A(t)|B > = < A^*(t)B > \equiv \frac{1}{\beta} \int_0^{\beta} d\lambda < A^*(t)B(i\hbar\lambda) >_{\text{eq}}, \tag{B.9}$$

where $A^*(t)$ is the complex conjugate of $A(t)$. The relaxation functions $C_{AB}(t)$ are defined by

$$C_{AB}(t) \equiv < A(t)|B > = < A|e^{-itL}|B >, \tag{B.10}$$

with the Liouville operator defined in eqn B.1. If we choose $A = B = n(\vec{q}, t)$ [see Table 2.1], then $S_{nn}^{\text{sym}}(q, \omega)$ (see eqn 2.9) is the Fourier transform of eqn B.10. We note that the $C_{AB}(t)$ satisfy, for all temperatures,

$$\frac{i}{2} \beta \partial_t C_{AB}(t) = \chi''_{AB}(t). \tag{B.11}$$

The fact that this relation holds for all temperatures is a direct consequence of the choice of the scalar product (if $A = B = n(\vec{q}, t)$, then eqn B.11 is equivalent to eqn 2.9). We have opted for this choice so as to obtain transparent sum rules later on (see Appendix B.3). The static susceptibility χ_{AB} is given by the $z = 0$ value (see eqn B.7) as

$$\chi_{AB} = \beta < A|B > = \int_{-\infty}^{\infty} \frac{d\omega}{\pi} \frac{\chi''_{AB}(\omega)}{\omega}. \tag{B.12}$$

Equation B.10 is Laplace transformed to yield

$$\frac{1}{2i}C_{AB}(z) = < A|\frac{i}{z-L}|B > = \int_{-\infty}^{\infty} \frac{d\omega}{\pi i} \frac{\chi"_{AB}(\omega)}{\omega(\omega-z)}. \tag{B.13}$$

Next, we define the projection operator P, which projects onto a subspace R of selected variables that will be those variables between which long-lived correlations are present. Thus,

$$P \equiv' \sum_{i,j} |A > \beta\chi_{ij}^{-1} < B| \equiv 1 - Q, \tag{B.14}$$

where $'\sum$ implies that the summation is restricted to the subspace R. Writing $L = LP + LQ$, and making use of the operator identity $\frac{1}{X+Y} = \frac{1}{X} - \frac{1}{X}Y\frac{1}{X+Y}$, we rewrite eqn. B.13 for $i, j \in R$ as

$$\frac{1}{2}C_{AB}(z) = < A|\frac{i}{z-LQ} + \frac{1}{z-LQ}LP\frac{i}{z-L}|B > . \tag{B.15}$$

Substituting eqn B.14 into eqn B.15 and rearranging the terms we then find

$$'\sum_C [z\delta_{AC} - \Omega_{AC} + i\Sigma_{AC}(z)]C_{CB}(z) = -\frac{2}{\beta}\chi_{AB}, \tag{B.16}$$

with the frequencies Ω_{AC} given by

$$\Omega_{AC} =' \sum_D < A|L|D > \beta\chi_{DC}^{-1}, \tag{B.17}$$

and the damping rates Σ_{AC} given by

$$\Sigma_{AC}(z) =' \sum_D < A|LQ\frac{i}{z-QLQ}QL|D > \beta\chi_{DC}^{-1}. \tag{B.18}$$

In matrix notation, eqn B.16 can simply be written as

$$(z1 + iH^a(z))C(z) = -\frac{2}{\beta}\chi, \tag{B.19}$$

with $H^a(z) = i\Omega + \Sigma(z)$ and 1 is the unit matrix. The aim of the effective eigenmode description is now to find a subspace R such that the z dependence of $\Sigma(z)$ can be ignored so that $\Sigma(z)$ can be replaced by its $z = 0$ value (which can obviously be done if R is chosen in such a way so as to span the full Hilbert space). Finally, using eqns B.6 and B.11–B.13, and assuming the z dependence has been dealt with ($\Sigma(z) \rightarrow \Sigma(0)$), we find for the 'basic' equation for $\chi(q, z)$ (we now explicitly include the q dependence)

$$\chi(q, z)\chi^{-1}(q) = \frac{iH^a(q)}{z1 + iH^a(q)} \tag{B.20}$$

for $\text{Im}[z] > 0$. Thus, $H^a(q)$ is an asymmetric matrix, independent of z, which describes all the interactions between a set of microscopic variables spanning a subspace of the full Hilbert space of observables. $\chi(q, \omega)$ is given by

$$\chi(q, \omega) = \lim_{\epsilon \to 0^+} \chi(q, z = \omega + i\epsilon) = \frac{iH^a(q)}{\omega 1 + iH^a(q)} \chi(q). \tag{B.21}$$

The poles of $\chi(q, \omega)$ are thus given by the eigenvalues of $H^a(q)$. The density–density response function $\chi_{nn}(q, \omega)$, accessible by means of inelastic neutron- and X-ray-scattering experiments, is given by the nn element of this matrix equation (as before, we refer to the Hilbert vector representing density fluctuations with the label n). The imaginary part of the dynamic susceptibility $\chi''_{nn}(q, \omega)$ is calculated straightforwardly from eqn B.21, using that the matrix elements $\chi_{AB}(q)$ are real, and is given by

$$\frac{\beta S_{nn}^{\text{sym}}(q, \omega)}{2} = \frac{\chi''_{nn}(q, \omega)}{\omega} = \text{Re} \left[\frac{\chi(q)}{i\omega 1 + H^a(q)} \right]_{nn}. \tag{B.22}$$

From observation of the above equations, it is clear that we have chosen to analyze the data in terms of $S_{nn}^{\text{sym}}(q, \omega)$ out of convenience, since we are mainly interested in the function $[i\omega 1 + H^a(q)]^{-1}$ that determines the poles of the response function $\chi(q, \omega)$ (see eqn B.21). One could of course equally well analyze the data in terms of $S_{nn}(q, \omega)$ or $\chi''_{nn}(q, \omega)$, *provided* the proper ω-dependent factors are incorporated in the description (see eqn 2.9).

In practice, it is more convenient to work with a symmetric matrix, which can be achieved by successive orthonormalization of the microscopic variables. For the density and momentum variables this implies a scaling with $\sqrt{\chi_{nn}(q)}$ and $(\sqrt{\beta m})^{-1}$, respectively. Following the procedure outlined in de Schepper *et al.* (1988), eqn B.22 is rewritten as

$$S_{nn}^{\text{sym}}(q, \omega) = \chi_{nn}(q) \frac{2}{\beta} \text{Re}[i\omega 1 + G(q)]_{nn}^{-1}, \tag{B.23}$$

with the symmetric matrix $G(q)$ given by $G(q) = U(q)H^a(q)U^{-1}(q)$. Here, $U(q)$ is a triangular matrix determined by $\chi_{AB}(q)$. Explicit expressions for the matrix elements of $U(q)$ can be found in terms of the equal-time correlation functions of $F_{AB}(q, t = 0)$ (de Schepper *et al.*, 1988). We do not need the explicit expressions here, rather we give the resulting expressions for the coupling constants that enter the symmetric matrix $G(q)$ in the following section.

B.2   Coupling constants

The four coupling constants of the 5×5 matrix $G(q)$ of the extended hydrodynamic description (Fig. 2.12) determine the behavior of the eigenmodes. These coupling constants can be expressed in terms of generalizations of basic thermodynamic quantities, as we show in this section.

The basic thermodynamic quantities that are generalized in the generalized hydrodynamics description are the following (de Schepper *et al.*, 1988). The isothermal compressibility κ_T related to the $q = 0$ value of the static structure factor $S(q)$

$$S(q = 0) = nk_{\text{B}}T\kappa_T = -nk_{\text{B}}T(\partial V/\partial p)_T/V; \tag{B.24}$$

the enthalpy h

$$h = (U + pV)/N; \qquad (B.25)$$

the volumetric expansion coefficient α

$$\alpha = (\partial V/\partial T)_p/V; \qquad (B.26)$$

the specific heat at constant volume c_V

$$c_V = (\partial U/\partial T)_V/N; \qquad (B.27)$$

and the specific heat at constant pressure c_p

$$c_p = T(\partial S/\partial T)_p/N. \qquad (B.28)$$

We also generalize the ratio of the specific heats $\gamma = c_p/c_V$. In the above, U and S are the total energy and entropy of the fluid, respectively.

Using the numbering of the microscopic quantities a_j given in Table 2.1 we generalize the thermodynamic quantities specified above to finite q values using the initial values $V_{ij}(q)$ of the intermediate scattering functions $F_{ij}(q,t)$: $V_{ij}(q) = F_{ij}(q, t = 0)$. We follow the generalization procedure outlined in de Schepper *et al.* (1988). The initial values $V_{ij}(q)$ are given by the following expressions that implicitly define $S(q)$, $h(q)$, $\alpha(q)$, $c_V(q)$, $c_p(q)$, and $\gamma(q)$.

$$
\begin{aligned}
V_{11}(q) &= S(q) \\
V_{22}(q) &= V_{14}(q) = \frac{k_B T}{m} \\
V_{13}(q) &= h(q)S(q) - k_B T^2 \alpha(q) \\
V_{33}(q) &= k_B T c_p(q) + h(q)^2 S(q) - 2k_B T^2 \alpha(q)h(q) \qquad (B.29)\\
V_{34}(q) &= V_{25}(q) = \frac{k_B T}{m} h(q) \\
V_{44}(q) &= -\frac{1}{q^2} \lim_{t \to 0} \frac{\partial^2}{\partial t^2} F_{22}(q, t) \\
V_{55}(q) &= -\frac{1}{q^2} \lim_{t \to 0} \frac{\partial^2}{\partial t^2} F_{33}(q, t).
\end{aligned}
$$

Here, we have made use of the fact that the Liouville operator (eqn B.1) acting on the microscopic quantities produces the following results:

$$
\begin{aligned}
La_1(q) &= -iqa_2(q) \\
La_2(q) &= -iqa_4(q) \qquad (B.30)\\
La_3(q) &= -iqa_5(q).
\end{aligned}
$$

All $V_{ij}(q)$'s that are not specifically listed are identical to zero because of the even or oddness of the microscopic variables in the microscopic velocity v.

We can now write the expressions for the coupling constants f that enter the matrix $G(q)$ of eqn 2.36. These coupling constants refer to the set of microscopic variables $\{b_j(q)\}$ that is obtained through successive orthonormalization of the set $\{a_j(q)\}$. The orthonormalization procedure itself (de Schepper *et al.*, 1988) is not very revealing and is not reproduced here. It yields the following results for the Liouville operator acting on the b_js:

$$Lb_1(q) = -if_{un}(q)b_2(q)$$
$$Lb_2(q) = -if_{un}(q)b_1(q) - if_{uT}(q)b_3(q) - if_{u\sigma}(q)b_4(q) \qquad \text{(B.31)}$$
$$Lb_3(q) = -if_{uT}(q)b_2(q) - if_{Tq}(q)b_5(q).$$

Here, the forces f are given by:

$$f_{un}(q) = \left[\frac{k_{\mathrm{B}}T}{mS(q)}\right]^{1/2} q$$

$$f_{uT}(q) = \left[\frac{(\gamma(q)-1)k_{\mathrm{B}}T}{mS(q)}\right]^{1/2} q$$

$$f_{u\sigma}(q) = \left[\frac{m^2 S(q)V_{44}(q) - k_{\mathrm{B}}^2 T^2 \gamma(q)}{mk_{\mathrm{B}}TS(q)}\right]^{1/2} q \qquad \text{(B.32)}$$

$$f_{Tq}(q) = \left[\frac{mV_{55}(q) - k_{\mathrm{B}}Th(q)^2}{mk_{\mathrm{B}}T^2 c_V(q)}\right]^{1/2} q.$$

These expressions are particularly useful when performing molecular dynamics computer simulations. In such simulations the $V_{ij}(q)$ can be evaluated directly from the simulations by evaluating the equal-time correlations functions $F_{ij}(q, t = 0)$ of the (full) set of microscopic variables $a_j(q)$, from which all coupling constant f can be determined without resorting to a fitting procedure.

Of particular interest is under what conditions these coupling constants become negligibly small, such as is the case with $f_{uT}(q)$ for very cold liquids (see Chapter 8). In order to make this clearer, we rewrite eqn B.32 directly in terms of the $V_{ij}(q)$s:

$$f_{un}(q) = \left[\frac{V_{22}}{V_{11}}\right]^{1/2} q$$

$$f_{uT}(q) = \left[\frac{[V_{13} - V_{11}V_{34}/V_{22}]^2 V_{22}}{[V_{33}V_{11} - V_{13}^2]V_{11}}\right]^{1/2} q = \sqrt{\gamma-1}\left[\frac{V_{22}}{V_{11}}\right]^{1/2} q$$

$$f_{u\sigma}(q) = \left[\frac{V_{44}}{V_{22}} - \frac{\gamma(q)V_{22}}{V_{11}}\right]^{1/2} q \qquad \text{(B.33)}$$

$$= \left[\frac{V_{44}}{V_{22}} + \frac{V_{22}}{V_{11}} - \frac{V_{22}}{V_{11}}\frac{[V_{13} - V_{11}V_{34}/V_{22}]^2}{[V_{33}V_{11} - V_{13}^2]}\right]^{1/2} q$$

$$f_{Tq}(q) = \left[\frac{V_{55} - V_{25}^2/V_{22}}{V_{33} - V_{13}^2/V_{11}}\right]^{1/2} q.$$

Here, we have dropped the q dependence of the $V_{ij}(q)$ since it is too late to return the book anyway. For instance, a value of $\gamma(q)$ close to 1 occurs if $V_{13} \approx V_{11}V_{34}/V_{22}$, or equivalently if $V_{13}/V_{11} \approx V_{25}/V_{22}$, or if $V_{25}/V_{13} \approx V_{22}/V_{11}$. Here, we have used that $V_{34}(q) = V_{25}(q)$. In words (using classical terminology) this requirement states that if the ratio of the 2nd and 0th frequency moments of the density–temperature correlation function (V_{25}/V_{13}) equals the same ratio of the density–density correlation function (V_{22}/V_{11}), then the generalized ratio of the specific heats will be close to one. Or, bearing in mind that the second frequency moment is related to the double time derivative of the intermediate scattering function $F_{ij}(q, t)$, it can be rephrased by saying that the initial time behavior of both correlation functions (the initial shape of the functions) should be identical up to the second derivative for $t = 0$. If we were to use quantum terminology, then all frequency moments drop by one, i.e. the second frequency moment becomes the first, etc.

Note that the (generalized) specific heat c_V is proportional to $V_{33} - V_{13}^2/V_{11}$. Since the specific heat is not close to zero this implies that the ratio of the 0th frequency moments of the temperature–temperature and the density–temperature correlation functions (V_{33}/V_{13}) is not the same as the ratio of the 0th frequency moments of the density–temperature and the density–density correlation functions (V_{13}/V_{11}). Thus, while the density–density $(F_{11}(q, t))$ and the density–temperature $(F_{13}(q, t))$ correlation function might have the same shape for short times in liquid metals and in very cold liquids, the same does not hold true for the density–temperature and the temperature–temperature $(F_{33}(q, t))$ correlation functions.

The coupling constant $f_{u\sigma}$ is very small, provided that $\gamma(q)$ is close to one as it is in very cold liquids and liquid metals, and provided that the excitations are sharp (see eqn B.33). Under these conditions we can expect the ratio of the fourth and second frequency moments of the density–density correlation function to be close to the ratio of the second and zeroth moments: $V_{44}/V_{22} \approx V_{22}/V_{11}$. The latter is the case for superfluid helium (see Chapter 9) when it comes to the single-phonon excitations. This situation can be achieved within the simple viscoelastic model when the damping rate of the momentum flux becomes very large (see Appendix E for details). Note that this is not the case in liquid metals.

Finally, the coupling constant $f_{Tq}(q)$ is proportional to the square root of $V_{55} - V_{25}^2/V_{22}$. Therefore, this coupling constant is expected to be small when the ratio of the second frequency moments of the temperature–temperature and the density–temperature correlation functions (V_{55}/V_{25}) is the same as the ratio of the second frequency moments of the density–temperature and the density–density correlation functions (V_{25}/V_{22}). Since such a relationship does not hold for the ratios of the zeroth frequency moments (see the discussion on c_V in the preceding paragraphs), we can safely assume that f_{Tq} will not be vanishingly small for ordinary fluids, or for liquid metals. The latter statement is pertinent to our discussion on the usage of a particular memory function in liquid metals (see Section 7.1.1).

B.3 Frequency-sum rules

The short-time behavior of the intermediate scattering function $F_{nn}(q, t)$ translates into frequency-sum rules for the dynamic structure factor (Balucani and Zoppi, 1994). In this section, we will only discuss the sum rules for the density–density correlation function, and we will drop the subscript 'nn' from our notation in this section. We have derived the effective eigenmode formalism for the symmetrized dynamic structure factor $S^{\text{sym}}(q,\omega)$, whereas the standard dynamic structure factor $S(q,\omega)$ is measured in experiments. Since the two are related through a pre-factor (eqn 2.9), the frequency sum rules for the two are related:[1]

$$\int_{-\infty}^{\infty} \omega^{2m-1} S^{\text{sym}}(q,\omega)\mathrm{d}\omega = 0, \tag{B.34}$$

and

$$\int_{-\infty}^{\infty} \omega^{2m} S^{\text{sym}}(q,\omega)\mathrm{d}\omega = \frac{2}{\hbar\beta}\int_{-\infty}^{\infty} \omega^{2m-1} S(q,\omega)\mathrm{d}\omega. \tag{B.35}$$

This implies that only the odd frequency moments of $S(q,\omega)$ play a role in the decay of density fluctuations. In particular, out of the three most frequently used sum rules of $S(q,\omega)$, namely

$$\int_{-\infty}^{\infty} \omega^{-1} S(q,\omega)\mathrm{d}\omega = \hbar\pi\chi(q) = \hbar\beta S_s(q)/2, \tag{B.36}$$

$$\int_{-\infty}^{\infty} S(q,\omega)\mathrm{d}\omega = S(q), \tag{B.37}$$

$$\int_{-\infty}^{\infty} \omega S(q,\omega)d\omega = \frac{\hbar q^2}{2m}, \tag{B.38}$$

the static structure factor $S(q)$ plays no role. Its role has been taken over by the static susceptibility $\chi(q)$. This reflects the fact that the effective eigenmode formalism is based on the poles of the dynamic susceptibility rather than on the peaks of the dynamic structure factor. Equation B.38 is frequently referred to as the f-sum rule, and this exact result is derived (Balucani and Zoppi, 1994; Hansen and McDonald, 2006) using the conservation of the number of particles.

Most textbooks discuss the higher-frequency-sum rules for the classical counterpart of $S(q,\omega)$. In the classical limit, $S(q,\omega)$ and $S^{\text{sym}}(q,\omega)$ are identical (see eqn 2.9) so that the second frequency moment in this limit is given by the f-sum rule. It may seem confusing how the exact result for the 1st frequency moment of $S(q,\omega)$ can show up as a second frequency moment in calculations that pertain to $S(q,\omega)$, albeit in the classical limit. This problem is closely related to efforts in trying to come up with the classical equivalent $S_{cl}(q,\omega)$ of the dynamic structure factor $S(q,\omega)$. First, the classical dynamic structure factor does not really exist, only the classical limit $(1/\beta \gg \hbar\omega)$ of the (quantum) dynamic structure factor $S(q,\omega)$ exists. This limit is reached (inadvertently) in molecular dynamics computer simulations where the classical equations of motions are solved. In order to compare the computer simulations directly to experiment, one recasts $S(q,\omega)$ into a symmetric (in ω) form. The inverse Fourier transform of this symmetric function will yield an $F(q,t)$ that is real-valued, and that can then be compared to computer simulations. Of course, there are many

---

[1] The detailed balance condition $S(q,-\omega) = \mathrm{e}^{-\beta\hbar\omega} S(q,\omega)$ is needed in the derivation of these equalities.

ways in which one can obtain a symmetric function out of $S(q, \omega)$. For instance, one can use $S^{\text{sym}}(q, \omega)$ of eqn 2.9, or one could use the detailed balance inspired choice of

$$S_{\text{cl}}(q, \omega) = \text{e}^{-\beta\hbar\omega/2} S(q, \omega). \tag{B.39}$$

Both are fine choices, but neither represents the true classical equivalent of $S(q, \omega)$ for the simple reason that there is no equivalent. This is just a fact of life and represents a limitation on the comparison between molecular dynamics computer simulations and scattering experiments. This also explains why it may be confusing to see the f-sum rule appear as the 2nd frequency moment; it is a consequence of assuming that there exists such a thing as the classical equivalent of $S(q, \omega)$.

In the preceding section we gave the coupling constants in terms of the initial values for the molecular dynamics case, that is, for the classical case (eqn B.33). This is the case for which the coupling constants can be determined directly without a fitting procedure to $F(q, t)$. Expressions for the values of the higher-order (higher than 1st) frequency moments given in the literature pertain to this case. Expressions for the coupling constants in the experimental (quantum) case have not been worked out, and in practice one determines parameters such as $f_{u\sigma}$ from a best fit to the data. From these fitted parameters one can then determine the higher-order frequency moments, and compare them to the classical predictions for these moments. In fact, even in experiments on classical liquids, the experimental data never seems to cover an extensive enough ω range in order to be able to determine any of the frequency moments directly from experiments. Add to this the fact that the scattering at high energy transfers tends to be so weak that the background scattering cannot be corrected for accurately enough in order to carry out the frequency integrals, and one quickly ends up in the situation where $S(q)$ is also determined from the data as a fitted parameter (through the f_{un} coupling parameter). Of course, this only holds for classical liquids, when one applies such a fitting procedure to real fluids then one actually determines $\chi_{nn}(q)$ through the fitting procedure.

Appendix C
(Almost) exact results
for hard-sphere fluids

In this appendix we provide a range of results valid for hard-sphere fluids; we do not present a derivation of those results, we simply enumerate them so that the reader can compare these predictions to the experimental results. The temperature only enters the equations as a secondary parameter in the sense that what happens during a collision is independent of temperature. The temperature merely determines the thermal velocities.

Static quantities:

The fraction of the volume V occupied by N hard spheres at close packing ($V = V_{cp}$) is calculated to be

$$V_0/V_{cp} = \pi/3\sqrt{2} = 0.741; n_{cp}\sigma^3 = 1.415, \tag{C.1}$$

with $V_0 = N4\pi(\sigma/2)^3/3$. The hard-sphere fluid actually solidifies at a lower density $n_{melt}\sigma^3 \approx 0.667 n_{cp}\sigma^3 = 0.943$ (Hoover and Ree, 1968), reflecting that the hard spheres can no longer stream past each other. The relationship that $n_{melt}\sigma^3 = $ constant goes under the name of Lindemann's law; in Fig. C.1 we reproduce the findings of van Loef (1974) who demonstrated that this law can be used to derive an equivalent hard-sphere diameter for real fluids, fitted to be $n_{melt}\sigma^3 = 0.93 \pm 0.02$. This relationship is seen to hold for a wide range of liquids, from simple liquids to liquid metals and diatomic fluids and it can be used in determining the equivalent hard-sphere diameter for the atoms in the fluid.

Results for hard-sphere theory are frequently represented as a function of packing fraction ϕ, which is defined for all densities $n = N/V$ as $\phi \equiv V_0/V = \pi n\sigma^3/6$. This is the same packing fraction that is used to denote the density of colloidal suspensions (Chapter 5). For instance, the freezing transition in colloids (Segrè et al., 1995a) is given as $\phi_{cryst} = 2\phi_{cp}/3 = 0.494$. Expressed in terms of ϕ the hard-sphere prediction for the pair correlation function at contact $\chi(\phi) = g(r = \sigma)$ reads

$$\chi(\phi) = \frac{1 - \phi/2}{(1 - \phi)^3}. \tag{C.2}$$

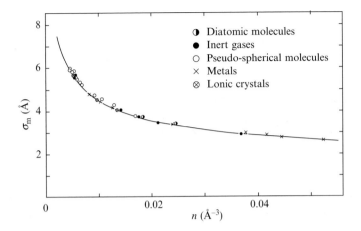

Fig. C.1 Figure adapted with permission from van Loef (1974). The effective hard-sphere diameter σ_m of a range of liquids at their melting temperatures T_m. The solid line is the prediction that freezing of a liquid is a packing problem, largely independent of the details of the interatomic potential: $n_{\mathrm{melt}}\sigma^3 = 0.93 \pm 0.02$.

A very good approximation to the static structure factor of a hard-sphere fluid is the Percus–Yevick approximation (Wertheim, 1963). The only input parameter for this approximation is the packing fraction ϕ:

$$\frac{1}{S(q)} = 1 + \frac{24\phi}{x^3}\{a[\sin(x) - x\cos(x)] + b[(\frac{2}{x^2} - 1)x\cos(x) + 2\sin(x) - \frac{2}{x}]$$
$$+ \frac{\phi a}{2}[\frac{24}{x^3} + 4(1 - \frac{6}{x^2})\sin(x) - (1 - \frac{12}{x^2} + \frac{24}{x^4})x\cos(x)]\}. \tag{C.3}$$

Here, $x = q\sigma/2$, and a and b are given by

$$a = \frac{(1 + 2\phi)^2}{(1 - \phi)^4}, \tag{C.4}$$

$$b = -\frac{3}{2}\phi\frac{(\phi + 2)^2}{(1 - \phi)^4}. \tag{C.5}$$

Transport-related quantities:

The Boltzmann coefficient of diffusion at low densities D_B is given by

$$D_B = \frac{3}{8n\sigma^2}\sqrt{\frac{k_B T}{\pi m}} = \frac{k_B T}{3\pi\sigma\eta_0}. \tag{C.6}$$

This equation also gives the expression for the concomitant viscosity η_0 for the dilute hard-sphere fluid. The Enskog diffusion coefficient D_E for all densities is given by

$$D_{\mathrm{E}} = D_{\mathrm{B}}/\chi(\phi) = \frac{1}{16}\left[\frac{\pi k_{\mathrm{B}}T}{m}\right]^{1/2}\left[\frac{6}{\pi n\phi^2}\right]^{1/2}\frac{(1-\phi)^3}{1-\phi/2}. \tag{C.7}$$

The mean-free path between collisions l_{E} in the Enskog theory is given by

$$l_{\mathrm{E}} = \frac{l_{\mathrm{Boltzmann}}}{\chi(\phi)} = \frac{1}{\sqrt{2}n\pi\sigma^2\chi(\phi)}. \tag{C.8}$$

Approach to ideal-gas behavior for a hard-sphere fluid:

The approach to ideal-gas behavior can be calculated for the incoherent dynamic structure factor – the self-correlation function $S_s(q,\omega)$ – in the high-q limit. These calculations are detailed in Montfrooij *et al.* (1986), here we merely reproduce the final results. The correction to the ideal-gas behavior of $S_s(q,\omega)$ for a hard-sphere liquid at temperature T is given by

$$S_{\mathrm{s}}(q,\omega) = \left[\frac{\beta m}{2\pi}\right]^{1/2}\frac{1}{q}\left[\mathrm{e}^{-\omega^{*2}/2} + \frac{s_1(\omega^*)}{ql_{\mathrm{free}}} + O(s_2(\omega^*/q^2))\right], \tag{C.9}$$

where $\omega^* = \omega(\beta m)^{1/2}/q$ and

$$s_1(\omega^*) = \frac{8}{9\pi}\sum_{j=0}^{\infty}\frac{(j+1)!}{(2j)!}{}_3F_2(\frac{1}{2},\frac{3}{2},j+2;\frac{5}{2},\frac{5}{2};\frac{1}{2})[-2(\omega^*)^2]^j. \tag{C.10}$$

This correction function is expressed using the generalized hypergeometric functions ${}_pF_q$:

$$_pF_q(\alpha_1,\alpha_2,...,\alpha_p;\beta_1,\beta_2,...,\beta_q;z) = \sum_{n=0}^{\infty}\frac{(\alpha_1)_n(\alpha_2)_n \times ... \times (\alpha_p)_n}{(\beta_1)_n(\beta_2)_n \times ... \times (\beta_q)_n}\frac{z^n}{n!}. \tag{C.11}$$

Here, $(\alpha)_n$ is the Pochhammer symbol

$$(\alpha)_n = (\alpha)(\alpha+1)\times...\times(\alpha+n-1) = \frac{\Gamma(\alpha+n)}{\Gamma(\alpha)}, \tag{C.12}$$

and $\Gamma(x)$ is the gamma function. The correction function $s_1(\omega^*)$ is shown in Fig. 2.7.

The expressions for the top values $S_s(q,\omega=0)$ and for the half-width $\omega_{\mathrm{H}}(q)$ follow from this:

$$S_{\mathrm{s}}(q,\omega=0) = \left[\frac{\beta m}{2\pi}\right]^{1/2}\frac{1}{q}\left[1 + \frac{s_1(0)}{ql_{\mathrm{free}}} + O(1/q^2)\right], \tag{C.13}$$

where

$$s_1(0) = \frac{8}{9\pi^3} F_2(\frac{1}{2}, \frac{3}{2}, 2; \frac{5}{2}, \frac{5}{2}; \frac{1}{2}) = 0.32808. \tag{C.14}$$

$$\omega_H(q) = \left[\frac{2\ln2}{\beta m}\right]^{1/2} q \left[1 - \frac{\xi}{q l_{\text{free}}} + O(1/q^2)\right], \tag{C.15}$$

with ξ given by

$$\xi = \frac{s_1(0) - 2s_1(\sqrt{2\ln2})}{2\ln2} = 0.4486. \tag{C.16}$$

Appendix D
Detailed behavior of the eigenmodes

In this appendix we illustrate the generic tendencies of the five effective eigenmodes corresponding to Fig. 2.12 for various combinations of the parameters f and z, including the hitherto largely ignored coefficient $z_{q\sigma}$. We do this by changing the value of one parameter while keeping the remainder of the seven parameters fixed.

For all figures we have used that the five off-diagonal elements of the matrix $G(q)$ (eqn 2.36) vary linearly with q for q values that are relatively small: $q < q_{max}/2$ with $q_{max}/2$ the peak position of the main peak of the static structure factor $S(q)$. We have kept the damping rates z_σ and z_q constant at values given in the figure captions. This reflects the fact that these damping rates only appear to be weakly q dependent and that they approach a constant limit for $q \to 0$. Thus, these figures should help to provide some insight into the transition from hydrodynamic behavior to the behavior relevant to the decay of density fluctuations at shorter wavelengths. The main purpose of this appendix is not to provide the reader with any particular set of parameters that will work for certain types of fluids, instead the aim is to provide some guidance into what the effects are of changes in the variables that make up the dynamic matrix $G(q)$ on the overall behavior of the effective eigenmodes.

The figures depict the speeds of propagation of the extended modes, their damping rates and their amplitudes in $S_{nn}^{\mathrm{sym}}(q,\omega)$. We also give a bird's eye view of $S_{nn}^{\mathrm{sym}}(q,\omega)$ that hopefully will slightly facilitate the interpretation of the many lines in the figures. By scrutinizing these figures the reader will find that there is a large shift in amplitude of the modes whenever modes merge. This shift takes place in a very narrow range of parameters, implying that when doing experiments is is quite possible to observe a mode that appears to evolve continuously but that in reality has changed its character by, for instance, mixing the σ channel into its decay tree. In the text we frequently refer to the character of a mode. This is a loose term but in essence what we mean by it is the following. The eigenvector of every mode has components on the various microscopic quantities. If one of these components is dominant, then we refer to that as the character of the mode. For instance, if the heat mode eigenvector has components of $(0.2, 0.2, 0.8, 0.1, 0.52)$ on ('n','u','T','σ','q'), respectively, then we would say that this mode has the character of a temperature fluctuation. It is visible in $S_{nn}(q,\omega)$ because it also has a component on the density 'n'. If the heat mode components are $(0.8, 0.2, 0.2, 0.1, 0.52)$ instead, then we say that the heat mode has the character of a density fluctuation. This latter character of the heat mode is frequently encountered when $q\sigma \approx 2\pi$ (Kirkpatrick, 1985).

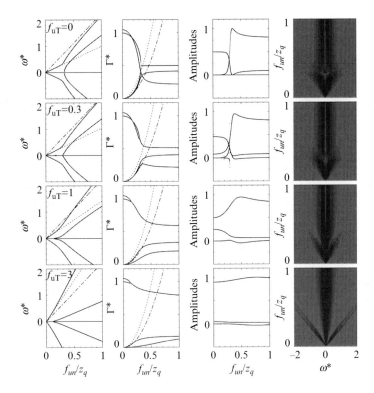

Fig. D.1 The dependence of the effective eigenmodes on the coupling parameter f_{uT}. The plots show, from left to right, the propagation frequencies ω^*, the damping rates Γ^* and the amplitudes of the five modes of the matrix $G(q)$ corresponding to Fig. 2.12. The rightmost panel depicts a bird's eye view of $S_{nn}^{\mathrm{sym}}(q, \omega^*)$. In this, as well as in the other plots in this appendix, the stars indicate that all frequencies are scaled since they depend upon dimensionless parameters f/z. This plot was generated with the following ratios for the coupling constants: $f_{un} = f_{u\sigma}/2 = f_{Tq}/1.5$. The decay rates are given by $z_\sigma = 1.05$, $z_q = 1$ and $z_{q\sigma} = 0$. The values for the parameter f_{uT} that is being varied are given in the figure. The notation $f_{uT} = 3$ is short for $f_{uT} = 3f_{un}$. The two damping rates were chosen to be slightly different from each other so as to be able to distinguish between the two kinetic modes at $q = 0$. In the figures depicting the propagation frequencies, the dotted line is given by $[f_{un}^2 + f_{uT}^2]^{1/2}$ (the adiabatic propagation frequency), the dashed line by $[f_{un}^2 + f_{u\sigma}^2]^{1/2}$. In the damping rate figures, the dotted curve is given by $f_{u\sigma}^2/z_\sigma$, the dashed curve by f_{Tq}^2/z_q. A value of $f_{uT} = 0$ corresponds to $\gamma = 0$ and hence an absence of the heat mode. This is visible in the amplitudes shown in the top line of the figure. However, note that once the kinetic modes mix in that there is a very abrupt change in the character of the modes (at $f_{un}/z_q = 0.25$, corresponding to $f_{u\sigma} \approx z_\sigma/2$), as well as in the spectra (top line of the figure). The top two panels show that should the observed damping of the modes in experiments be very small compared to the hydrodynamics predictions, that it is likely that the lowest q value in the measured spectra corresponds to $f_{un}/z_q > 0.3$, rendering the interpretation of the character of the modes exceedingly difficult.

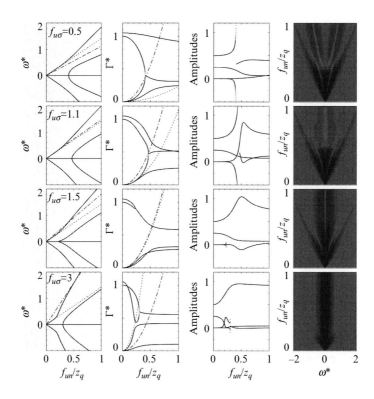

Fig. D.2 Same as Fig. D.1 except now the parameter $f_{u\sigma}$ is being varied. The notation $f_{u\sigma} = 3$ is short for $f_{u\sigma} = 3f_{un}$. The ratios for the other coupling constants have been kept fixed at $f_{un} = f_{uT} = f_{Tq}/1.5$, $z_\sigma = 1.05$, while $z_q = 1$ and $z_{q\sigma} = 0$. The spectra for the intermediate values of $f_{u\sigma}$ appear to be very close to what is actually observed in experiments. Note the very limited ranged of values that this coupling parameter can have before drastic changes are observed, such as the appearance of another set of propagating modes (top line), or the very rapid demise of the extended sound mode (bottom panel). In fact, the results for $f_{u\sigma} = 1.5f_{un}$ seem to be most consistent with what is observed in experiments. Note, however, that this case corresponds to a non-mixing in of the kinetic modes, something that we know from computer simulations to be incorrect (de Schepper *et al.*, 1988) for simple liquids, but perhaps it does describe the dynamics of liquid metals. If this were the case, then the observed smallness of the damping of the extended sound and heat modes would imply that the lowest q value probed in scattering experiments already corresponds to $f_{un}/z_q > 0.5$.

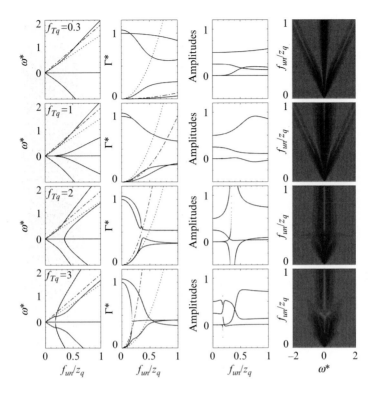

Fig. D.3 Same as Fig. D.1 except now the parameter f_{Tq} is being varied. The notation $f_{Tq} = 3$ is short for $f_{Tq} = 3f_{un}$. The ratios for the other coupling constants have been kept fixed at $f_{un} = f_{uT} = f_{u\sigma}/1.5$, $z_\sigma = 1.05$, while $z_q = 1$ and $z_{q\sigma} = 0$. This figure illustrates how the change in value of a coupling parameter deep down in the decay tree can completely change the complexion of the scattering spectra. The parameters are very similar to the previous figure for the case of $f_{u\sigma} = 1.5$. Note the very different interpretation of the scattering spectra when we change the value of f_{Tq} from 1 (second line), to 1.5 (third line of the preceding figure) to 2 (third line of this figure). Also note the craziness that takes place for the largest value of f_{Tq} shown in this figure. We do not believe that this corresponds to any measured spectra in any type of liquid; however, we are not aware of any rule that would prevent the ratio of f_{Tq}/f_{un} taking on this value, but apparently such a rule does exist. The spectra for $f_{Tq} = 2f_{un}$ and $f_{un}/z_q = 0.4$ are a good example of how modes can change their character over a very small q range, and of how two modes can essentially cancel each other by having almost identical damping rates and propagation frequencies, while having the opposite amplitude.

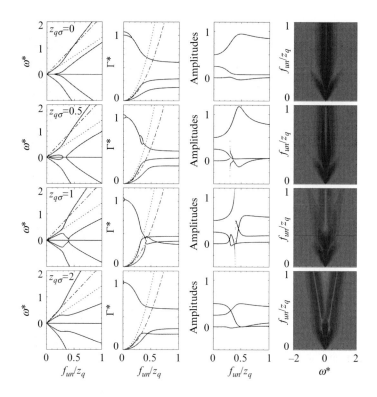

Fig. D.4 Same as Fig. D.1 except now the parameter $z_{q\sigma}$ is being varied. The notation $z_{q\sigma} = 3$ is short for $z_{q\sigma} = 3f_{un}$. The ratios for the coupling constants have been kept fixed at $f_{un} = f_{uT} = f_{u\sigma}/2 = f_{Tq}/1.5$, while $z_\sigma = 1.05$, and $z_q = 1$. The first line of this figure corresponds to the third line of Fig. D.1. In this line we observe pretty standard behavior, where we see an increase in the propagation frequency of the extended sound modes once (in this case) $f_{u\sigma}$ becomes comparable to $z_\sigma/2$ at $f_{un}/z_q = 0.25$. The amplitude of the extended sound modes diminishes when this happens, but note that the amplitude does not go to zero. The extended heat mode becomes more prominent at this transition. The other propagating mode that appears around $f_{un}/z_q = 0.2$ has such a small amplitude in $S_{nn}(q,\omega)$ that it is very unlikely to be observed, should this set of parameters describe a real fluid. With increasing $z_{q\sigma}$ we at first observe only very subtle changes (second line of this figure) in $S_{nn}(q,\omega)$, even though it is not possible to tell from the amplitude figure what the character is of the propagating mode; the changes are too rapid, and too intertwined. This would be a situation where only computer simulations can identify the character of the modes. When $z_{q\sigma} = 1$ we observe a very complicated q dependence of the mode, however, for $z_{q\sigma} = 2$ the situation appears to resemble that of real liquids much more closely. In the latter case, we have two sets of propagating modes at the lowest q values, one set (with the lowest speed of propagation) that has an amplitude in $S_{nn}(q,\omega)$, and one set that is entirely absent in $S_{nn}(q,\omega)$. With increasing q, this lower set becomes an even more prominent feature of $S_{nn}(q,\omega)$, absorbing the heat mode intensity. Also note that the behavior of this mode is what would be identified as a change to isothermal propagation should this be observed in a fluid such a liquid nickel (see Section 7.2). However, this change would not be brought about through any of the mechanisms discussed in Section 7.2. This illustrates once more that an unambiguous interpretation of scattering experiments can only be successful in most cases with the aid of computer simulations.

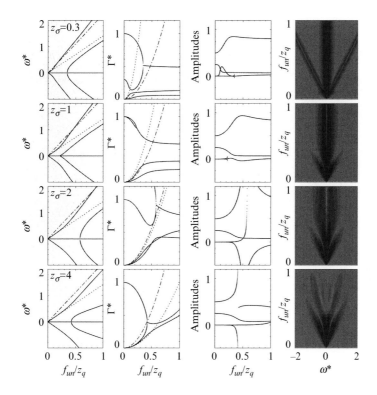

Fig. D.5 Same as Fig. D.1 except now the damping rate z_σ is being varied. The ratios for the coupling parameters have been kept fixed at $f_{un} = f_{uT} = f_{u\sigma}/2 = f_{Tq}/1.5$, while $z_q = 1$ and $z_{q\sigma} = 0$. This, and the following figure, are intended to illustrate the change in the modes when the two damping rates, z_σ and z_q, are no longer equal. The second line of the figure can be compared directly to the preceding figures. First, when we increase the damping rate z_σ from 1 (second line) to 2 (third line), the bird's eye view of $S_{nn}(q,\omega)$ appears to undergo only subtle changes. Studying the behavior of the amplitudes and damping rates of the modes shows that the interpretation of the modes in $S_{nn}(q,\omega)$ is instead very different between the two cases when it comes to the heat mode (the behavior of the extended sound mode is largely unaffected, albeit that the damping rate differs between the two cases). If we lower the damping rate z_σ to 0.3 we observe a deceptively simple structure in $S_{nn}(q,\omega)$, namely an extended heat mode and two extended sound modes over the entire q region. However, even in this case appearances are deceiving. When we look at the behavior of the heat mode near $f_{un}/z_q \approx 0.15$, we observe that the heat mode actually consists of two separate contributions with damping rates that differ by less than 50%. With increasing q, this additional central mode disappears again from $S_{nn}(q,\omega)$. Thus, while it would seem to be possible to do an extrapolation from finite q to $q = 0$, in practice this would not represent an allowed procedure. When we look at the spectra for $z_\sigma = 4$ (lower line) we observe a particular interesting behavior with the appearance (in $S_{nn}(q,\omega)$ for $f_{un}/z_q > 0.4$) of propagating temperature fluctuations in the absence of a heat mode. Only at $f_{un}/z_q = 1$, corresponding to $f_{u\sigma} = z_\sigma/2$ do we see the reappearance of the heat mode as indicated by the upward curvature in the graph depicting the amplitudes of the modes.

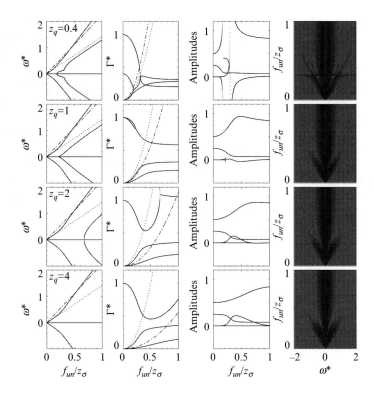

Fig. D.6 Same as Fig. D.1 except now the damping rate z_q is being varied. The ratios for the coupling parameters have been kept fixed at $f_{un} = f_{uT} = f_{u\sigma}/2 = f_{Tq}/1.5$, while $z_\sigma = 1$ and $z_{q\sigma} = 0$. The results have now been plotted as a function of f_{un}/z_σ since we are varying z_q. The second line in this figure can be compared directly to the preceding figures in this appendix. The behavior of the modes with increasing z_q is similar to the first line of Fig. D.5, where z_σ was lowered instead. The behavior of the extended sound modes is the same in all cases, however for $z_q = 2,4$ there is a region in q space (near $f_{un}/z_\sigma \approx 0.4$) where the central part of $S_{nn}(q,\omega)$ consists of two non-propagating modes with very similar damping rates. The exact location and the extent of this region depends on the damping rates z_q and z_σ. Note that a central feature consisting of two modes has not been observed in liquids, with the caveat that it would be very difficult to identify two such similar modes as being separate entities. Having said that, it would appear once again that there is only a very limited range of values that the parameters that make up the dynamic matrix $G(q)$ can take on in real liquids. For the case of the two damping rates, it seems that the two have to be of comparable value for the description to reflect the decay mechanisms in real fluids. Finally, the top line of the figure for $z_q = 0.4$ can be compared to the bottom line in Fig. D.5; both figures represent a very similar ratio of z_σ/z_q, albeit that the ratio in the present case is still low enough to prevent the occurrence of propagating temperature fluctuations.

Is some of these figures the term heat mode is a bit of a misnomer as it does not necessarily imply that a density fluctuation relaxes through coupling with the microscopic temperature; central lines also appear in these spectra even in the absence of any f_{uT} coupling. The term heat mode merely refers to the fact that this mode appears to be a continuous extension of the hydrodynamic heat mode. We stress that it appears to be a continuous extension; as we showed in the preceding figures what appears to be continuous is in some cases far from continuous.

In the figures displaying the damping rates of the various modes, we also display the equivalents of the decay rates z_T and z_u of the 3×3 hydrodynamic matrix $G(q)$. These elements are given by f_{Tq}^2/z_q and $f_{u\sigma}^2/z_\sigma$, respectively, and they are shown as dashed and dotted quadratic curves in the figures. These decay rates can be compared to the strength hydrodynamic coupling parameters f_{un} and f_{uT}. Also note that while $f_{u\sigma}^2/z_\sigma$ is equivalent to the hydrodynamic damping rate of the sound modes ϕq^2, f_{Tq}^2/z_q is equivalent to the hydrodynamic damping rate aq^2 of the heat mode when divided by the ratio of the specific heats: $aq^2 = f_{Tq}^2/\gamma z_q$.

Appendix E
Memory functions and effective eigenmodes

There exists a close connection between the effective eigenmode formalism presented in this book, and the memory function formalism. In this appendix we provide a quick guide so as to make it easier for the reader to alternate between the two formalisms. In the memory formalism, the dynamic susceptibility for density–density correlations is given by (in Laplace functions)

$$\frac{\chi_{nn}(q,z)}{\chi_{nn}(q)} = \left[z + \frac{f_{un}^2}{z + M(q,z)} \right]^{-1}, \tag{E.1}$$

where $M(q,z)$ is the memory kernel, the Laplace transform of the memory function $M(q,t)$. When we compare it to the effective eigenmode expression for (the imaginary part of) the dynamic susceptibility

$$\frac{S_{nn}^{\text{sym}}(q,\omega)}{\pi S_{\text{sym}}(q)} = \frac{\chi''_{nn}(q,\omega)}{\omega\chi_{nn}(q)} = \text{Re} \left[\frac{1}{i\omega 1 + G(q)} \right]_{11} \tag{E.2}$$

then we can give the equivalent expressions for $M(q,\omega)$ for the models representing the various decay trees mentioned in this book and in the literature. In Table E.1 we show the right-hand side of eqn E.2 obtained by taking the '11' elements of the matrix $[i\omega 1 + G(q)]^{-1}$ (but not the real part), and in Table E.2 we show the memory functions in the time domain, where possible.

In addition, some functions are used particularly often in fitting routines, so for sake of ease of use we write out the full expressions for $S_{nn}^{\text{sym}}(q,\omega)$. In the hydrodynamic model we have (see Fig. 2.11)

$$\frac{\pi S_{nn}^{\text{sym}}(q,\omega)}{S_{\text{sym}}(q)} = \frac{f_{un}^2[\omega^2 z_u + (f_{uT}^2 + z_u z_T)z_T]}{\omega^2[f_{un}^2 + f_{u\sigma}^2 + z_u z_T - \omega^2]^2 + [f_{un}^2 z_T - \omega^2(z_u + z_T)]^2}. \tag{E.3}$$

The simple viscoelastic model can be expressed as (see Fig. 2.14)

$$\frac{\pi S_{nn}^{\text{sym}}(q,\omega)}{S_{\text{sym}}(q)} = \frac{z_\sigma f_{un}^2 f_{u\sigma}^2}{\omega^2[f_{un}^2 + f_{u\sigma}^2 - \omega^2]^2 + z_\sigma^2[f_{un}^2 - \omega^2]^2}. \tag{E.4}$$

And the damped harmonic oscillator (Fig. 2.16) is written as

$$\frac{\pi S_{nn}^{\text{sym}}(q,\omega)}{S_{\text{sym}}(q)} = \frac{f_{un}^2 z_u}{(f_{un}^2 - \omega^2)^2 + (\omega z_u)^2}. \tag{E.5}$$

Table E.1 The equivalent expressions for the effective eigenmode formalism in terms of continued fractions H. The conversion from sketch to mathematical expression is straightforward to carry out with the following recipe. Every variable (circle with a letter in it in the sketches of the decay tree) gives rise to a term $i\omega + \ldots$ where the dots will be replaced with the remainder of the tree branch. Every spring (coupling constant) between the variables a and b will give rise to a term $\dfrac{f_{ab}^2}{\ldots}$, where the dots will be replaced with the remainder of the tree branch. Every arrow (decay rate) will give rise to a term z_c, if the decay rate refers to the variable c. For brevity, the q dependences of the quantities have not been displayed.

model name	decay tree	H
ordinary hydrodynamics		$\left[i\omega + \dfrac{f_{un}^2}{i\omega + z_u + \dfrac{f_{uT}^2}{i\omega + z_T}} \right]^{-1}$
simple viscoelastic		$\left[i\omega + \dfrac{f_{un}^2}{i\omega + \dfrac{f_{u\sigma}^2}{i\omega + z_\sigma}} \right]^{-1}$
extended viscoelastic		$\left[i\omega + \dfrac{f_{un}^2}{i\omega + \dfrac{f_{uT}^2}{i\omega + z_T} + \dfrac{f_{u\sigma}^2}{i\omega + z_\sigma}} \right]^{-1}$
damped harmonic oscillator		$\left[i\omega + \dfrac{f_{un}^2}{i\omega + z_u} \right]^{-1}$
extended eigenmodes ($z_{q\sigma} = 0$)		$\left[i\omega + \dfrac{f_{un}^2}{i\omega + \dfrac{f_{uT}^2}{i\omega + \dfrac{f_{Tq}^2}{i\omega + z_q}} + \dfrac{f_{u\sigma}^2}{i\omega + z_\sigma}} \right]^{-1}$

Note that the damped harmonic oscillator is the limiting case of the viscoelastic model when z_σ becomes large. This can be seen by rewriting eqn E.4 as

$$
\frac{\pi S_{nn}^{\mathrm{sym}}(q,\omega)}{S_{\mathrm{sym}}(q)} = \frac{f_{un}^2 (f_{u\sigma}^2/z_\sigma)}{\omega^2 [f_{un}^2/z_\sigma + f_{u\sigma}^2/z_\sigma - \omega^2/z_\sigma]^2 + [f_{un}^2 - \omega^2]^2}
$$
$$
\approx \frac{f_{un}^2 (f_{u\sigma}^2/z_\sigma)}{\omega^2 [f_{u\sigma}^2/z_\sigma]^2 + [f_{un}^2 - \omega^2]^2}.
\tag{E.6}
$$

Table E.2 The equivalent memory functions $M(q,t)$ for the decay trees of the effective eigenmode formalism. For brevity, the q dependences of the quantities have not been displayed.

model name	decay tree	$M(q,t)$
ordinary hydrodynamics		$2z_u\delta(t) + f_{uT}^2 e^{-z_T t}$
simple viscoelastic		$f_{u\sigma}^2 e^{-z_\sigma t}$
extended viscoelastic		$f_{u\sigma}^2 e^{-z_\sigma t} + f_{uT}^2 e^{-z_T t}$
two-time viscoelastic	See discussion near eqn 2.57	$f_{u\sigma}^2[(1-\alpha)e^{-z_\sigma t} + \alpha e^{-z_{new}t}] + f_{uT}^2 e^{-z_T t}$
damped harmonic oscillator		$2z_u\delta(t)$
extended eigenmodes $(z_{q\sigma} = 0)$		$f_{uT}^2[se^{-rt} - re^{-st}]/(s-r) + f_{u\sigma}^2 e^{-z_\sigma t};$ $s = z_q/2 + [z_q^2/4 - f_{Tq}^2]^{1/2},$ $r = z_q/2 - [z_q^2/4 - f_{Tq}^2]^{1/2}$

Here, we have used, given the dominance of z_σ, that $\omega^2 \approx f_{un}^2$ in the expression $f_{un}^2/z_\sigma + f_{u\sigma}^2/z_\sigma - \omega^2/z_\sigma$ for those frequencies, where S_{nn}^{sym} is significantly different from zero. Identifying $f_{u\sigma}^2/z_\sigma$ as z_u reduces the above equation to eqn E.5. We show this transition from viscoelastic behavior to that of a damped harmonic oscillator with increased damping in Fig. E.1.

This transition nicely illustrates how the extended eigenmode formalism almost automatically selects the appropriate cut-off in the decay tree. In this particular case, when the damping of the momentum flux becomes so large that all power put into this channel is lost instantaneously, then we should simply ignore this channel and replace it with an effective decay constant $z_u = f_{u\sigma}^2/z_\sigma$. It is also intuitively clear why for very large damping z_σ the coupling constant $f_{u\sigma}$ cannot play a role in the propagation frequency of extended sound modes. If all power is lost immediately, then there is simply no way for fluctuations to propagate through this channel. In the extended

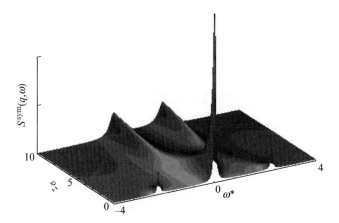

Fig. E.1 The transition from viscoelastic behavior to that of an harmonic oscillator is determined by the damping rate z_σ. For very low damping rates $S^{\mathrm{sym}}(q,\omega)$ displays three modes: the central heat mode and two propagating sound modes with propagation frequency $(f_{un}^2 + f_{u\sigma}^2)^{1/2}$ and damping rate $\frac{z_\sigma}{2}\frac{f_{u\sigma}^2}{f_{un}^2 + f_{u\sigma}^2}$. With increased damping the propagation frequency decreases and the amplitude of the heat mode diminishes. For very large damping rates z_σ the heat mode has disappeared, while the propagation frequencies are now given by f_{un} and their damping rates by $f_{u\sigma}^2/2z_\sigma$. This figure has been plotted in reduced units where $f_{un} = 1$; $f_{u\sigma}$ has been kept fixed at $f_{u\sigma} = 2$.

eigenmode formalism, we then find that the propagation frequencies are given by the other coupling constants in the problem (f_{un} in this case). For the viscoelastic model we see a progression from a propagation frequency of $(f_{un}^2 + f_{u\sigma}^2)^{1/2}$ for low decay rates z_σ to f_{un} for very high z_σ. Note that this generic type of behavior of when a decay channel becomes irrelevant in terms of determining propagation frequencies holds true for all types of decay trees.

Appendix F
Effective eigenmode formalism for mixtures

In order to be able to describe mixtures in such a way that it incorporates the hydrodynamic variables and can account for the fast- and slow-sound modes, we need a minimum of five microscopic quantities. We list these quantities in Table F.1. For our effective eigenmode formalism we take linear combinations of these quantities in order to arrive at an orthonormal set. We reproduce the procedure here from Westerhuijs *et al.* (1992) including all the details that show how the partial structure factors $S_{11}(q)$, $S_{12}(q)$, and $S_{22}(q)$ enter the formalism. Again, the following notation is for the classical case; the generalization to the quantum-mechanical case involves replacing the static structure factors by the static susceptibilities and the symmetrization of the expression for the microscopic variables.

The microscopic quantities listed in Table F.1 satisfy the following normalizations:[1]

$$< \delta n_i(q)|\delta n_j(q) >= S_{ij}(q)$$

$$< \delta u_i(q)|\delta u_j(q) >= \frac{\delta_{ij}}{\beta m_j} \tag{F.1}$$

$$< \delta n_i(q)|\delta u_j(q) >= < \delta e(q)|\delta u_j(q) >= 0.$$

With the aid of eqn F.1 the following five variables b_j can be constructed: b_c, b_N, b_P, b_v, and b_T. These names refer to fluctuations in the mutual number concentration ('c'), the total number density ('N'), the total longitudinal momentum ('P'), the mutual relative velocity ('v'), and the total temperature ('T'). In terms of the relative concentrations x_j of the two species ($x_1 + x_2 = 1$) the expressions for b_j are given by:

$$b_c = \frac{1}{A_c(q)}[\sqrt{x_2}\delta n_1(q) - \sqrt{x_1}\delta n_2(q)],$$

$$b_N = \frac{1}{A_N(q)}\{[\sqrt{x_1}S_{22}(q) - \sqrt{x_2}S_{12}(q)]\delta n_1(q) + [\sqrt{x_2}S_{11}(q) - \sqrt{x_1}S_{12}(q)]\delta n_2(q)\}, \tag{F.2}$$

---

[1] As before, we do not explicitly write out the equilibrium values such as $n_{1,eq}$ in the definitions and normalizations of the microscopic quantities.

Table F.1 The five microscopic quantities relevant to mixtures from which an orthonormal set is constructed. The quantities are written out for the classical case. Here, $\vec{r}_p^{(s)}$ denotes the position of particle p of species s.

microscopic quantity	expression
$\delta n_1(q)$	$\dfrac{1}{\sqrt{N_1}}\sum_{p=1}^{N_1} e^{i\vec{q}\cdot\vec{r}_p^{(1)}}$
$\delta n_2(q)$	$\dfrac{1}{\sqrt{N_2}}\sum_{p=1}^{N_2} e^{i\vec{q}\cdot\vec{r}_p^{(2)}}$
$\delta u_1(q)$	$\dfrac{1}{\sqrt{N_1}}\sum_{p=1}^{N_1} (\vec{v}_p^{(1)}\cdot\vec{q}/q)e^{i\vec{q}\cdot\vec{r}_p^{(1)}}$
$\delta u_2(q)$	$\dfrac{1}{\sqrt{N_2}}\sum_{p=1}^{N_2} (\vec{v}_p^{(2)}\cdot\vec{q}/q)e^{i\vec{q}\cdot\vec{r}_p^{(2)}}$
$\delta e(q)$	$\dfrac{1}{\sqrt{N_1+N_2}}\sum_{s=1}^{2}\sum_{p=1}^{N_s} \epsilon_p^{(s)} e^{i\vec{q}\cdot\vec{r}_p^{(s)}}$
	with $\epsilon_p^{(s)} = \dfrac{1}{2}m_s v_p^{(s)2} + \dfrac{1}{2}\sum_{w=1}^{2}\sum_{q=1}^{N_w} \phi_{ws}(|\vec{r}_q^{(w)} - \vec{r}_p^{(s)}|)$

$$b_P = \sqrt{\frac{\beta}{x_1 m_1 + x_2 m_2}}[\sqrt{x_1}m_1\delta u_1(q) + \sqrt{x_2}m_2\delta u_2(q)],$$

$$b_v = \sqrt{\frac{\beta m_1 m_2}{x_1 m_1 + x_2 m_2}}[\sqrt{x_2}\delta u_1(q) + \sqrt{x_1}\delta u_2(q)], \qquad \text{(F.3)}$$

$$b_T = \frac{1}{A_T(q)}[\delta e(q) - <b_c(q)|\delta e(q)> b_c(q) - <b_N(q)|\delta e(q)> b_N(q)].$$

Of the three normalization factors $A_c(q)$, $A_N(q)$ and $A_T(q)$ we only require explicit expression for the first two:

$$A_c(q) = [x_1 S_{22}(q) + x_2 S_{11}(q) - 2\sqrt{x_1 x_2}S_{12}(q)]^{1/2},$$

$$A_N(q) = A_c[S_{11}(q)S_{22}(q) - S_{12}(q)^2]^{1/2}. \qquad \text{(F.4)}$$

In order to make the connection between the matrix $G(q)$ and the neutron-scattering results (eqn 6.1) we also need the inverse transformations:

$$\delta n_j(q) = \sum_{i=c,N} s_i^{(j)}(q)b_i(q), \qquad \text{(F.5)}$$

where $j = 1$ or 2 and

$$s_c^{(1)}(q) = \frac{1}{A_c(q)}[\sqrt{x_2}S_{11}(q) - \sqrt{x_1}S_{12}(q)]$$

$$s_c^{(2)}(q) = \frac{1}{A_c(q)}[\sqrt{x_2}S_{12}(q) - \sqrt{x_1}S_{22}(q)] \tag{F.6}$$

$$s_N^{(1)}(q) = \sqrt{x_1}A_N(q)/A_c(q)^2$$

$$s_N^{(2)}(q) = \sqrt{x_2}A_N(q)/A_c(q)^2.$$

Using the above equations, the $S_{ij}(q,\omega)$ can be expressed in terms of $S_{cc}(q,\omega)$, $S_{cN}(q,\omega)$ and $S_{NN}(q,\omega)$ that are in turn determined by

$$S_{\alpha\beta}(q,\omega) = \frac{1}{\pi}\left[\frac{1}{i\omega\mathbf{1} + G(q)}\right]_{\alpha\beta}. \tag{F.7}$$

Note that the above quantities simplify significantly if we are to employ the ideal-mixing rules from eqn 6.4. In that case, $A_c(q) = 1$ and $A_N(q) = \sqrt{S_T(q)}$. Moreover, eqn F.6 reduces to (with S_T given in eqn 6.5)

$$s_c^{(1)}(q) = \sqrt{x_2}$$

$$s_c^{(2)}(q) = -\sqrt{x_1}$$

$$s_N^{(1)}(q) = \sqrt{x_1 S_T(q)} \tag{F.8}$$

$$s_N^{(2)}(q) = \sqrt{x_2 S_T(q)}.$$

As an example of how the above procedure works, if we were to use the ideal mixing rules then the measured $S(q,\omega)$ would be given by

$$S(q,\omega) = S_{cc}(q,\omega)\,[x_1 x_2(b_1^{*2} - b_2^{*2})]$$

$$+S_{NN}(q,\omega)\,[(x_1 b_1^* + x_2 b_2^*)^2 S_T(q)] + \tag{F.9}$$

$$+S_{cN}(q,\omega)\,2\sqrt{x_1 x_2}(b_1^* - b_2^*)(x_1 b_1^* + x_2 b_2^*)\sqrt{S_T(q)}.$$

Without using the ideal mixing rules, the above expression would contain the partial structure factors.

The three coupling constants $f_{cv}(q)$, $f_{NP}(q)$, and $f_{Nv}(q)$ are given (Westerhuijs *et al.*, 1992) by the ratio of sum rules (as is $f_{un}(q)$ in the case of monoatomic liquids). These coupling constants are given by (without using the ideal mixing rules)

$$f_{cv}(q) = \frac{q}{\sqrt{\beta\overline{m}}}\frac{1}{A_c(q)}\,[\overline{m}^2/(m_1 m_2)]^{1/2}$$

$$f_{NP}(q) = \frac{q}{\sqrt{\beta\overline{m}}}\frac{A_c(q)^2}{A_N(q)} \tag{F.10}$$

$$f_{Nv}(q) = \frac{q}{\sqrt{\beta\overline{m}}}\frac{1}{A_N(q)}\{\sqrt{x_1 x_2 m_2/m_1}S_{22}(q) - \sqrt{x_1 x_2 m_1/m_2}S_{11}(q)$$

$$+[x_1\sqrt{m_1/m_2} - x_2\sqrt{m_2/m_1}]S_{12}(q)\},$$

where $\overline{m} = x_1 m_1 + x_2 m_2$. Finally, if the ideal mixing rules are used, then the expressions for these coupling constants reduce to

$$
\begin{aligned}
f_{cv}(q) &= \frac{q}{\sqrt{\beta \overline{m}}} [\overline{m}^2/(m_1 m_2)]^{1/2} \\
f_{NP}(q) &= \frac{q}{\sqrt{\beta \overline{m}}} \frac{1}{\sqrt{S_T(q)}} \\
f_{Nv}(q) &= \frac{q}{\sqrt{\beta \overline{m}}} \frac{1}{\sqrt{S_T(q)}} \sqrt{x_1 x_2} \frac{m_2 - m_1}{\sqrt{m_1 m_2}}.
\end{aligned}
\tag{F.11}
$$

We did not provide expressions for the two remaining coupling constants of the matrix $G(q)$ or for its three damping rates. The procedure on how to get somewhat meaningful expressions for these numbers is outlined by Westerhuijs *et al.* (1992); in practice, they are determined from a fitting procedure.

Appendix G
Glory oscillations

In order to solve the collision between two atoms one needs to take into account the wave nature of the particles and solve the Schrödinger wave equation for the sum of the incoming and scattered waves where the incoming wave is represented by a plane wave and where the scattered wave is spherically symmetric:

$$\Psi(\vec{r}) \sim e^{i\vec{q}\cdot\vec{r}} + f(\theta)\frac{e^{iqr}}{r}. \tag{G.1}$$

The interference pattern between the incoming and scattered waves results in an angle-dependent scattering length $f(\theta)$. The parameters that enter the problem are the reduced mass of the two particles μ, their relative velocity v along the line of collision, and the two-particle interaction $V(r)$, which is assumed to be spherically symmetric. The quantum-mechanical problem is solved through standard separation of variables, resulting in a radial equation that depends on the angular momentum quantum number l:

$$\frac{\mathrm{d}^2 G_l(r)}{\mathrm{d}r^2} + (k^2 - U(r) - \frac{l(l+1)}{r})G_l(r) = 0. \tag{G.2}$$

Here, $U(r) = (2\mu/\hbar)V(r)$ and the relative velocity enters through the wave number k as $k = \sqrt{2\mu E_i/\hbar^2}$ where E_i is the initial energy. The full wavefunction can then be reconstructed from the solutions for each of the $G_l(r)$ by summing over all values of l. This standard procedure goes by the name of the partial-wave method.

The solution of eqn G.2 for a given value of l can be captured in one parameter, namely the phase shift η_l. This number represents by what phase the wavefunction has been altered due to the scattering event once the scattered particle is so far from the scattering center that its wavefunction is that of a free particle once more. This is sketched in Fig. G.1 for a typical value of l as a function of relative velocity v (or equivalently, kinetic energy K). Since the collection of $\{\eta_l\}$ fully describes the solution to the scattering problem it is possible to rewrite the quantity of interest in terms of these phase shifts (Mott and Massey, 1971):

$$f(\theta) = \frac{1}{q}\sum_{l=0}^{\infty}(2l+1)e^{i\eta_l}\sin\eta_l P_l(\cos\theta), \tag{G.3}$$

with P_l the Legendre polynominals.

The details of the scattering problem, mostly being the range of the interparticle potential, determine how many l terms one needs to calculate in practice in order

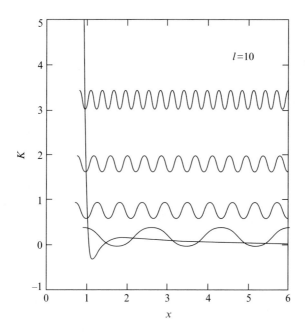

Fig. G.1 This figure has been redrawn with permission from Bernstein (1960). Shown are the radial part of the wavefunctions for the scattered particle (in this case H_2 being scattered by a mercury atom according to the effective potential shown in the figure) for the partial wave corresponding to $l = 10$. Depending on the initial kinetic energy K (relative velocity) of the particle, the scattered wave will resemble that of a free particle once it is a certain distance x away from the scattering center. Both the kinetic and potential energy are shown in reduced units where energies are scaled to the well depth of the potential and $x = r/\sigma$ (with σ the characteristic length scale of the potential). The effective potential shown consists of the bare potential plus the term $\sim l(l+1)/r$ (eqn G.2).

to satisfactorily describe the scattering problem. The solutions to eqn G.2 involve (Bessel) functions that behave asymptotically as $\sim (qr)^l$ for small qr. Thus, the larger the range of the potential (loosely speaking the size of the particles), the more l terms are required for the full description of the scattering problem. Since $\hbar q = mv$, the number of l terms also depends on the mass and the relative velocity between the particles in the scattering problem. As an example, in the description of a neutron being scattering by the nucleus of an atom one only needs to retain the $l = 0$ terms given the very short range of the strong force. Thus, neutron scattering is s-wave scattering. In contrast, for two Ga atoms colliding one needs to retain hundreds of l terms, depending on the relative velocity v. Given the power of computers, the latter does not represent an insurmountable problem.

When one solves the scattering problem and when one calculates the scattering length $f(\theta)$ as a function of relative velocity v, one finds pronounced oscillations in this quantity as well as in the resulting total cross-section $\sigma(v)$:

$$\sigma(v) = 2\pi \int |f(\theta)|^2 \sin\theta \mathrm{d}\theta = \frac{4\pi}{q^2} \sum_{l=0}^{\infty} (2l+1)\sin^2\eta_l(v). \tag{G.4}$$

As an example of the outcome of the quantum-mechanical scattering problem we show the calculated $f(\theta)$ for the collision between an H_2 molecule and a Hg atom in Fig. G.2 as evaluated by Bernstein, while the total cross-section for the collision between two ^{4}He atoms is shown in Fig. 10.1.

The actual numerical solutions to such two-body scattering problems are straight-forward, one just has to take care to properly (anti)symmetrize the wavefunction of the two particles for the case of fermions and bosons. We refer the reader to the details presented in Mott and Massey (1971) and in the paper by Bernstein (1960). The origin of the oscillations can be captured by looking at the scattering problem from a classical point of view in the geometry shown in Fig. G.3. This classical scattering problem is detailed in textbooks (see, e.g., Mott and Massey 1971; Goldstein 1980).

Classically, the scattering problem is phrased (for a stationary scattering center) in terms of impact parameter $s = L/(mv) = L/\sqrt{2mE_i}$, where L stands for classical angular momentum rather than quantum number l. Classically, an impact parameter $s = 0$ implies a head-on collision resulting in a deflecting angle $\theta = \pi$ of the incoming particle (see Fig. G.3). A very large impact parameter results in zero deflection angle since this obviously corresponds to the particles missing each other. Also, since all

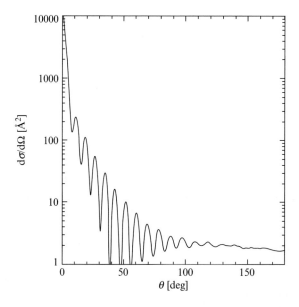

Fig. G.2 Shown are the results for the calculation of the differential cross-section $|f(\theta)|^2$ for hydrogen in mercury for a relative velocity determined by the restriction that $q\sigma = 18$. This figure has been recalculated with permission from Bernstein (1960). The total cross-section for this process, which involves integration over all angles weighted by the factor $\sin(\theta)$ consists mostly of processes that involve deflection angles of less than 90 degrees.

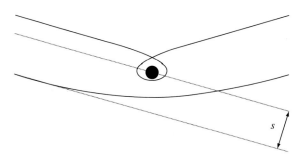

Fig. G.3 Shown are two trajectories (solid lines) for a scattered particle, while the scattering center (big dot) is kept fixed. In this figure, the particles' trajectories start at the left. The impact parameter is defined as $s = L/mv$, which corresponds to the distance between the incoming particles trajectory and the line of collision, as shown in the figure. The two scattering events shown lead to the same scattering angle θ, even though they have very different angles ϕ over which the incoming particle is deflected during the scattering process. The deflection angle and the scattering angle are related by $\theta = \pm\phi - 2m\pi$ with the $\pm$ sign and the integer m chosen in such a way that the scattering angle is between 0 and π.

incoming particles must end up somewhere, one quickly arrives at the basic equation (Goldstein, 1980) for the scattering cross-section related to a flux I of incoming particles:

$$I 2\pi s |\mathrm{d}s| = I\sigma(\theta) 2\pi \sin\theta |\mathrm{d}\theta|; \tag{G.5}$$

hence

$$s|\mathrm{d}s| = \sigma(\theta) \sin\theta |\mathrm{d}\theta|, \tag{G.6}$$

and

$$\sigma(\theta) = |f(\theta)|^2 = \frac{s}{\sin\theta} \frac{|\mathrm{d}s|}{|\mathrm{d}\theta|}. \tag{G.7}$$

The cross-section $\sigma(\theta)$ will show strong maxima whenever the deflection angle is zero, or when $\mathrm{d}\theta/\mathrm{d}s$ is zero. Both can occur. As can be seen from inspection of eqn G.2, the solution to the scattering problem for a given value of l (or L) is determined by the effective potential $U(r) + \frac{l(l+1)}{r}$ (with $l(l+1)$ replaced by L^2 in the classical treatment). This effective potential that determines the radial part of the solution is sketched in Fig. G.4 for typical s values, which for a fixed relative velocity v (or energy E_i) is the same sketch as for typical l values. Depending on the particle energy (that is, on the relative velocity), the particle can make it (or not) over the hump that determines the distance of closest approach. A particle with just a little more than the required amount of energy to make it over the hump caused by the centrifugal term in the effective potential will take a relatively long time to actually make it up the hump; meanwhile, its angular velocity will not go to zero, resulting in a diverging deflection angle by the time the particle rolls back down the hump (see middle panel

Deflection angle

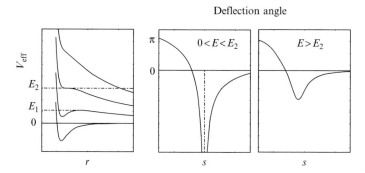

Fig. G.4 Figure adapted from Goldstein (Goldstein, 1980). The left panel depicts the effective potential $V_{\text{eff}}(r) = U(r) + L^2/r$ (eqn G.2) for various L values. With increasing L, the attractive part of the effective potential becomes less pronounced. For a small, but non-zero value of L there exists an energy E_1 that determines the distance of closest approach. A particle with energy E_1 will take a very long time to reach this distance of closest approach. During this approach, the particle's deflection angle increases without limit. This is referred to as orbiting and is shown in the middle panel, where the deflection angle has been plotted as a function of impact parameter $s = L/mv$. The divergence is given by the vertical dashed line, and the $s = 0$ value corresponds to a head-on collision in which the incoming particle reverses course, whereas the $s = \infty$ limit corresponds to zero deflection angle since the particles miss each other. Also note that there is an impact parameter for which the deflection angle is zero (just to the left of the vertical dashed line). Repeating the above reasoning for increasing L values tells us that a critical L value must exist (yielding an energy E_2 shown in the left panel) that would signify the last orbiting possibility. For larger L values, orbiting is no longer possible, even though large (non-diverging) deflection angles will still be found as long as the slope of the effective potential is fairly flat (for $E > E_2$). This yields a deflection angle pattern as shown in the right panel. Note that now there is a region where the deflection angle depends only very weakly on the impact parameter, leading to a divergence in the cross-section $|f(\theta)|^2 \sim \mathrm{d}s/\mathrm{d}\theta$.

of Fig. G.4). This type of behavior is called orbiting or spiralling, and the deflection angle diverges when the orbiting condition is (almost) satisfied. This divergence of the deflecting angle corresponds to $\mathrm{d}s/\mathrm{d}\theta = 0$, hence a minimum in the cross section. Essentially, an incident particle meeting the orbiting condition will be scattered at an almost random angle. In addition, given the continuity of the $\theta(s)$ curve, there must exist an impact parameter (or l value) that results in a deflection angle of zero (or of multiples of π). Classically speaking, the particle goes around the other particle, to continue its way in the forward (backward) direction.

For larger values of impact parameter s (or equivalently, L) the hump will no longer be as important (Fig. G.4). The well will be filled in, and the effective potential starts to transform into a repulsive potential. What we will find is that the divergence will get smeared out and rounded (Goldstein, 1980) once a critical value is reached (see Fig. G.4). This rounding will give rise to rainbow glories, as there now exists a range of impact parameters s resulting in the same deflection angle. In this case, we find that $ds/d\theta = \infty$ and the cross-section will diverge, or at the very least reach a strong maximum. As in the case of a diverging deflection angle, we will also still find an impact parameter for a given energy where the deflection angle will be zero. And depending on the depth of the rounded divergence, we can also find deflection angles of $-\pi$ and so forth. Thus, the combination of an attractive part to the potential and the repulsive centrifugal contribution to the effective potential leads to the occurrence of forward, backward and rainbow glories. For very large values of s (L), nothing much of interest in terms of glories occurs (classically speaking). When we deal with waves, as we do in optics and quantum mechanics, then these rays, characterized by large l, are important because they interfere with the rays that suffer zero deflection angle or a 180-deg deflection angle.

When determining the cross-section $|f(\theta)|^2$ in the quantum-mechanical case we need to sum over all l values, and hence we will run through all the eventualities of glory scattering. In fact, in the quantum-mechanical case we will also encounter higher-order glory scattering. We refer the reader to the paper by Helbing (1969) for the mechanism behind these glories. Exactly where all the glories occur in $f(\theta)$ depends on the relative velocity v. When summing over all l and integrating over all scattering angles, we find remnant oscillations in the total cross-section that originate from the various glories in the partial wave equations. It is these remnant oscillations, such as the ones shown in Fig. 10.1 that are important in interpreting the characteristic widths of scattering spectra.

References

Albergamo, F., Verbeni, R., Huotari, S., Vankó, G., and Monaco, G. (2007). *Phys. Rev. Lett.*, **99**, 205301.

Albergamo, F., Verbeni, R., Huotari, S., Vankó, G., and Monaco, G. (2008). *Phys. Rev. Lett.*, **100**, 239602.

Alley, W.E. and Alder, B.J. (1983). *Phys. Rev. A*, **27**, 3158.

Andersen, K.H., Bossy, J., Cook, J.C., Randl, O.G., and Ragazzoni, J.-L. (1996). *Phys. Rev. Lett.*, **77**, 4043.

Andersen, K.H., Stirling, W.G., and Glyde, H.R. (1997). *Phys. Rev. B*, **56**, 8978.

Anderson, M.H., Ensher, J.R., Matthews, M.R., Wieman, C.E., and Cornell, E.A. (1995). *Science*, **269**, 198.

Anento, N. and Padró, J.A. (2001). *Phys. Rev. E*, **64**, 021202.

Badyal, Y.S., Bafile, U., Miyazaki, K., de Schepper, I.M., and Montfrooij, W. (2003). *Phys. Rev. E*, **68**, 061208.

Bafile, U., Barocchi, F., and Guarini, E. (2008). *Condens. Matter Phys.*, **11**, 107.

Bafile, U., Guarini, E., and Barocchi, F. (2006). *Phys. Rev. E*, **73**, 061203.

Bafile, U., Verkerk, P., Barocchi, F., de Graaf, L.A., Suck, J.-B., and Mutka, H. (1990). *Phys. Rev. Lett.*, **65**, 2394.

Bafile, U., Verkerk, P., Guarini, E., and Barocchi, F. (2001). *Phys. Rev. Lett.*, **86**, 1019.

Balucani, U. and Zoppi, M. (1994). *Dynamics of the Liquid State*. Oxford University Press, Oxford.

Baskes, M.I., Chen, S.P., and Cherne, F.J. (2002). *Phys. Rev. B*, **66**, 104107.

Bedeaux, D. and Mazur, P. (1974). *Physica*, **73**, 431.

Bedell, K., Pines, D., and Zawadowski, A. (1984). *Phys. Rev. B*, **29**, 102.

Beenakker, C.W.J. and Mazur, P. (1984). *Physica A (Amsterdam)*, **126**, 349.

Behrend, O.P. (1995). *Phys. Rev. E*, **52**, 1164.

Bermejo, F.J., Bustinduy, I., Cabrillo, C., Levett, S., and Taylor, J. (2005a). *Phys. Rev. Lett.*, **95**, 269601.

Bermejo, F.J., Bustinduy, I., Levett, S.J., Taylor, J.W., Fernández-Perea, R., and Cabrillo, C. (2005b). *Phys. Rev. B*, **72**, 104103.

Bermejo, F.J., Fernández-Perea, R., Cabrillo, C., Jiménez-Ruiz, M., Maira-Vidal, A., Ruiz-Martin, M.D., and Stunault, A. (2007). *J. Non-Cryst. Solids*, **353**, 3129.

Bermejo, F.J., Saboungi, M.L., Price, D.L., Alvarez, M., Roessli, B., Cabrillo, C., and Ivanov, A. (2000). *Phys. Rev. Lett.*, **85**, 106.

Berne, B.J. and Pecora, R. (2000). *Dynamic Light Scattering*. Dover Publishing, Dover.

Bernstein, R.B. (1960). *J. Chem. Phys.*, **33**, 795.

Bodensteiner, T., Morkel, Chr., Gläser, W., and Dorner, B. (1992). *Phys. Rev. A*, **45**, 5709.

Bogoliubov, N.N. (1947). *J. Phys. (USSR)*, **11**, 23.

Bosse, J., Jacucci, G., Ronchetti, M., and Schirmacher, W. (1986). *Phys. Rev. Lett.*, **57**, 3277.

Bove, L.E., Dorner, B., Petrillo, C., Sacchetti, F., and Suck, J.-B. (2003). *Phys. Rev. B*, **68**, 024208.

Bove, L.E., Formisano, F., Sacchetti, F., Petrillo, C., Ivanov, A., Dorner, B., and Barocchi, F. (2005). *Phys. Rev. B*, **71**, 014207.

Bove, L.E., Petrillo, C., and Sacchetti, F. (2008). *Condens. Matter Phys.*, **11**, 119.

Bove, L.E., Sacchetti, F., Perillo, C., Dorner, B., Formisano, F., and Barocchi, F. (2001). *Phys. Rev. Lett.*, **87**, 215504.

Bove, L.E., Sacchetti, F., Petrillo, C., Dorner, B., Formisano, F., Sampoli, M., and Barocchi, F. (2002). *Philos. Mag. B*, **82**, 365.

Bruin, C., Michels, J.P.J., van Rijs, J.C., de Graaf, L.A., and de Schepper, I.M. (1985). *Phys. Lett. A*, **110**, 40.

Bruin, C., van Rijs, J.C., de Graaf, L.A., and de Schepper, I.M. (1986). *Phys. Rev. A*, **34**, 3196.

Brush, S.G. (1976). *The Kind of Motion We Call Heat*. North-Holland, Amsterdam. Books 1 and 2.

Bryk, T. and Mryglod, I. (2002). *J. Phys.: Condens. Matter*, **14**, L445.

Bryk, T. and Mryglod, I. (2005). *J. Mol. Liq.*, **120**, 83.

Bryk, T. and Wax, J.-F. (2009). *Phys. Rev. B*, **80**, 184206.

Cabrillo, C., Bermejo, F.J., Alvarez, M., Verkerk, P., Maira-Vidal, A., Bennington, S.M., and Martín, D. (2002). *Phys. Rev. Lett.*, **89**, 075508.

Calderin, L., González, D.J., González, L.E., and López, J.M. (2008). *J. Chem. Phys.*, **129**, 194506.

Calderin, L., González, L.E., and González, D.J. (2009). *J. Chem. Phys.*, **130**, 194505.

Campa, A. and Cohen, E.G.D. (1988). *Phys. Rev. Lett.*, **61**, 853.

Campa, A. and Cohen, E.G.D. (1989). *Phys. Rev. A*, **39**, 4909.

Campa, A. and Cohen, E.G.D. (1990). *Phys. Rev. A*, **41**, 5451.

Cazzato, S., Scopigno, T., Hosokawa, S., Inui, M., Pligrim, W.-C., and Ruocco, G. (2008). *J. Chem. Phys.*, **128**, 234502.

Ceperley, D.M. (1995). *Rev. Mod. Phys.*, **67**, 279.

Ceperley, D.M. and Pollock, E.L. (1986). *Phys. Rev. Lett.*, **56**, 351.

Chang, C.C. and Campbell, C.E. (1976). *Phys. Rev. B*, **13**, 3779.

Chapman, S. and Cowling, T.G. (1970). *The Mathematical Theory of Nonuniform Gases*. Cambridge University Press, London.

Chester, G.V. (1967). In *Liquid Helium* (ed. G. Careri), Academic Press Inc., New York, p. 51.

Clouter, M.J., Luo, H., Kiefte, H., and Zollweg, J.A. (1990). *Phys. Rev. A*, **41**, 2239.

Cohen, E.G.D. (1966). In *Lectures in Theoretical Physics* (ed. W. Britten), Volume VIIIA, University of Colorado Press, Boulder, Colorado, pp. 145-181.

Cohen, E.G.D. (1967). In *Lectures in Theoretical Physics* (ed. W. Britten), Volume IXC, Gordon and Breach, New York, pp. 279-333.

Cohen, E.G.D. (1993*a*). *Physica A*, **194**, 229.

Cohen, E.G.D. (1993*b*). *Am. J. Phys.*, **61**, 524.

Cohen, E.G.D. and de Schepper, I.M. (1990). *Il Nuovo Cimento D*, **12**, 521.

Cohen, E.G.D., de Schepper, I.M., and Zuilhof, M.J. (1984). *Physica B*, **127**(1-3), 282.

Cohen, E.G.D., Kamgar-Parsi, B., and de Schepper, I.M. (1986). *Phys. Lett. A*, **114**, 241.

Cohen, E.G.D., Westerhuijs, P., and de Schepper, I.M. (1987). *Phys. Rev. Lett.*, **59**, 2872.

Cowley, R.A. and Woods, A.D.B. (1971). *Can. J. Phys.*, **49**, 177.

Crevecoeur, R.M., Smorenburg, H.E., and de Schepper, I.M. (1996). *J. Low Temp. Phys.*, **105**, 149.

Crevecoeur, R.M., Verberg, R., de Schepper, I.M., de Graaf, L.A., and Montfrooij, W. (1995). *Phys. Rev. Lett.*, **74**, 5052.

Dalberg, P.S., Boe, A., Strand, K.A., and Sikkeland, T. (1978). *J. Chem. Phys.*, **69**, 5473.

Davis, K.B., Mewes, M.-O., Andrews, M.R., van Druten, N.J., Durfee, D.S., Kurn, D.M., and Ketterle, W. (1995). *Phys. Rev. Lett.*, **75**, 3969.

de Gennes, P.G. (1959). *Physica (Utrecht)*, **25**, 825.

de Schepper, I.M. and Cohen, E.G.D. (1982). *J. Stat. Phys.*, **27**, 223.

de Schepper, I.M., Cohen, E.G.D., Bruin, C., van Rijs, J.C., Montfrooij, W., and de Graaf, L.A. (1988). *Phys. Rev. A*, **38**, 271.

de Schepper, I.M., Cohen, E.G.D., Pusey, P.N., and Lekkerkerker, H.N.W. (1989). *J. Phys.: Condens. Matter*, **1**, 6503.

de Schepper, I.M., Cohen, E.G.D., and Zuilhof, M.J. (1984*a*). *Phys. Lett. A*, **101**, 399.

de Schepper, I. and Montfrooij, W. (1989). *Phys. Rev. A*, **39**, 5807.

de Schepper, I.M., van Beijeren, H., and Ernst, M.H. (1974). *Physica*, **75**, 1.

de Schepper, I.M., van Rijs, J.C., van Well, A.A., Verkerk, P., de Graaf, L.A., and Bruin, C. (1984*b*). *Phys. Rev. A*, **29**, 1602.

de Schepper, I.M., Verkerk, P., van Well, A.A., and de Graaf, L.A. (1983). *Phys. Rev. Lett.*, **50**, 974.

de Schepper, I.M., Verkerk, P., van Well, A.A., and de Graaf, L.A. (1984*c*). *Phys. Lett. A*, **104**, 29.

Demmel, F., Hosokawa, S., Lorenzen, M., and Pilgrim, W.-C. (2004). *Phys. Rev. B*, **69**, 012203.

Dierker, S.B., Pindak, R., Fleming, R.M., Robinson, I.K., and Berman, L. (1995). *Phys. Rev. Lett.*, **75**, 449.

Dietrich, O.W., Graf, E.H., Huang, C.H., and Passell, L. (1972). *Phys. Rev. A*, **5**, 1377.

Donnely, R.J., Donnely, J.A., and Hills, R.N. (1981). *J. Low Temp. Phys.*, **44**, 471.

Drewel, M., Ahrens, J., and Podschus, U. (1990). *J. Opt. Soc. Am.*, **7**, 206.

Dzugutov, M., Larsson, K.-E., and Ebbsjo, I. (1988). *Phys. Rev. A*, **38**, 3609.

Egelstaff, P.A., Gläser, W., Litchinsky, D., Schneider, E., and Suck, J.B. (1983). *Phys. Rev. A*, **27**, 1106.

Enciso, E., Almarza, N.G., Dominguez, P., Gonzalez, M.A., and Bermejo, F.J. (1995). *Phys. Rev. Lett.*, **74**, 4233.

Fåk, B., Gückelsberger, G., Scherm, R., and Stunault, A. (1994). *J. Low. Temp. Phys.*, **97**, 445.

Feynman, R.P. (1953a). *Phys. Rev.*, **91**, 1291.

Feynman, R.P. (1953b). *Phys. Rev.*, **91**, 1301.

Feynman, R.P. (1954). *Phys. Rev.*, **94**, 262.

Feynman, R.P. (1955). In *Progress in Low Temperature Physics* (ed. C. Gorter). North-Holland, Amsterdam.

Feynman, R.P. (1972). *Statistical Mechanics: A Set of Lectures*. Benjamin, Reading, Massachusetts.

Feynman, R.P. and Cohen, M. (1956). *Phys. Rev.*, **102**, 1189.

Forster, D. (1975). *Hydrodynamic Fluctuations, Broken Symmetry, and Correlation Functions*. W.A. Benjamin Inc, Reading, Massachusetts.

Gibbs, M.R., Andersen, K.H., Stirling, W.G., and Schober, H. (1999). *J. Phys.: Condens. Matter*, **11**, 603.

Gibbs, M.R. and Stirling, W.G. (1996). *J. Low Temp. Phys.*, **102**, 249.

Gläser, W. and Morkel, Ch. (1984). *J. Non-Cryst. Solids*, **61**, 309.

Glyde, H.R. (1992). *Phys. Rev. B*, **45**, 7321.

Glyde, H.R. (1994). *Excitations in Liquid and Solid Helium*. Oxford University Press, Oxford.

Glyde, H.R., Azuah, R.T., and Stirling, W.G. (2000a). *Phys. Rev. B*, **62**, 14337.

Glyde, H.R., Fåk, B., van Dijk, N.H., Godfrin, H., Guckelsberger, K., and Scherm, R. (2000b). *Phys. Rev. E*, **61**, 1421.

Glyde, H.R., Gibbs, M.R., Stirling, W.G., and Adams, M.A. (1998). *Europhys. Lett.*, **43**, 422.

Glyde, H.R. and Griffin, A. (1990). *Phys. Rev. Lett.*, **65**, 1454.

Goldstein, H. (1980). *Classical Mechanics* (2nd edn). Addison-Wesley Publishing Company, Reading, Massachusetts.

González, L.E., González, D.J., Calderín, L., and Şengül, S. (2008). *J. Chem. Phys.*, **129**, 171103.

Griffin, A. (1993). *Excitations in a Bose-Condensed Liquid*. Cambridge University Press, New York.

Grimm, R. (2005). *Nature*, **435**, 1035.

Grüner, F. and Lehman, W.P. (1982). *J. Phys. A: Math. Gen.*, **15**, 2847.

Hansen, J-.P. and McDonald, I.R. (2006). *Theory of Simple Liquids* (3rd edn). Academic Press, London.

Helbing, R.K.B. (1969). *J. Chem. Phys.*, **50**, 493.

Henshaw, D.G. (1958). *Phys. Rev. Lett.*, **1**, 127.

Hohenberg, P.C. and Martin, P.C. (1965). *Ann. Phys.*, **34**, 291.

Hoover, W.G. and Ree, F.H. (1968). *J. Chem. Phys.*, **49**, 3609.

Hosokawa, S., Pilgrim, W.-C., Kawakita, Y., Ohshima, K., Takeda, S., Ishikawa, D., Tsutsui, S., Tanaka, Y., and Baron, A.Q.R. (2003). *J. Phys.: Condens. Matter*, **15**, L623.

Hunter, R.J. (1989). *Foundations of Colloid Science*. Oxford University Press, Oxford.

Jackson, H.W. and Feenberg, E. (1962). *Rev. Mod. Phys.*, **34**, 686.

Juge, K.J. and Griffin, A. (1994). *J. Low Temp. Phys.*, **97**, 105.

Kamgar-Parsi, B., Cohen, E.G.D., and de Schepper, I.M (1987). *Phys. Rev. A*, **35**, 4781.

Kapitza, P.L. (1938). *Nature*, **141**, 71.

Ketterle, W. and Zwierlein, M.W. (2008). In *Ultracold Fermi Gases* (ed. M. Inguscio, W. Ketterle, and C. Salomon), IOS Press, Amsterdam, pp. 95–287.

Kirkpatrick, T.R. (1984). *Phys. Rev. B*, **29**, 3966.

Kirkpatrick, T.R. (1985). *Phys. Rev. A*, **32**, 3130.

Ladd, A.C. (1994*a*). *J. Fluid Mech.*, **271**, 285.

Ladd, A.C. (1994*b*). *J. Fluid Mech.*, **271**, 311.

Landau, L.D. (1941). *J. Phys. USSR*, **5**, 71.

Landau, L.D. (1947). *J. Phys. (USSR)*, **11**, 91.

Landau, L.D. and Khalatnikov, I.M. (1949). *Zh. Eksp. Teor. Fiz.*, **19**, 637.

Lee, D.K. and Lee, F.J. (1975). *Phys. Rev. B*, **11**, 4318.

Leegwater, J.A. (1991). *J. Chem. Phys.*, **94**, 7402.

Leggett, A.J. (1965). *Phys. Rev.*, **140**, 1869.

Leggett, A.J. (1972). *Phys. Rev. Lett.*, **29**, 1227.

London, F. (1938). *Nature*, **141**, 643.

Lovesey, S.E. (1971). *Theory of Neutron Scattering*, Oxford University Press, Oxford.

Lurio, L.B., Lumma, D., Sandy, A.R., Borthwick, M.A., Falus, P., Mochrie, S.G.J, Pelletier, J.F., Sutton, M., Regan, L., Malik, A., and Stephenson, G.B. (2000). *Phys. Rev. Lett.*, **84**, 785.

Manousakis, E. and Pandharipande, V.R. (1984). *Phys. Rev. B*, **30**, 5062.

Manousakis, E. and Pandharipande, V.R. (1986). *Phys. Rev. B*, **33**, 150.

Maxwell, J.C. (1902). *Theory of Heat.* Longmans, Green & Co., London.

Mayers, J. (2006). *Phys. Rev. B*, **74**, 014516.

McGreevy, R.L. (2001). *J. Phys.: Condens. Matter*, **13**, R877.

McGreevy, R.L. and Mitchell, E.W.J. (1985). *Phys. Rev. Lett.*, **55**, 398.

Mezei, F. (1980). *Phys. Rev. Lett.*, **44**, 1601.

Mezei, F. and Stirling, W.G. (1983). In *75th Jubilee Conference on Helium-4* (ed. J. Armitage), World Scientific, Singapore, p. 111.

Miceli, P.F., Montfrooij, W., Taub, H., Schoen, K., Worcester, D.L., and Winholtz, R.A. (2005). *Trans. Am. Nucl. Soc.*, **92**, 174.

Miyazaki, K. and de Schepper, I.M. (2001). *Phys. Rev. Lett.*, **87**, 214502.

Miyazaki, K., Srinivas, G., and Bagchi, B. (2001). *J. Chem. Phys.*, **114**, 6276.

Montfrooij, W., de Graaf, L.A., and de Schepper, I.M. (1991). *Phys. Rev. A*, **44**, 6559.

Montfrooij, W., de Graaf, L.A., and de Schepper, I.M. (1992*a*). *Phys. Rev. B*, **45**, 3111.

Montfrooij, W. and de Schepper, I. (1989). *Phys. Rev. A*, **39**, 2731.

Montfrooij, W. and de Schepper, I. (1995). *Phys. Rev. B*, **51**, 15607.

Montfrooij, W. and Svensson, E.C. (1994). *Physica B*, **194-196**, 521.

Montfrooij, W. and Svensson, E.C. (1996). *Czech. Phys.*, **46 S5**, 2559.

Montfrooij, W. and Svensson, E.C. (1997). *Physica B:*, **241-243**, 924.

Montfrooij, W. and Svensson, E.C. (2000). *J. Low Temp. Phys.*, **121**, 293.

Montfrooij, W., Svensson, E.C., and de Schepper, I.M. (1992*b*). *J. Low Temp. Phys.*, **89**, 437.

Montfrooij, W., Svensson, E.C., de Schepper, I.M., and Cohen, E.G.D. (1997). *J. Low Temp. Phys.*, **109**, 577.

Montfrooij, W., Verkerk, P., and de Schepper, I. (1986). *Phys. Rev. A*, **33**, 540.

Montfrooij, W., Westerhuijs, P., de Haan, V.O., and de Schepper, I.M. (1989). *Phys. Rev. Lett.*, **63**, 544.

Montfrooij, W., Westerhuijs, P., and de Schepper, I. (1988). *Phys. Rev. Lett.*, **61**, 2155.

Montfrooy, W., de Schepper, I., Bosse, J., Gläser, W., and Morkel, Ch. (1986). *Phys. Rev. A*, **33**, 1405.

Moon, R.M., Riste, T., and Koehler, W.C. (1969). *Phys. Rev.*, **181**, 920.

Moraldi, M., Celli, M., and Barocchi, F. (1992). *Phys. Rev. A*, **46**, 7561.

Mori, H. (1965). *Prog. Theor. Phys.*, **33**, 423.

Morkel, Chr., Gronemeijer, Chr., Gläser, W., and Bosse, J. (1987). *Phys. Rev. Lett.*, **58**, 1873.

Mott, N.F. and Massey, H.S.W. (1971). *The Theory of Atomic Collisions* (3rd edn). Clarendon Press, Oxford.

Murray, C.A., Woerner, R.L., and Greytak, T.J. (1975). *J. Phys. C: Solid State Physics*, **8**, L90.

Ngai, K.L. (2007). *J. Non-Cryst. Solids*, **353**, 709.

Ohbayashi, K. (1991). In *Excitations in 2-Dimensional and 3-Dimensional Quantum, Fluids* (ed. A. Wyatt and H. Lauter), Plenum Press, New York, p. 77.

Osheroff, D.D. (1997). *Rev. Mod. Phys.*, **69**, 667.

Palevsky, H., Otnes, K., and Larsson, K.E. (1958). *Phys. Rev.*, **112**, 11.

Palevsky, H., Otnes, K., Larsson, K.E., Pauli, R., and Stedman, R. (1957). *Phys. Rev.*, **108**, 1346.

Patty, M. (2009). Ph.D. thesis.

Patty, M., Schoen, K., and Montfrooij, W. (2006). *Phys. Rev. E*, **73**, 021202.

Patty, M., Schoen, K., Montfrooij, W., and Yarmani, Z. (2009). Unpublished.

Penrose, O. and Onsager, L. (1956). *Phys. Rev.*, **104**, 576.

Pilgrim, W.C. and Morkel, C. (2002). *J. Non-Cryst. Solids*, **312-314**, 128.

Pusey, P.N. (1978). *J. Phys. A: Math. Gen.*, **11**, 119.

Reijers, H.T.J., van der Lugt, W., van Dijk, C., and Saboungi, M.-L. (1989). *J. Phys.: Condens. Matter*, **1**, 5229.

Résibois, P. and de Leener, M. (1977). *Classical Kinetic Theory of Fluids*. Wiley, New York.

Robkoff, H.N. and Hallock, R.B. (1981). *Phys. Rev. B*, **24**, 159.

Root, J.H. and Svensson, E.C. (1991). *Physica B*, **169**, 505.

Ross, M. (1969). *Phys. Rev.*, **184**, 233.

Ruiz-Martín, M.D., Jiménez-Ruiz, M., Stunnault, A., Bermejo, F.J., Fernandéz-Perea, R., and Cabrillo, C. (2007). *Phys. Rev. B*, **76**, 174201.

Ruocco, G. and Sette, F. (2008). *Condens. Matter Phys.*, **11**, 29.

Ruvalds, J. and Zawadowski, A. (1970). *Phys. Rev. Lett.*, **25**, 333.

Sacchetti, F., Suck, J.-B., Petrillo, C., and Dorner, B. (2004). *Phys. Rev. E*, **69**, 061203.

Sampoli, M., Bafile, U., Guarini, E., and Barocchi, F. (2002). *Phys. Rev. Lett.*, **88**, 085502.

Santucci, S.C., Fioretto, D., Comez, L., Gessini, A., and Masciovecchio, C. (2006). *Phys. Rev. Lett.*, **97**, 225701.

Scherm, R., Guckelsberger, K., Fåk, B., Dianoux, A.J., Godfrin, H., and Stirling, W.G. (1987). *Phys. Rev. Lett.*, **59**, 217.

Schimmel, H.G., Montfrooij, W., Verhoeven, V.W.J., and de Schepper, I.M. (2002). *Europhys. Lett.*, **60**, 868.

Schmets, A.J.M. and Montfrooij, W. (2008). *Phys. Rev. Lett.*, **100**, 239601.

Scopigno, T., Balucani, U., Ruocco, G., and Sette, F. (2000*a*). *Phys. Rev. Lett.*, **85**, 4076.

Scopigno, T., Balucani, U., Ruocco, G., and Sette, F. (2000*b*). *Phys. Rev. E*, **63**, 011210.

Scopigno, T., Balucani, U., Ruocco, G., and Sette, F. (2002). *Phys. Rev. E*, **65**, 031205.

Scopigno, T., di Leonardo, R., Cornez, L., Baron, A.Q.R., Fioretto, D., and Ruocco, G. (2005*a*). *Phys. Rev. Lett.*, **94**, 155301.

Scopigno, T., di Leonardo, R., Cornez, L., Baron, A.Q.R., Fioretto, D., Ruocco, G., and Montfrooij, W. (2005*b*). *Phys. Rev. Lett.*, **95**, 269602.

Scopigno, T. and Ruocco, G. (2005). *Rev. Mod. Phys.*, **77**, 881.

Sears, V.F. (1973). *Phys. Rev. A*, **7**, 340.

Sears, V.F. (1975). *Nucl. Instr., Methods*, **123**, 521.

Sears, V.F. and Svensson, E.C. (1979). *Phys. Rev. Lett.*, **43**, 2009.

Sears, V.F., Svensson, E.C., Martel, P., and Woods, A.D.B. (1982). *Phys. Rev. Lett.*, **49**, 279.

Segrè, P.N., Behrend, O.P., and Pusey, P.N. (1995*a*). *Phys. Rev. E*, **52**, 5070.

Segrè, P.N. and Pusey, P.N. (1996). *Phys. Rev. Lett.*, **77**, 771.

Segrè, P.N., van Megen, W., Pusey, P.N., Schätzel, K., and Peters, W. (1995*b*). *J. Mod. Opt.*, **42**, 1929.

Sharma, R.K., Tankeshwar, K., and Pathak, K.N. (1998). *J. Chem. Phys.*, **108**, 2919.

Silver, R.N., Sivia, D.S., and Gubernatis, J.E. (1990). *Phys. Rev. B*, **41**, 2380.

Sjögren, L. and Sjölander, A. (1979). *J. Phys. C*, **12**, 4369.

Sokol, P.E. (1995). In *Bose-Einstein Condensation* (ed. A. Griggin, D. Snoke, and S. Stringari), p. 51. University Press, New York.

Squires, G.L. (1994). *Thermal Neutron Scattering*. Oxford University Press, Oxford.

Stirling, W.G. (1983). In *75th Jubilee Conference on Helium-4* (ed. J. Armitage). World Scientific, Singapore, p. 109.

Stirling, W.G. (1991). In *Excitations in 2-dimensional and 3-dimensional Quantum Fluids* (ed. A. Wyatt and H. Lauter), Plenum Press, New York, p. 47.

Sumi, T., Miyoshi, E., and Tanaka, K. (1999). *Phys. Rev. B*, **59**, 6153.

Svensson, E.C. (1989). In *Elementary Excitations in Quantum Fluids* (ed. K. Ohbayashi and M. Watabe), Springer-Verlag, Heidelberg, p. 59.

Svensson, E.C. (1991). In *Excitations in 2-dimensional and 3-dimensional Quantum Fluids* (ed. A. Wyatt and H. Lauter), Plenum Press, New York, p. 59.

Svensson, E.C., Martel, P., Sears, V.F., and Woods, A.D.B. (1976). *Can. J. Phys.*, **54**, 2178.

Svensson, E.C., Montfrooij, W., and de Schepper, I.M. (1996). *Phys. Rev. Lett.*, **77**, 4398.

Sychev, V.V., Vasserman, A.A., Kozlov, A.D., Spiridonov, G.A., and Tsymarny, V.A. (1987). *Thermodynamic Properties of Helium*, Hemisphere Publishing Corp., New York.

Talbot, E.F., Glyde, H.R., Stirling, W.G., and Svensson, E.C. (1988). *Phys. Rev. B*, **38**, 11229.

Torcini, A., Balucani, U., de Jong, P.H.K., and Verkerk, P. (1995). *Phys. Rev. E*, **51**, 3126.

Tsai, K.H., Wu, T.-M., Tsay, S.-F., and Yang, T.-J. (2007). *J. Phys.: Condens. Matter*, **19**, 205141.

van Beijeren, H. and Ernst, M.H. (1973). *Physica*, **68**, 437.

van Hove, L. (1954). *Phys. Rev.*, **95**, 249.

van Loef, J.J. (1974). *J. Chem. Phys.*, **61**, 1605.

van Loef, J.J. and Cohen, E.G.D. (1989). *Phys. Rev. B*, **39**, 4715.

van Well, A.A. and de Graaf, L.A. (1985). *Phys. Rev. A*, **32**, 2396.

van Well, A.A., Verkerk, P., de Graaf, L.A., Suck, J.-B., and Copley, J.R.D. (1985). *Phys. Rev. A*, **31**, 3391.

Varoquaux, E., Ihas, G.G., Avenel, O., and Aarts, R. (1993). *Phys. Rev. Lett.*, **70**, 2114.

Varoquaux, E., Jr, W. Zimmermann, and Avenel, O. (1991). In *Excitations in 2-dimensional and 3-dimensional Quantum Fluids* (ed. A. Wyatt and H. Lauter), Plenum Press, New York, p. 343.

Verberg, R., de Schepper, I.M., and Cohen, E.G.D. (1999). *Europhys. Lett.*, **48**, 397.

Verkerk, P., van Well, A.A., and de Schepper, I.M (1987). *J. Phys. C: Solid State Phys.*, **20**, L979.

Verkerk, P., Westerweel, J., Bafile, U., de Graaf, L.A., Montfrooij, W., and de Schepper, I.M. (1989). *Phys. Rev. A*, **40**.

Wegdam, G.H., Bot, A., Schram, R.P.C., and Schaink, H.M. (1989). *Phys. Rev. Lett.*, **63**, 2697.

Wertheim, M.S. (1963). *Phys. Rev. Lett.*, **10**, 321.

Westerhuijs, P., Montfrooij, W., de Graaf, L.A., and de Schepper, I.M. (1992). *Phys. Rev. A*, **45**, 3749.

Williams, G.A. (1992). *Phys. Rev. Lett.*, **68**, 2054.

Woods, A.D.B. and Cowley, R. A. (1973). *Rep. Prog. Phys.*, **36**, 1135.

Woods, A.D.B. and Svensson, E.C. (1978). *Phys. Rev. Lett.*, **41**, 974.

Woods, A.D.B., Svensson, E.C., and Martel, P. (1976*a*). In *Proceedings of the Conference on Neutron Scattering* (ed. R. Moon), Volume 2, National Technical Information Service, Springfield, p. 1010.

Woods, A.D.B., Svensson, E.C., and Martel, P. (1976*b*). *Phys. Lett. A*, **57**, 439.

Woods, A.D.B., Svensson, E.C., and Martel, P. (1978). *Can. J. Phys.*, **56**, 302.

Xu, R. (2000). *Particle Characterization: Light Scattering Methods*. Springer, Kluwer Academic Publishers, Dordrecht, The Netherlands.

Yarnell, J.L., Arnold, G.P., Bendt, P.J., and Kerr, E.C. (1958). *Phys. Rev. Lett.*, **1**, 9.

Yarnell, J.L., Arnold, G.P., Bendt, P.J., and Kerr, E.C. (1959). *Phys. Rev.*, **113**, 1379.

Zuilhof, M.J., Cohen, E.G.D., and de Schepper, I.M. (1984). *Phys. Lett. A*, **103**, 120.

Zwanzig, R. (1961). In *Lectures in Theoretical Physics*, Volume 3, Interscience, New York, p. 106.

Zwierlein, M.W., Abo-Shaefer, J.R., Schirotzek, A., Schunk, C.H., and Ketterle, W. (2005). *Nature*, **435**, 1047.

Index

Aluminum, 146
Argon, 93, 96
Atomic separation
  simple model, 101

Bose-condensate, 4
  elementary excitations, 185
  lithium, 186
  sodium, 186
  superfluid, 185
Bragg scattering, 66
Brillouin lines, 4
Brillouin scattering, 91

Cage diffusion, 110
  amplitude, 59, 60
    memory function formalism, 57
  background, 59
  binary collisions, 61
  colloids, 116
  correlation loss, 59, 60
  form factor, 60
  graphical representation, 29
  heat flux, 61
  long-time behavior, 20
  momentum flux, 61
  short time behavior
    visibility in experiments, 111
  short-time behavior, 111
    colloids, 116
Cesium, 151, 152
Close packing, 101, 236
  exceptions, 162
Collisional transfer, 107
Colloids, 115
  background, 115
  charged, 118
  neutral, 122
Computer simulations, 81
  colloids, 116, 123
  fit functions, 84
  gallium, 164
  graphical representation, 82
  hard spheres, 89
  helium, 86, 87
  helium–neon, 133, 136, 137
  hydrogen, 21
  krypton–argon, 136, 137
  Lennard-Jones fluid, 98, 100
  Li–Pb, 127
  limitations, 81, 84, 85

mercury, 83
method, 82
microscopic variables, 83
mixtures, 126
molecular dynamics, 82
Monte Carlo, 84
needs for, 62, 94, 119, 150
path-integral Monte Carlo, 86
quantum fluids, 86
reverse Monte Carlo, 85
Condensate fraction, 24, 185
Conversion, 5, 226
Convolution, 8
Correlation functions, 9, 227
  classical versus quantum, 9
  density–temperature, 41, 43, 89
  dynamic structure factor, 11
  dynamic susceptibility, 12, 176
  eigenmode formalism, 41
  equal-time correlation functions, 13
  Fourier transform, 10
  intermediate scattering function, 116
  pair correlation function, 13
  relaxation function, 12
  response function, 227
  table, 12
  temperature–temperature, 41, 89
  van Hove, 10, 227
    generalized, 10
  velocity auto-correlation, 18
    measurement of, 22
    mode-coupling corrections, 21
Coupling parameters, 230
  background, 2
  definition
    density–velocity, 34
    velocity–temperature, 34
  definitions, 232
  express as correlation functions,
    232
  generalized, 231
  hydrodynamic limit, 92
  mixtures, 131, 254
  q dependence, 49
    computer simulation, 98
  temperature dependence, 198
  velocity–momentum flux, 45

Damping
  background, 49
  eigenmodes, 47, 156

Damping (*cont.*)
  hydrodynamics, 15, 16
  Landau–Khalatnikov, 213
  origin, 50
  particle-hole excitations, 183
  propagation gap, 97
  q dependence, 49
  sound modes
    simple model, 105
  superfluid, 197
  temperature dependence
    superfluid, 198, 200
Data analysis
  errors in, 8, 78
    determination propagation frequency,
      78, 147, 176
    imposing restrictions, 80, 81
    interpretation of fit results, 80
    number of fit parameters, 81
    symmetrization, 114
de Gennes minimum, 102, 107
Debye sphere, 118
Decay rates
  heat flux, 47
  hydrodynamic limit, 92
  hydrodynamics, 15, 16
  momentum flux, 47
  q dependence, 3, 47
Decay time
  definition, 15
  exponentials, 89
Decay tree, 2
  diffusion-dominated, 108
  extended hydrodynamics, 37
  extent, 49, 62
    metals, 148
  harmonic oscillator, 46
  hydrodynamics, 2, 34
  mixtures, 130
  q range, 90
  table, 91, 249, 250
  viscoelastic model, 44
  viscoelastic model (extended),
    45, 58
Detailed balance, 11, 41
Diffusion, 92
  background, 10
  Boltzmann, 119
  Brownian motion, 119
  colloids, 116
  Enskog, 107
  heat, 16
  mixtures, 136
  scattering experiments, 69
  self-diffusion, 10, 111
Dynamic structure factor
  definition, 11
  generalized, 11
Dynamic susceptibility, 13, 228

Effective eigenmodes
  number of visible modes, 40
  amplitudes, 39, 43, 90
    absence of heat mode, 177
    metals, 159
    mixtures, 137
    numerical examples, 51, 160
  background, 39, 40, 88
  character of the modes, 40, 41, 89, 94,
    99, 240
  connection to heat and sound modes, 39
  damping, 47, 49
  detailed balance, 41
  experimental determination, 39, 40
  exponentials, 41
  heat mode, 39, 88
  hydrodynamics transition, 92
    computer simulation, 99
    mathematical representation, 48
    measurement of, 93
  kinetic modes, 99
  Lorentzian lines, 41, 43
  metals
    numerical examples, 157–160
  number of visible modes, 4, 7, 39, 90
    metals, 157
    more than three, 113
    numerical examples, 51, 157
  numerical examples, 240
  propagating versus overdamped, 49, 53, 59,
    113
  sound modes, 39, 88
  sum rules, 40
Eigenmode formalism, 32
  (eigen)modes, 38
  adjustable parameters, 8, 33, 49
  background, 1, 32, 36, 40
  classical limit, 14
  classical versus quantum, 14
  continued fraction expressions, 249
  decay tree, 32, 35
    extended hydrodynamics, 37
    extent, 36, 49, 138
    harmonic oscillator, 46
    hydrodynamics, 34
    mixtures, 130
    table, 249
    viscoelastic model, 44, 45, 58
  dynamic matrix, 35, 36
  eigenvalues, 38, 42, 99
  eigenvectors, 38, 42, 99
  extent of decay tree, 47
  fit function, 7, 230
  function to fit to, 13, 248
  hard spheres, 38
  hydrodynamics transition, 47, 94
    numerical examples, 50–53
  Lorentzian lines, 39, 43
  memory function comparison, 149, 248

memory function formalism equivalence, 55, 248

mixtures, 129, 254
  metals as binary mixture, 163

poles of susceptibility, 42

projection formalism, 227

transport coefficients, 35

Elementary excitations, 4, 185
  measurement of, 187

Energy gap, 186
  Bose statistics, 190
  superfluidity, 189

Ensemble average, 227

Enskog theory, 28
  applied to colloids, 121
  applied to metals, 152, 161
  applied to simple fluids, 100
  background, 28
  cage diffusion, 29
  heat mode, 99
  mixtures, 127
  predictions for $q\sigma \simeq 2\pi$, 107
  propagation gap, 97
  sound modes, 99
  vortex diffusion, 29

European Synchrotron Radiation Facility, 143–146

Excitations
  background, 4, 9
  excitations with $\lambda = d$, 101
  new types, 162, 165, 171

Excited state, 192, 224
  graphical representation, 192, 225
  perturbation theory, 202

Experimental results
  aluminum, 146
  argon, 93, 96
  cesium, 151, 152
  gallium, 75, 161, 167, 169
  helium, 25, 173–176, 187, 194, 214
  helium–neon, 128, 133
  helium-3, 182
  hydrogen–argon, 132
  lead, 168
  lithium, 145
  mercury, 112, 167
  monitor contamination, 73
  neon, 95
  nickel, 154
  polymethyl-methacrylate spheres, 123
  polystyrene spheres, 120, 124
  PPMA, 117, 120
  silicon, 153
  sodium, 19, 27
  sodium–chlorine, 139
  water, 140

Fast sound, 126
  helium–neon, 128, 133
  hydrogen–argon, 132

krypton–argon, 136
  sodium–chlorine, 138, 139
  speed of propagation, 128
  water, 140

Fick's law, 15

Fit functions, 7, 248
  convolution, 8
  molecular dynamics, 84

Fit parameters
  number of free parameters, 7

Fluctuation–dissipation theorem, 13, 228

Fluctuations, 1
  density
    graphical representation, 104
  graphical representation, 2
  relaxation, 10
  temperature, 113, 157
    propagating, 193

Fourier transform, 228

Free-streaming, 90

Friedel oscillations, 162

Gallium, 161, 167, 169, 222

Glory oscillations, 220, 256

Ground state, 191, 225
  graphical representation, 191

Hard spheres, 236
  colloids, 120
  comparison to experiment
    sodium, 27
  effective diameter, 31, 119, 236
  eigenmode formalism, 38
  Enskog theory, 28
  equivalent diameter, 31, 119, 236
  ideal-gas limit, 26
  Lindemann law, 31, 236
  metals, 152, 161
  mixtures, 127, 132
  propagation gap, 97

Harmonic oscillator, 25, 224, 248
  Fermi liquid, 182
  fit function, 80
  graphical representation, 46
  helium, 173

Heat mode, 4, 39, 88
  amplitude, 39
    hydrodynamics, 43
  character, 105
  colloids, 120
  dominance at $q\sigma \simeq 2\pi$, 99, 108, 120
  graphical representation, 104
  metals, 152
  mixtures, 131
  q dependence, 177
  simple model, 102, 105

Helium, 25, 86, 173–176, 187, 194, 198, 214

Helium–neon, 128

Helium-3, 182

Hybridization gap, 195
  temperature dependence, 196, 213
Hydrodynamics, 15
  decay rate, 15, 16
  diffusion, 15
  extrapolation from experiment, 54
  length scale, 91
  table, 16
  transition, 47, 92
    metals, 147, 151, 156, 158
    mixtures, 128, 132
    numerical examples, 240
    water, 141
Hydrogen, 21
Hydrogen–argon, 132

Ideal gas, 22
  Bose gas, 185
  classical limit, 26
  quantum limit, 24, 195
Ideal mixing rules, 130
Institut Laue-Langevin, 93
ISIS, 112, 167, 188, 194, 195, 202

Kinetic modes, 50, 99

Laplace transform, 228
Lead, 168
Lineshape distortion, 114, 174
Liouville operator, 227
  definition, 14
Lithium, 145
Lorentzian lines
  lineshape distortion, 14
  quantum effects, 14

Maxon, 199
  superfluid transition, 199, 200
Mean free path, 105
  density expansion, 6
Memory function formalism, 54, 248
  adjustable parameters, 55
    restrictions, 59
  decay rates, 56
  effective eigenmodes comparison, 149
  eigenmode formalism equivalence, 55,
    248
  hydrodynamics, 56
  sum rules, 55
  table, 250
  viscoelastic model, 57
  viscoelastic model (extended), 57
  what function to fit to, 55
Mercury, 83, 112, 167
Metals, 143
  electron–ionic coupling, 143, 149, 160,
    163
  fluctuating magnetic moments, 163
  heat mode
    weakness, 143

  sound modes
    prominance, 143
    propagation speed, 147
  unique excitations, 162
Microscopic density, 9
  quantum-mechanical expression,
    33
Microscopic theory, 6
  cage diffusion, 7
  importance of cage diffusion, 62
  outlook, 218
  vortex diffusion, 7
Microscopic variables
  mixtures, 130, 252
  orthonormalization, 33, 230, 252
  table, 11, 253
Microscopic velocity
  quantum-mechanical expression, 33
Mixtures, 126
  coupling parameters, 254
  hydrodynamics transition, 128, 131
  ideal mixing rules, 130, 255
  optical modes, 129, 134
Mode-coupling, 16
  background, 17
  colloids, 124
  comparison to experiment, 19, 21,
    22
  corrections
    diffusion, 17
    velocity autocorrelation, 21
  temperature dependence, 179
  validity range versus density, 17, 19
Multiphonon, 193
  graphical representation, 76, 211
  measurement, 195
  temperature dependence, 199, 215

Neon, 95
Nickel, 154
Notation, 5, 226

Open questions, 149, 160,
    163, 178, 218
Optical modes, 129, 134
Orbiting, 260

Pair correlation function, 13, 86
  at contact, 119
Partial waves expansion, 221, 256
Particle–hole excitations, 181
Percus–Yevick approximation, 237
Perturbation theory, 202
  comparison to experiment,
    203, 207, 209
  excited states, 206
  graphical representation, 205
  hydrodynamic limit, 92
  resonances, 206
  self-energy, 205

Probing wavelength, 65
Projection formalism, 32, 227
Propagation gap, 96
  density dependence, 97

Quantum effects, 33
  absence of heat mode, 178
  diffraction, 219
  increase of decay rate, 179
  lineshape distortion, 174
  oscillations in cross-section, 220–222, 258
Quantum fluids, 4

Rainbow glories, 220
Rayleigh line, 4, 89
Recoil, 24, 191, 195, 202
Relaxation function, 13
  experimental determination, 42
Relaxation of fluctuations
  background, 10
Response function, 227
Roton, 187
  decay rate, 214
  superfluid transition, 198
  temperature dependence, 214

Scalar product, 42, 228
Scattering, 64
  Bragg, 65
  Brillouin, 91, 93
  coherent, 66
  cross-section, 68
    gallium–gallium, 222
    graphical representation, 65
    helium–helium, 220
    mixtures, 127
    neutron, 70
    oscillations, 258
    resonances, 259
  data corrections, 71
    attenuation, 72, 77
    container, 71
    detectors, 71
    monitor contamination, 72
    multiple scattering, 75, 122, 170
  dynamic light scattering, 115
  graphical representation, 65
  incoherent, 67
  incoherent limit, 67
  inelastic, 68
    graphical representation, 70
  interference, 64
  magnetic, 68, 69, 166
  magnetic form factor, 68
  multiple scattering
    graphical representation, 76, 77
  neutron versus xray, 68, 69, 138, 143
  polarized, 167
  spin-flip (neutron), 68, 167

Second sound, 193
Self-energy, 205
Silicon, 153
Simple models, 101
  atomic separation, 101
  excitations in superfluid helium, 208
  heat mode, 102
  sound modes, 105
Slow sound, 126
  helium–neon, 133
  hydrogen–argon, 132
  krypton–argon, 136
Sodium, 19, 27
Sound modes, 4, 39, 88
  amplitude, 39
    hydrodynamics, 43
  background, 10
  cold liquids, 172
  coupling to particle–hole excitations, 181
  graphical representation, 104
  high versus low temperature, 178
  metals, 144
  propagating versus overdamped, 46, 96, 214
  propagation speed, 96
    adiabatic, 50, 51
    isothermal, 50, 155, 244
    metals, 153
  simple model, 97, 103
  zero sound, 47
SPring-8, 153
Static structure factor, 13
  background, 66
  partial structure factors, 127
  Percus–Yevick approximation, 237
Static susceptibility, 13
Stokes line(s), 89
Sum rules, 42, 233
  eigenmodes, 40
  f-sum rule, 42
  higher order, 234
  memory function formalism, 55
Superfluid
  Bose-condensate, 185
  energy gap, 187
  fermionic, 186
  Feynman explanation of superfluidity, 190
  graphical overview, 212
  graphical representation, 190
  transition to normal fluid, 175, 176, 196
    changes to maxon, 199
    changes to roton, 197
    graphical overview, 213
  vortices, 186, 189

Transforms
  Fourier, 228
  Laplace, 228
Transition
  hydrodynamics, 47

Transport coefficients, 92
    determination from experiment,
        93
    Enskog, 237
    generalized, 94, 110
    link to binary collisions, 218

Unit conversion, 5, 226

Viscoelastic model, 44, 249
    background, 44
    extentions thereof, 45
    graphical representation, 44, 45,
        58

memory function, 44
metals, 145
Viscosity, 92
    density expansion, 6
Vortex diffusion
    Enskog theory, 29
    graphical representation,
        30

Water, 140

Zero-point motion, 172
    freezing, 186
Zero sound, 47